Mechanical Engineering Series

Series Editor

Francis A. Kulacki, Department of Mechanical Engineering,
University of Minnesota, Minneapolis, MN, USA

The Mechanical Engineering Series presents advanced level treatment of topics on the cutting edge of mechanical engineering. Designed for use by students, researchers and practicing engineers, the series presents modern developments in mechanical engineering and its innovative applications in applied mechanics, bioengineering, dynamic systems and control, energy, energy conversion and energy systems, fluid mechanics and fluid machinery, heat and mass transfer, manufacturing science and technology, mechanical design, mechanics of materials, micro- and nano-science technology, thermal physics, tribology, and vibration and acoustics. The series features graduate-level texts, professional books, and research monographs in key engineering science concentrations.

More information about this series at http://www.springer.com/series/1161

Lin-Shu Wang

A Treatise of Heat and Energy

Springer

Lin-Shu Wang
Stony Brook University
Stony Brook, NY, USA

ISSN 0941-5122 ISSN 2192-063X (electronic)
Mechanical Engineering Series
ISBN 978-3-030-05745-9 ISBN 978-3-030-05746-6 (eBook)
https://doi.org/10.1007/978-3-030-05746-6

This Springer imprint is published by the registered company Springer Nature Switzerland AG
The registered company address is: Gewerbestrasse 11, 6330 Cham, Switzerland

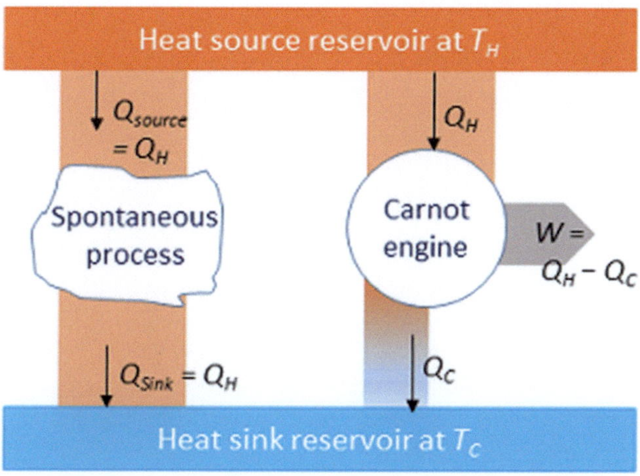

Heat and work flow for a Carnot cycle, which is an example of extracting from the heat sink reservoir "heat of the amount, $Q_H - Q_C$", (that would have been lost in a spontaneous heat transfer process) for the production of work, $W = Q_H - Q_C$.

To Ming

Preface

Thermodynamic understanding of heat and energy is based on the mechanical theory of heat (MTH), which resulted from the synthesis, by Kelvin and Clausius, of Carnot's theory of heat and the Mayer–Joule principle. Yet, there are no good definitions for heat or energy in the current literature on thermodynamics. It is noted that the advent of the entropy principle created the scientific stream of thermodynamics (a new stream branched off from its original source, the engineering stream) and led to, in quick succession, the successful formulation of equilibrium thermodynamics. Here, I make the case that the impression of the Kelvin–Clausius synthesis' success is formed from its success in producing a coherent system of equilibrium thermodynamics, not in resulting in a coherent system of engineering stream of thermodynamics—the failure of which is reflected in the fact that engineering thermodynamics cannot even talk about heat and energy without self-contradictions as well as fail to provide students of thermodynamics real grasp on reversibility. This disquisition–essay makes the case that the uneven achievement of Joule, Kelvin, and Clausius is because they made the classic error of equating correlation between heat and work to causality between heat and work, and, as a result, prevented the (later) formulation of the entropy principle from realizing its full power. While this error has been pointed out in a number of papers, the authors of those papers advocated, for removing the error, a return to Carnot's theory as a caloric theory of heat. That was clearly a mistake: it is argued here that Carnot's theory is a relational theory of heat not an ontological theory and, in fact, it can be made to incorporate with, ontologically, either the caloric theory or MTH. This disquisition essay presents a relational, i.e., predicative, theory of heat embracing fully MTH's ontology for an updated understanding of heat, spontaneous energy conversion, and reversible-like processes.

Stony Brook, USA — Lin-Shu Wang

Contents

Symbols and Abbreviations

A Area, m^2
A_H Helmholtz function, $U - TS$, kJ
AF Air–fuel ratio
c Speed of sound, m/s
c_p Constant pressure specific heat, kJ/kg-K
c_p Constant pressure molar specific heat, kJ/kmol-K
c_v Constant volume specific heat, kJ/kg-K
c_v Constant volume molar specific heat, kJ/kmol-K
C_p Constant pressure heat capacity, kJ/K
C_v Constant volume heat capacity, kJ/K
e Specific total energy, kJ/kg
e Molar specific total energy, kJ/kmol
E Total energy, $E = U + KE + PE$, kJ
Ex Exergy, kJ
Ex^{MTL} Material exergy
ex Mass specific exergy, kJ/kg
EGP Entropy growth potential, kJ/K
F Force, kN
FA Fuel–air ratio
fe Flow specific exergy, defined by (120)
g Specific Gibbs function, kJ/kg
G Gibbs function, $G = H - TS$, kJ
h Specific enthalpy, kJ/kg
H Enthalpy, $H = U + pV$, kJ
$\bar{h}_f$ Enthalpy of formation, kJ/kmol
$\bar{h}_f^0$ Enthalpy of formation at 25 °C and 1 atm
HHV Higher heating value, kJ/kg fuel
J Joule's constant (the MEH constant), $joule/calorie_{15}$
J_Q Heat flow
J_S Entropy flow
k Specific heat ratio, c_p/c_v
K_p Equilibrium constant
KE Kinetic energy, $\frac{1}{2}mV^2$, kJ

LHV	Lower heating value, kJ/kg fuel
m	Mass, kg
$\dot{m}$	Mass flow rate, kg/s
M	Molar mass, kg/kmol
MEH	Mechanical equivalent of heat
MTH	Mechanical theory of heat
n	Normal unit vector
N	Number of moles, kmol
p	Pressure, kPa
p_i	Partial pressure, kPa
p_0	Surroundings pressure, kPa
p_0	Pressure at 1 atm (101.325 kPa)
p^r	Under constant reservoir pressure
PE	Potential energy, kJ
PETH	Predicative entropic theory of heat
q	Heat transfer per unit mass, kJ/kg
q	Heat transfer per unit molar mass, kJ/kmol
Q	Heat transfer, kJ
$\dot{Q}$	Heat transfer rate, kW
Q_H	Heat transfer with high-temperature body
Q_A	Heat transfer with high-temperature body
Q_L	Heat transfer with low-temperature body
Q_C	Heat transfer with low-temperature, i.e., cold body
Q_B	Heat transfer with low-temperature body
Q_{rev}	System heat exchange of reversible event
Q_{spon}	System heat exchange of spontaneous event
$\Delta\hat{Q}$	Additional heat extracted for the reversible event, $= Q_{rev} - Q_{spon}$
R	Universal gas constant, 8.31447 kJ/kmol-K
R_i	Gas constant of *i*th specie, kJ/kg-K
s	Specific entropy, kJ/kg-K
s	Molar specific entropy, kJ/kmol-K
S	Entropy, kJ/K
t	Time, s
T	Temperature, °C or K
T_H	Temperature of high-temperature body, K
T_A	Temperature of high-temperature body, K
T_L	Temperature of low-temperature body, K
T_C	Temperature of low-temperature cold body, K
T_B	Temperature of low-temperature body, K
T_0	Surroundings temperature, K
T_0	Standard temperature at 298.15 K
T^r	Under constant reservoir temperature
u	Specific internal energy, kJ/kg
u	Molar specific internal energy, kJ/kmol

U	Internal energy, kJ
v	Specific volume, m^3/kg
v	Molar specific volume, $m^3/kmol$
V	Volume, m^3
$\dot{V}$	Volume flow rate, m^3/s
V	Velocity, m/s
W	Work, kJ
W_{rev}	Reversible, or maximum useful, work
$\dot{W}$	Power, kW
x	Quality
x_i	Mole fraction
y_i	Mass fraction

Greek Letters

β	Thermal expansion coefficient, K^{-1}
Δ	Finite change in quantity
η_{th}	Thermal efficiency
θ	Absolute temperature, defined by Eq. (46)
λ	Integrating denominator
μ	Chemical potential
ν	Stoichiometric coefficient
ξ	Extent of reaction, kmol
ρ	Density, kg/m^3

Subscripts

a	Air
A	High temperature (as in T_A)
atm	Atmospheric
B	Low temperature (as in T_B)
C	Cold body
cv	Control volume
cs	Surface of control volume
e	Exit conditions
f	Saturated liquid
fg	Difference in property between saturated liquid and saturated vapor
g	Saturated vapor
G	Growth
H	High temperature or hot body

i	Inlet conditions
i	*i*th component
L	Low temperature
P	Potential
rev	Reversible
s	Isentropic
sat	Saturated
spon	Spontaneous event
surr	Surroundings
sys	System
0	Standard temperature/pressure at 298.15 K and 1 atm
0	Reference surroundings
1	Initial state
2	Final state

Superscripts

r	Reservoir: process under constant reservoir temperature or pressure condition
0	Standard temperature/pressure at 298.15 K and 1 atm

Introduction: Temperature and Some Comment on Work

1

Just as Newton first conclusively showed that this is a world of masses, so [Sadi Carnot and] Willard Gibbs first revealed it as a world of systems.
—L.J. Henderson in *The Order of Nature* [1] [p. 126 (1917)].

Abstract

This introductory chapter informs the reader that this book is a disquisition of heat and energy. Understanding of heat and energy has two general requirements: the reader must achieve mastery of both the first law of thermodynamics and the second law of thermodynamics, and the reader must appreciate the difference between thermodynamic objects as systems and mechanical objects as mere mass bodies. Treatment of heat begins, in this chapter, with the introduction of the intensity of heat, i.e., temperature, and equation of state for ideal gases and ideal-gas mixtures.

Keywords

Heat · Thermodynamic system · Thermodynamic equilibrium · Temperature · Ideal gases · Thermal equation of state for ideal gases · Mixtures of ideal gases · Dalton's law · Work

Thermodynamics is mainly concerned with relation between heat and mechanical work. This relation is prescribed by two premises, one of which is *equivalence between heat and mechanical work* (equivalence principle) established by Mayer and Joule (the other premise is Carnot's principle). A central question addressed by this title is the meaning of this "equivalence".

Thermodynamics is a theory of heat. At the present time, the prevailing theory is the mechanical theory of heat (MTH). One remarkable thing about MTH is that there are no suitable definitions on heat and energy in MTH; the most commonly used ones are "heat is energy in transit," and "energy is the capacity for doing

L.-S. Wang, *A Treatise of Heat and Energy*, Mechanical Engineering Series,
https://doi.org/10.1007/978-3-030-05746-6_1

work," both of which are problematic. Without good definitions, a theory of heat and energy is destined to be a disappointment. A new theory of heat is developed here with the formulation of acceptable definitions to be its minimum requirement. I present the definitions of work, heat, energy, and a new term (to be introduced), entropy growth potential, here first without explanation (which will be fully accounted for by the end of Chap. 8):

- Work *is energy in transit.*
- Heat, *as denoted by Q, is energy and* entropy *in transit; waste* heat *or heat in a body (i.e.,* heat *used as short for thermal internal energy) is high-entropy form of energy.*
- Energy *is a conserved quantity that can be neither created nor destroyed; an important characteristic of the quantity is its exergetic content, which is measured in terms of energy system's capacity for doing work; though energy of a system and its interacting surroundings cannot be destroyed, the system-surroundings' capacity for doing* work *is being lost incessantly.*
- Entropy growth potential (EGP) *is the driver of every event in a given Poincare range; in the mechanical theory of heat (MTH, i.e., standard thermodynamics), the "consumption" of* heat *and the "consumption" of energy are treated as proxies of EGP, which is the real driver—the correct identification of the real driver leads to, in departure from MTH, how we classify "energy conversion" processes.*

This essay will succeed in its goal if a reader in going through it has a good grasp on the meaning of heat and energy as well as the two laws of thermodynamics.

1.1 Heat, Its Two Laws

When our ancestors learned to manipulate fire, they crossed the threshold from beast to human and started off on the long journey seeking answers to those simple questions that continue to haunt us: What is warmth and what is heat? How is heat created? What heat can do for us or what we can do with heat? This essay tells a story of heat, which is a long fascinating one, one that is still profoundly relevant to today's technologically ultra-modern society.

The key to this story of heat (or the science of thermodynamics) is two laws, the first law and the second law: the first law of thermodynamics (the principle of conservation of energy) asserts that energy can neither be created nor destroyed; the second law of thermodynamics (the entropy principle) posits the *unidirectionality* of the universe (in the sense that entropy always increases), or that the universe is irreversible. The idea of energy conservation is an extremely useful concept with the broadest application, and, generally, well understood by students of thermodynamics—with the singular exception of the aforementioned "precise meaning of equivalence." The idea of entropy growth is equally useful and profoundly powerful, yet it is not well understood by students of thermodynamics. As a result, its powerful significance is not as fully appreciated as that of energy conservation is: for instance, the meaning of equivalence between heat and work is universally considered to be a matter of the first law without the appreciation that, in fact, one cannot have a true comprehension of which without the second law.

For an additional instance, the economic activity of man is often expressed in terms of production. But, we are unclear about *what* is really being produced: "For man is not a producer but only a converter, and for every job of conversion he needs primary products [matter]. In particular, his power to convert depends on primary energy…" noted Schumacher in *Small is Beautiful* [2]. Man can neither produce (i.e., create) primary matter nor create primary energy: what is produced is not the matter or energy of a product, but the function of the product in its new form—the *value* in the new product system. When secondary energy (or energy carrier) is "produced" from primary energy, primary energy is merely converted or transformed into secondary energy, a different form of energy of higher value. When an energy carrier is "consumed" for "producing" a material product, there is no destruction of energy or creation of mass, but the conversion of the energy carrier into heat and residual energy in the product and the conversion of raw materials into the material product of higher value.

What makes any conversion possible is the result of *unidirectionality* in nature (i.e., the entropy *principle*_see Chap. 5), or *unidirectionality* in energy of various forms (i.e., the *energy principle*_see Chap. 4). It should be noted that the energy principle is subsumed, as will be shown in Chap. 8, under the entropy principle, thus, not a principle independent from the entropy principle; the energy principle derives its validity from the principle of universal entropy growth. Without the entropy principle, no economic activity, no value creation, and no life are possible. All these and everything else on Earth are the result of the entropy growth.

The essay will suggest that the reason for the second law being not correctly understood even today is the outcome of the way it was formulated. The law had its genesis in Carnot's 1824 paper while the genesis for the first law was the discovery of the "equivalence principle" by Rumford, Mayer, and Joule over the period from 1798 to 1850. The two competing premises were thought to be competing theories of heat. Both laws were then simultaneously formulated by Kelvin and by Clausius when each realized that the apparent conflict between Carnot and Rumford–Mayer–Joule could be resolved not by determining which theory-of-heat was right and the other wrong but by according the theory of equivalence principle the status of being ontologically correct, while finding in Carnot's treatment (even though the treatment had been based on wrong ontology) an important element that must be incorporated into the theory. That element concerns the impossibility question, "what kinds of machines, e.g., PMM of the second kind, are impossible and what types of thermodynamic processes are 'forbidden' in nature?" That theory is, of course, the mechanical theory of heat (MTH)—the cornerstone of which is the equivalence principle which became the first law. The second law was the formalization of the necessary element into restrictions in the transformation of heat into work.

But, this outcome was not Carnot's take on his principle (see Sects. 4.2, p. 64 and 4.5), which he viewed to be the cornerstone of understanding transformation of heat into work rather than being merely a necessary *restriction* on possible types of heat→work transformations. In other words, the second law as it was formulated in MTH is incomplete and, as a result, the law does not enjoy equal status with the first law and the understanding of heat is correspondingly incoherent.

Ultimately, the goal of this essay is to accord equal status to the first law and the second law in its complete form.

1.2 Thermal Equilibrium and Temperature

Joseph Fourier (1768-1830)

At the beginning of the nineteenth century, there were two competing ontological conceptions of heat: the caloric theory and the mechanical theory. Both Fourier and Carnot, the creators of this science (Fourier treated heat transfer as a workless dissipation, while, for Carnot, heat gradient was the source of dissipation-less work), used caloric theory productively to develop their theses.[1] Only later in the middle of the nineteenth century, the caloric theory was overthrown and superseded by the mechanical theory of heat (Mechanische Wärmetheorie, or MTH) which

[1]Though Carnot was not a member of the French Laplacian School, the center of the caloric theory, and Carnot's theory of heat is not a caloric theory of heat as this essay will argue, he did explicitly use the concept of "caloric" for studying the relation between "caloric" and power. (Though, it is important to note as discussed in Chaps. 4 and 8 that his use of the term was fundamentally different from the Laplacian School.) Fourier, on the other hand, did not, strictly speaking, assume the existence of caloric. In fact, he called his theory *Théorie analytique de la chaleur* stressing the analytical treatment without speculating on the nature of heat. Even though, since he only focused on the study of heat flow his mathematical theory was completely consistent with the central premise of the caloric theory that heat is conserved.

evolved, with further clarification of the second law of "thermodynamics," into modern thermodynamics and statistical mechanics.

At this early point of discourse, we need not to be concerned with the precise nature of heat, only with what we can do with heat. Planck described the goal of his treatise [3] being to offer a uniform viewpoint for the entire field and, for that purpose, his method was "distinct from the other two, in that it does not advance the mechanical theory of heat, but, keeping aloof from definite assumptions as to its nature, starts direct[ly] from a few very general empirical facts, mainly the two fundamental principles of Thermodynamics." I share with Planck the goal in seeking a uniform point of view for the entire field: rather than aiming for a unity of science (i.e., *reductionism*), the goal is for a unity of knowledge (*consilience* as originally defined by Whewell), which is a less lofty goal but one reachable and still deeply satisfactory. By having a good idea of what we can do with heat, we shall have a good understanding of heat, even in absence of a precisely worded definition.

The conception of heat arises from the sensation of warmth or coldness, which is immediately experienced upon touching the surface of a body. This direct sensation, however, furnishes no quantitative scientific measure of a body's state with regard to heat; it yields only qualitative impressions of warmth or coldness, which vary according to external circumstances and subjective perceptions. For quantitative purposes, we utilize the change of volume which takes place in all bodies when heated under constant pressure, for this admits of exact measurement. Heating produces in most substances and under most conditions an increase of volume, and thus we can tell whether a body gets hotter or colder with a quantitative yardstick, a purely mechanical observation affording a much higher degree of precision.

If two bodies, one of which feels colder than the other, are brought together (for example, a good size ice cube and warm water), it is invariably found that the warmer (or hotter) body is cooled to, and the colder one heated up to a certain point, and then change ceases. The two bodies are then said to be in *thermal equilibrium.* Experience shows that such a state of general equilibrium sets in, not only when two, but also when any number of bodies initially at different degrees of warmth (or coldness) are brought into mutual contact. From this follows the proposition, known as the *zeroth law of thermodynamics* (see Fig. 1.1):

> If a body, A, be in thermal equilibrium with another body, B, and with a second different body C, then B and C are in thermal equilibrium with one another.

For, if we bring *A*, *B*, and *C* together so that each touches the other two, then, according to our supposition of thermal equilibrium, there will be equilibrium at the points of contact *AB* and *AC*, and, therefore, also at the contact *BC*. If it were not so, no general thermal equilibrium would be possible, which is contrary to experience.

These facts enable us to compare the degree of heat of two bodies, *B* and *C*, without having to bring them into direct contact but by bringing each body into contact of an arbitrarily selected third body, *A*. This third standard body, for example, can be a column of mercury enclosed in a vessel of capillary tube and bulb. By observing the volume of *A* (height of mercury in the case of capillary tube mercury thermometer) in each case, it is possible to tell whether *B* and *C* are in

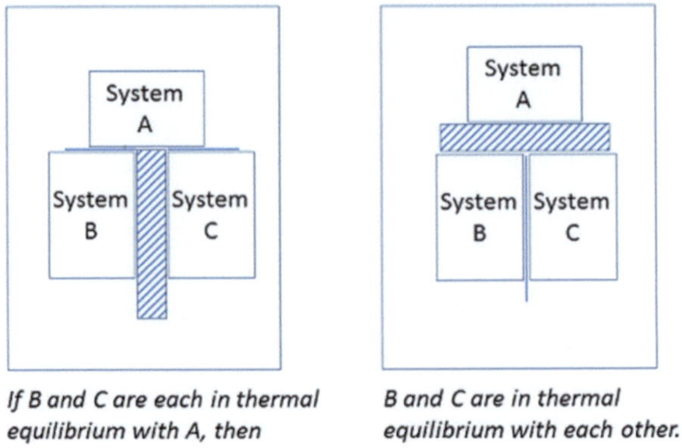

Fig. 1.1 The zeroth law of thermodynamics (cross-shading boxes designate adiabatic walls; heavy lines diathermic walls)

thermal equilibrium and if not, which of the two is the hotter body. The degree of heat of *A*, or of a body in thermal equilibrium with *A*, can thus be defined by either the volume of *A* or volume of *A* in reference to some arbitrarily selected *normal volume*.

One established practice is the selection of normal volumes when *A* is in thermal equilibrium with melting ice under atmospheric pressure and with steam under atmospheric pressure as the two reference points, or *fixed points* (the ice point and the steam point) and the division of the volume difference (or height difference) between the normal volume at the steam point, X_1, and at the ice point, X_2, by 100. Reading based on this choice of units is called *temperature*, *t*, in degrees Centigrade,

$$t(X) = 100 \times \frac{X - X_2}{X_1 - X_2}$$

which ranges from 0 °C at the ice point to 100 °C at the steam point. Temperature is the quantitative measure of the degree of heat. Two bodies of equal temperature are, therefore, in thermal equilibrium and vice versa.

The volume of a thermometric substance is one example of thermometric properties; other substance properties (pressure, electric resistance, thermal EMF, etc.) can also be used as thermometric properties. The temperature readings of *no* two thermometric substances agree in general except at 0 and 100 °C. The definition of temperature given above is, therefore, somewhat arbitrary. This will be rectified in terms of its numerical accuracy in Sect. 1.5. The definition of *thermodynamic* temperature (or absolute temperature) in terms of its theoretical/conceptual meaning will be given in Sect. 4.3.

1.3 Thermodynamic Systems and the General Concept of Equilibrium

A common practice in physics starts with the separation of a finite portion of matter or a restricted region of space from its surroundings. The portion/region that is set aside and on which the investigation is focused is called the *system*, and the totality of everything outside the system that has a direct bearing on its behavior is known as the *surroundings*.

Equilibrium in simple systems In microscopic physics—take the Newtonian mechanics, for example, the state of a system composed of N mass points requires the knowledge of $6N$ time-dependent variables (3 variables for the position and 3 for the momentum for each of the N mass points). Thermodynamics deals with the macroscopic set of the N particles of the corresponding system and, in this case, it is unnecessary to know the motion of each particle individually in order to represent the thermodynamic state, i.e., the macroscopic properties, of the system. When a *simple* system is in a state of thermodynamic *equilibrium*, the average properties of the system can be described in terms of time-independent *thermodynamic coordinates*. Thermodynamic coordinates are also known as state variables or properties (or property functions). Examples of thermodynamic coordinates are volume and mass, pressure and temperature.

Imagine that an experiment is performed on a constant mass of gas in a vessel equipped so that the pressure p, volume V, and temperature t can be measured. (Note that t is temperature in one of the empirical scales; later we shall use T to denote temperature defined by a gas thermometer or at the absolute scale [Sect. 4.3].) The empirical finding of such an experiment is that two of the three variables of a simple system in equilibrium can be set arbitrarily; once the two variables are set, the value of the third variable at equilibrium will be determined. For example, once V and t are chosen, the value of p is determined. That is, the three variables satisfy a functional relationship, $f(p, V, t) = 0$. We shall call the *concrete expression* of a functional relationship of state variables the EQUATION OF STATE.

When a system is not in a state of equilibrium, no single-valued thermodynamic coordinates are defined for the system *as a whole*. The description of the nonequilibrium system is still possible based on the concept of LOCAL THERMODYNAMIC EQUILIBRIUM,

> The assumption that the same equilibrium thermodynamic relations, $f(p, V, t) = 0$, which is originally determined for a whole equilibrium system, for instance, remain valid for the state variables assigned to every elemental volume of the nonequilibrium system.

(Local thermodynamic equilibrium is a central concept in the *modern formalism of thermodynamics*, as it will be presented in Chap. 6, Sect. 6.4.) Nonequilibrium systems are in general time-dependent as a result of the flows of various kinds driven by their corresponding "thermodynamic forces" in the interior of the nonequilibrium systems, e.g., a temperature gradient drives heat flow; a pressure gradient drives fluid motion; a chemical affinity drives chemical reaction (see discussion in Chaps. 7 and 9).

Composite systems Composite systems are one variety of nonequilibrium systems that are time independent. A composite system maintains a *time-independent* disequilibrium existence by its internal constraints. Examples of internal constraints are "semi-permeable walls" (introduced by van't Hoff), "impermeable diathermic walls", "adiabatic partitions", "restricted adiabatic partitions" (which allow transmission of mechanical energy), "chemical reaction barriers". A composite system is made of several simple subsystems, each of which is defined by the corresponding internal constraints. Within each simple subsystem, which is either defined by a physical wall, a movable partition, or by a non-spatially identifiable chemical barrier, a state of "meta-stable" equilibrium prevails.

Once a constraint is removed, the composite system will undergo change toward a meta-stable equilibrium that is defined by the remaining constraints; if all the constraints are removed the resulting change to the system will eventually bring the system into its internal stable equilibrium state. The determination of the internal equilibrium state is considered by Callen to be "the core of thermodynamic theory":

> The single, all-encompassing problem of thermodynamics is the determination of the equilibrium state that eventually results after the removal of internal constraints in a closed, composite system. [4]

There is merit in the identification of thermodynamics as the science of tendency toward equilibrium: the clear focus resulted in treatment as one reader put it, "The best treatment of thermodynamics I have seen." But, with this identification, Callen represents the revised view of physicists with regarding to "what thermodynamics is" by leaving out the engineering heritage of thermodynamics, i.e., *engineering thermodynamics*, which originated with studying the relation between heat and mechanical work. Instead, Callen focuses on the determination of equilibrium state in the narrow sense of *equilibrium thermodynamics* (exception is found in Callen's Chap. 4, in which obligatory treatment of maximum useful work is given; the case can be made, however, that Callen's treatment in [4] represents a paragon of equilibrium thermodynamics, not engineering thermodynamics).

1.3.1 Nonequilibrium and Irreversibility

Whereas the concept of equilibrium is fundamental to the edifice of thermodynamic theory, the theory derives its significance from its applications to systems at equilibrium or near equilibrium as well as to systems *far from equilibrium* [5] or at nonequilibrium states. In the former cases, thermodynamic systems of interest are material objects that EITHER move toward equilibrium states as the natural end states (which are Callen's problem) OR are prevented from reaching equilibrium states due to externally imposed boundary conditions.

The more important applications are the latter cases: applications to systems at nonequilibrium states (nonequilibrium thermodynamics, NET). As will be discussed in Chap. 4 (see Footnote 1) and afterward, there is no irreversibility if the world is at equilibrium. It is also impossible to be away from equilibrium without any manifestation of irreversibility. We study thermodynamics and NET because of the irreversibility and interconnectedness of nature, and the driving forces resulting from irreversibility manifested in the tendency toward equilibrium.

1.4 Dimension and Unit of Temperature

A macroscopic description of a system may be expressed in terms of its *extensive state variables*—mass and volume—and its *intensive state variables*—pressure and temperature. An intensive state variable is the macroscopic manifestation of microscopic processes of individual atoms and molecules. Pressure is the macroscopic manifestation of particle momentum transfer; its dimension is *force* per unit *area* (FL^{-2}, or $ML^{-1}T^{-2}$, with unit N/m^2 or kN/m^2). Temperature is the macroscopic manifestation of particle kinetic energy (the average particle kinetic energy). This might suggest that temperature could also be defined in terms of energy. However, temperature is not only the "quantitative" sum of single particles' energy but also the "qualitative" characterization of the collective distribution of particle energies. Therefore, such a suggestion would have been mistaken.[2]

The general concept of temperature logically follows from the postulate of thermal equilibrium, and following R. H. Fowler, the postulate is known as the *zeroth law of thermodynamics* (Sect. 1.2). Temperature, therefore, is strictly a macroscopic concept. The thermodynamic concept of temperature, in fact, was rigorously developed from the concept of *Carnot's temperature function* by Carnot and Kelvin. Unlike pressure, which is a derived dimension in terms of the three fundamental dimensions of *mechanics*, the dimension of temperature is a new fundamental dimension, which together with the three fundamental dimensions of mechanics defines the four fundamental dimensions of *thermodynamics*. Consequently, as we shall see in the following chapters, heat and thermal energy (which is a function of temperature)—though quantitatively equivalent to—are qualitatively different from work and mechanical energy (both of which are independent of temperature), respectively.

[2]This discussion of temperature follows the conventional approach, which preceded the introduction of entropy. That is, once the dimension of temperature was determined, the dimension of entropy was linked via its definition to the dimension of temperature. Reassessment of the meaning of temperature and entropy in recent years has led to the suggestion that entropy should be dimensionless, and temperature, correspondingly, is then redefined as *tempergy*—which has the dimension of energy. [H. S. Leff (1999) *Am. J. Phys.* ***67***(12):1114–1122].

This is an interesting reinterpretation. A dimensionless reduced entropy and an energy-dimension tempergy, however, do not infer that there is no need for a new dimension for the conjugate pair of tempergy-reduced entropy or temperature-entropy. What is significant is the existence of a new FUNDAMENTAL DIMENSION with the introduction of the conjugate pair.

1.4.1 Universal Constants: Dimensionless Conversion Factors and Dimensional Universal Constants

At this point, it is useful to point out that the formulation of physical theories involves a number of *universal constants*. A partial list includes g_c (the constant in Newton's second law of motion relating force to mass and acceleration), G (gravitational constant), R (universal molar gas constant), and k (Boltzmann constant), $\mathcal{N}_A$ (Avogadro number), h (Planck constant), and J (the constant relating heat to mechanical energy). Note that the Boltzmann constant is $k = 1.38054 \times 10^{-23} \frac{joule}{K-molecule}$ and the Avogadro number is $\mathcal{N}_A = 6.02252 \times 10^{26}$molecules/kmol. The universal gas constant R is related to the Boltzmann constant k: $R = \mathcal{N}_A k = 8314.47$ j/kmol K.

Progress in physics has been accompanied by increasing precision in the determination of these constants as well as the evolving understanding of the theoretical nature of universal constants, which is open to two different interpretations (see paragraphs below). The choice made on each universal constant is, in some cases, viewed to be a matter of whether a particular choice is capable of yielding useful applications, such as *dimensional analysis* in engineering. However, instead of suggesting possible arbitrariness in the choice, a choice that yields useful practical applications reflects a correct decipher of Nature.

Some of the universal constants such as g_c and J underwent a transformation from erstwhile dimensional universal constants to become *dimensionless* conversion factors: for instance, g_c became Newton's conversion factor relating *force* to *mass*; J became Joule's constant relating thermal energy to mechanical energy. When this happens, a new understanding of the constants emerged. For instance, with an arbitrary set of units, the Newton second law is written

$$F = \frac{1}{g_c} m \frac{d^2x}{dt^2}$$

With the choice of newton, kg, meter, and second as the units of F, m, x, and t, respectively,

$$g_c = 1 \frac{\text{kg} - \text{meter}}{\text{newton} - \text{sec}^2}$$

and the Newton second law assumes the form

$$F(\text{newton}) = \text{m(kg)} \frac{d^2x(\text{meter})}{dt(\text{sec})^2}$$

By becoming *unity*, g_c disappears from the physical formula. Similarly, J disappears from the physical formula with the adoption of the same mechanical energy unit of "joule" for the unit of heat and thermal energy (see Sect. 3.3 in Chap. 3).

Although the quantity of heat is determined by the same measure of that of mechanical energy in Chap. 3, heat (more on that in Sect. 5.6) as a physical concept remains distinctive from mechanical energy. Its full characterization requires a new variable, temperature. Had the Boltzmann constant undergone the same transformation of g_c and J, temperature too *would have assumed* the unit of energy (see Footnote 2) and the distinction in the characterization of heat and thermal energy versus work and mechanical energy would have disappeared completely—quantitatively as well as qualitatively. This emphatically is not the case, as we shall discuss in detail in Chaps. 4–8. The Boltzmann constant and the universal gas constant are considered to be *dimensional* universal constants. Thus, temperature is a new fundamental dimension, Θ[K], which is added to the MKS system of length L [m], mass M [kg], and time T [s] to form the SI system.

To recapitulate: We can identify universal constants into two groups: (1) dimensionless conversion factors (g_c and J), and (2) dimensional universal constants (k and R). The fundamental dimension set for the *science of motion and heat* is the set of length L, mass M, time T, and temperature Θ (the corresponding SI units are meter, kg, s, K). The complete SI system also includes the units of ampere and candela.

The history of temperature both in terms of the formation of the concept (see Sects. 4.3 and 4.4) and in terms of the evolution of its measurement and the standardization of temperature scales (based on the proposal by W. F. Giauque [6]:18,347]) is a rich one. The use of *gas thermometer* (see Sect. 1.5) and the formulation of the second law (see Chaps. 4 and 5) played crucial roles in establishing our current state of the art.

1.5 Thermal Equation of State for Ideal Gases

The temperature scales in use before 1954 required for their specification two fixed points, the ice point and the steam point. Hundreds of attempts were made to measure temperature t with high precision without much success. In 1939, W. F. Giauque proposed the use of a single fixed point, the triple point of water, set arbitrarily at 273.16 K (corresponding to the ice point of 273.15 K), as the thermometric reference point, i.e., temperature t' is in terms of

$$t' = t'(X_3)\frac{X}{X_3} = 273.16 \cdot \frac{X}{X_3}$$

where X is a thermodynamic property. At the 10th Conference on Weights and Measures in Paris during the summer of 1954, the Giauque proposal was passed.

Consider the property of ideal gases as the possible choice of thermometric property X. This determination is possible as a result of the way in which the products pv of gases depend on p. Zemansky ([6]:14, 111–116]) wrote

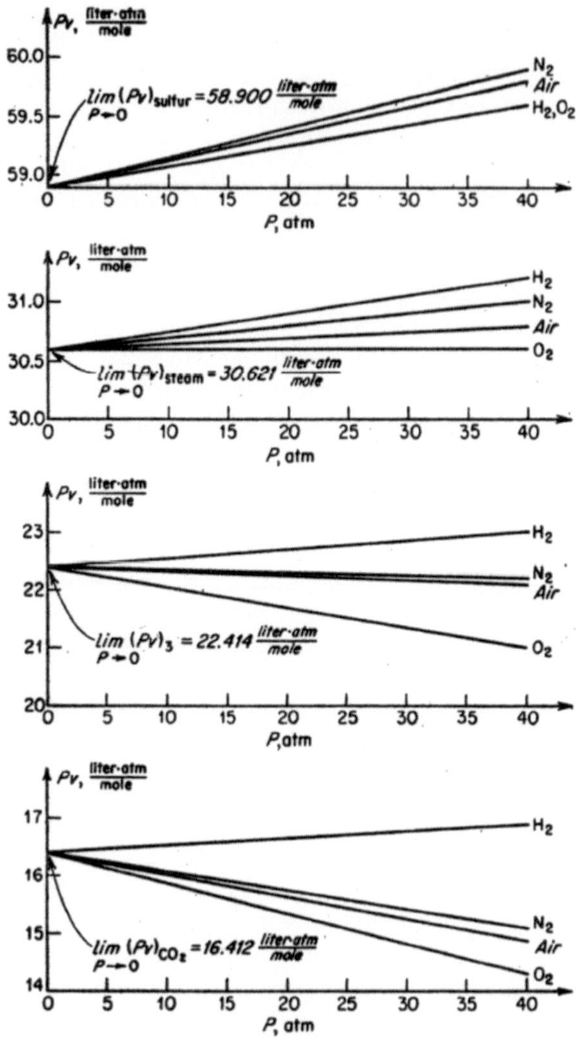

Fig. 1.2 The fundamental property of gases is that the product of pv in the limit of p approaching zero is independent of the nature of the gas and depends only on temperature

The remarkable property of gases that makes them so valuable in thermometry is displayed in Fig. 6.2 [reproduced here as Fig. 1.2] where the product of *pv* is plotted against *p* for four different gases, all at the temperature of boiling sulfur in the top graph, all at the temperature of boiling water in the one beneath, all at the triple point of water in the next lower graph, and all at the temperature of solid CO_2 in the lowest. In each case it is seen that, as the pressure approaches zero, the product of *pv* approaches the same value for all [four] gases at the same temperature.

That is, the pv readings of the gas thermometer at a given temperature, but filled with different thermometric gases in the limit of very low pressure, $p \to 0$, are found to be independent of the nature of the specific gases

$$\lim_{p\to 0}(pv)_{gas\_A} = \lim_{p\to 0}(pv)_{gas\_B} = \ldots = CONSTANT$$

or

$$\lim_{p\to 0}(pv)_{ideal\,gases} = A \tag{1}$$

where the constant $A(t')$ depends only on temperature t', independent of the nature of specific gases. If the choice of t' to be proportional to A is made, then one finds, for a thermometric substance of low-pressure ideal gas, the reading of temperature t' to be in terms of

$$t'(p) = 273.16 \cdot \frac{p}{p_3} \quad (constant\,V) \tag{2a}$$

$$t'(V) = 273.16 \cdot \frac{V}{V_3} \quad (constant\,p) \tag{2b}$$

In a similar manner, a choice can be made by convention that $A = RT$, i.e.,

$$\lim_{p\to 0} pv = RT$$

or,

$$pv = RT \tag{3}$$

Correspondingly,

$$pV = NRT \tag{4}$$

where N is the mole number, v is the molar specific volume, $v = \frac{V}{N}$. Note that $N = \frac{m}{M}$, where m is mass of an ideal gas and M is its molecular weight. Let R_i ($\equiv R/M_i$) be the gas constant for a specific gas, Eq. (4) becomes

$$pV\left(= \frac{m}{M_i} RT\right) = mR_iT \tag{5}$$

We shall see later (Sects. 4.3 and 4.5) that it is possible to define the same scale of temperature T by the general second law consideration. The advantage of that approach will be that the definition of the temperature is independent of not only the nature of specific gases but also the specific nature of the *gas* substance.

Because of the simple form of the equation of state of ideal gases, Eqs. (3)–(5), ideal gases will be broadly used in the following both as working fluids in engineering analyses and in *thought experiments* for theoretical demonstration/inference.

1.6 Mixtures of Ideal Gases

When two or more ideal gases are mixed, the behavior of a molecule normally is not influenced by the presence of other similar or dissimilar molecules and, therefore, a (non-reacting) mixture of ideal gases also behaves as an ideal gas. Air, a mixture of oxygen, nitrogen, and a trace of other gases, for example, is conveniently treated as an ideal gas. The prediction of the *p-V-T* of the gaseous mixture is based on two models: *Dalton's law* and *Amagat's law.*

Consider the following definitions. The composition of a mixture is specified by the number of moles of each of its components, N_i, or by the mass of each component, m_i. Let the mixture be composed of r components and the mole number N and mass m of the mixture be the sum of the respective values of their individual components

$$N = \sum_{i=1}^{n} N_i \quad \text{and} \quad m = \sum_{i=1}^{n} m_i$$

Note again $N_i = m_i/M_i$. Introducing mole fraction x_i and mass fraction y_i

$$x_i \equiv \frac{N_i}{N} \quad \text{and} \quad y_i \equiv \frac{m_i}{m}$$

the apparent molecular weight of the mixture is

$$M_{mixture} = \frac{m}{N} = \frac{\sum_i m_i}{N} = \frac{\sum_i N_i M_i}{N} = \sum_i x_i M_i \tag{6}$$

where M_i is the molecular weight of ith component. The apparent specific gas constant of the mixture is then $R_{mixture} = R/M_{mixture}$.

The *p-V-T* behavior of a mixture of ideal gases (which [the mixture] also behaves as an ideal gas) is based on Dalton's law

> Dalton's law of additive pressures: *The pressure of an ideal gas mixture is equal to the sum of the pressures* [known as partial pressures, p_i] each gas would exert if it existed alone at the mixture temperature *and volume.*

$$p_{mixture}(T, V) = \sum_i p_i(T, V) \tag{7}$$

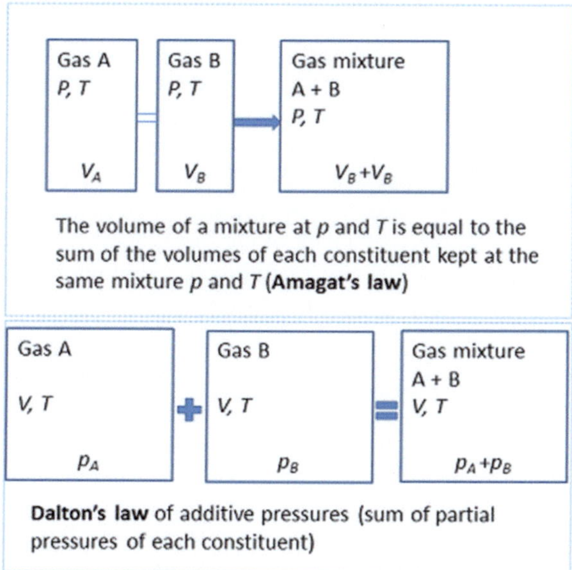

Fig. 1.3 Amagat's law and Dalton's law

The *p*-*V*-*T* behavior of a mixture of ideal gases can also be based on Amagat's law

> Amagat's law of additive volumes: The volume of an ideal gas mixture is equal to the sum of the volumes each gas would occupy if it existed alone at the mixture temperature and pressure (Fig. 1.3)

Since both the mixture and each of its component obey the ideal gas equation of state

$$pV = NRT$$

$$p_iV = N_iRT$$

where p_i is the partial pressure of the *i*th component. It follows

$$\frac{p_i}{p} = \frac{N_i}{N} = x_i \tag{8}$$

1.7 Work

If a system undergoes a displacement or a volume change under the action of a force, the amount of work is equal to the product of *the force* and *the component of the displacement parallel to the force*. For instance, we consider a gas enclosed in a cylinder having a movable piston of area S (see Fig. 1.4). If p is the pressure of the body against the walls and the piston of the cylinder, then pS is the force exerted by the body on the piston, which, as a result of the force, is moved from the solid-line position to the dotted-line position corresponding to an infinitesimal distance dz (dz denotes, in the general case, the movement of a point on a surface element dS). An infinitesimal amount of work

$$\delta W = pSdz$$

is performed by the system. But Sdz is equal to the increase, dV, in the volume of the system. Thus, we may write:

$$\delta W = pdV = p\int_S dS \cdot dn \tag{9}$$

where dn is the unit vector normal to dS. Equation (9) is known as the "quasi-static work" ([4]: p.19).

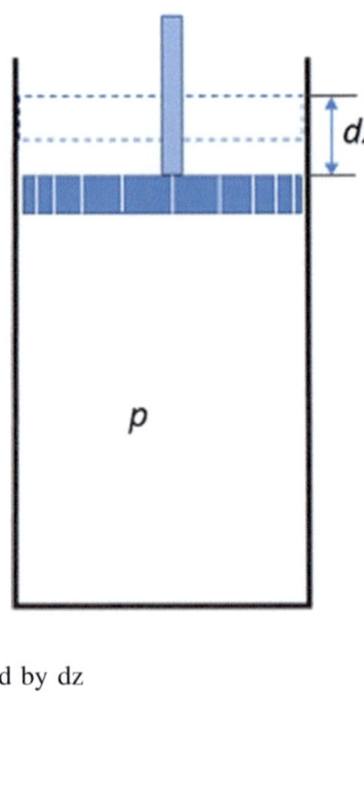

Fig. 1.4 Piston moves upward by dz

For a finite transformation, the work done by the system is obtained by integrating δW

$$W = \int \delta W = \int_1^2 pdV = \int_1^2 p \int_S dS \cdot dn$$

where the integral is taken over the whole process from the initial state '1' to the final state '2'.

The above example of work, the work done by the system as a *whole* on its "surroundings" (or the corresponding work on the system by its surroundings), is an example of *external work*. Implicit in Fig. 1.4, a work *reservoir* is assumed and the piston is connected through a piston rod and additional transmission means to the work reservoir. A system external work is made up of, in the case of expansion external work, (1) *expansion* work against a surrounding thermal-and-pressure reservoir and (2) *useful work* which is transmitted and *stored* in a work reservoir. Opposite kind of useful work exchange, work *extracted* from a work reservoir by the system, is also possible.

Examples of useful work exchange between a system and a mechanical work reservoir are the rising or lowering of a suspended weight; the winding or unwinding of a spring; the input or the output of power of a flywheel. (In the case of electrical work, the charging or discharging of a battery is an example.)

1.8 Calculation of $\int pdV$ for "Quasi-static Processes"

Work (as well as heat in Chaps. 2 and 3) is not a state variable. Change in a system's state variable is determined by the initial and final states of the change independent of the path. In contrast, work done by a system depends not only on the initial state and the final states but also on the intermediate states, i.e., on the path connecting the initial state and the final states.

A simple example of work process is the so-called "quasi-static process", defined as a process that the path of which can be determined. There are, however, two different ways of determining the path: One way is, as shown in the figure (Fig. 1.5), the result of balancing forces acting on a moving piston.

Another way of determining the path of a quasi-static process is through the removal of closely spaced mechanical constraints as shown in Fig. 1.6. We shall make a preliminary comment on the two processes in Sect. 1.9 and more detailed discussion on them in Chap. 6. At this point, it suffices to point out that they are fundamentally different kinds of quasi-static processes and only the first one (Fig. 1.5) may be called an *internally reversible, quasi-static process* (defined in Chap. 6), an example of such kind is given here.

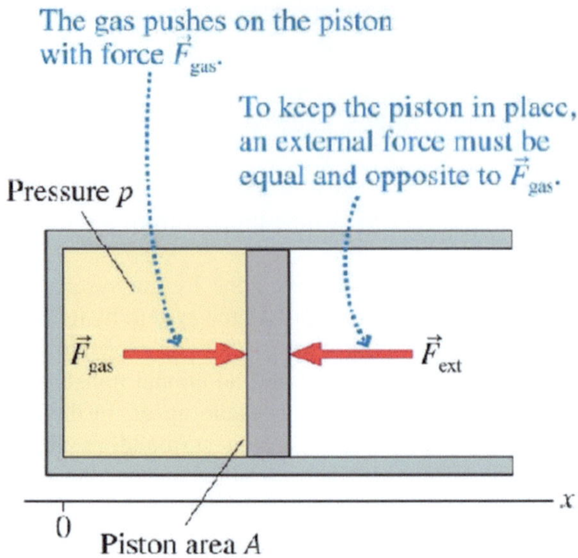

Fig. 1.5 An example of quasi-static process: it can be an expansion process if $\vec{F}_{gas}$ (= pA_{piston}) is slightly greater than $\vec{F}_{ext}$ with piston moving right, or a compression process in the opposite case. Note that the figure implies the existence of a mechanical arrangement for maintaining the required $\vec{F}_{ext}$—including the existence of a work reservoir (see discussion in Sect. 1.9)

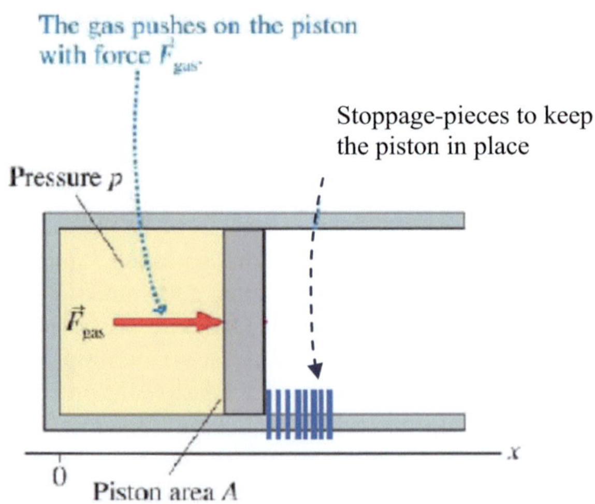

Fig. 1.6 Another example of a quasi-static process, resulting from the removal of closely spaced constraints: As it will be discussed in Chap. 6, with the proper arrangement, such kind of quasi-static process can be made to follow the same path as the above one in Fig. 1.5. In other words, not only work is path dependent but also two work processes along exactly the same path (as shown in Fig. 1.5 and here) can be different from each other

Consider the example of an internally reversible, quasi-static isothermal expansion or compression of an ideal gas. The "quasi-static" work is

$$W = \int_{V_i}^{V_f} pdV$$

The pressure p of an ideal gas is related to its volume V through the equation of state

$$pV = NRT (= \text{constant})$$

Substituting p by using the equation of state, we get

$$W = \int_{V_i}^{V_f} \frac{NRT}{V} dV = NRT \int_{V_i}^{V_f} \frac{dV}{V} = NRT ln \frac{V_f}{V_i}$$

If there is 0.1 kmol of gas kept at a constant temperature of 0°C or 273.15 K and the gas is compressed from a volume of 4 m^3 to 1 m^3, then

$$\begin{aligned} W &= 0.1 \cdot 8.3143 \cdot 273.15 \cdot \ln \frac{1}{4} \text{kJ} \\ &= -314.83\, kJ \end{aligned}$$

The minus sign indicates that work is done on the gas.

1.9 Difference Between a Mass Body and a Thermodynamic System

The nominal object of our study is a thermodynamic system. How does a thermodynamic system differ from a mass body? The obvious answer is that a mass body is subject to force interaction or work, while a thermodynamic system is subject to both work interaction and heat interaction. Though both subject to work interaction, work interaction involved in a thermodynamic system is fundamentally different from that in a mass body. A generally accepted thermodynamic work concept is quasi-static work (see Figs. 1.5 and 1.6, and Chap. 6 for definition). This work concept originated from dynamics, the study of mass bodies.

1.9.1 Quasi-static Process and Work Reservoir

Work in mass body dynamics is a straightforward concept: work done on a mass body equals mass body mechanical energy gain; work performed by a system is equal to the system's energy loss (see Sect. 3.4). Things become complicated in thermodynamics and it may thus be constructive to make a brief note of preparatory purpose here on two particular points before we divulge into details in Chap. 6.

The first point regards the "quasi-static" heat Q and work W formula,

$$\delta Q = TdS$$

(see Eq. [83] in Chapter 6)
and

$$\delta W = pdV$$

(see Eq. [81] in Chapter 6)
(S denotes the entropy, a state variable that will be introduced in Chap. 5). The usual assumption that they are applicable under the condition of quasi-staticity or quasi-equilibrium is mistaken—the argument will be laid out in Chap. 6. This is the first point of misunderstanding.

The second point concerns the role of a work reservoir in the consideration of the work of a thermodynamic system. Consider a case of expansion external work. System external work is made up of (1) expansion work against a surrounding thermal-and-pressure reservoir and (2) *useful* work, which is transmitted and stored in a work reservoir. Both reservoirs are necessary parts of the whole picture—explicitly or implicitly (as in the case of Figs. 1.5 and 1.6)—in the consideration of system work. Consider for instance the classical *Joule free expansion* [Sects. 3.8 and 6.3]: as it will be discussed below in Sect. 6.3 and Problems 6.2 and 6.3, the system work of Joule free expansion being zero is because reservoirs of both kinds are explicitly absent. (So is in Fig. 1.6, in which the work reservoir is explicitly absent.) The critical role of reservoirs is a central point in the irreversible universe, which will be fully discussed in Chap. 6:

> Comprehension of a system in the irreversible universe necessitates the consideration of the system and its reservoir(s) as an interconnected whole.

That is, though the object of our study is nominally a system, it is inadequate to study the system in itself. The thermodynamic literature, unfortunately, continues the tradition of focusing solely on systems, their interaction with work reservoirs is rarely discussed in explicit terms. This is the second mistake.

The two mistakes are at the bottom inseparable. The underlying root of both is the presupposition of *mechanistic* scientific knowledge (see below).

1.9.2 A Mass Body and a Thermodynamic System: No Thermodynamic System is an Island

The doctrine of *mechanism* presupposes that change in a particle is determined by local force-driven interactions alone, independent of everything else there is in the world. This presupposition was overthrown in the twentieth century by quantum mechanics, which discovered that nature can be described completely only as an *entangled* whole. While the mechanistic presupposition remains applicable to the study of a macroscopic mass body (as the classical limit of quantum system, see Fig. 1.7a), its validity in all macroscopic systems should have been called into question (even before quantum mechanics) in the science of heat: a macroscopic thermodynamic system can only be *fully* comprehended as a part of the interconnected world in terms of work reservoir and how it interacts with reservoirs and the rest of its surroundings (see Fig. 1.7b).

Throughout the course of this disquisition, the full implications of the difference between a mass body and a thermodynamic system will be developed. The thread of the discussion is irreversibility (or spontaneity) and interconnectedness—manifested in the phenomena of *heat* (see Sect. 3.5, *heat* is used as a broader term of heat: while the definition of heat will be given, no attempt will be made in defining the broader term *heat* precisely). By treating thermodynamics as more than a mere theory of energy understood mechanically, I make the case for identifying *heat* phenomena not just in terms of physical forces or energy (in the classical realm of Newton), but of a new "driving force" (in the statistical realm of Maxwell, Boltzmann, and Gibbs). Just as the dynamical forces can be quantitatively determined, a quantitative measure of the new driving force, the entropic "forces" or *entropy growth potential* will be given in Sects. 8.3–8.5.

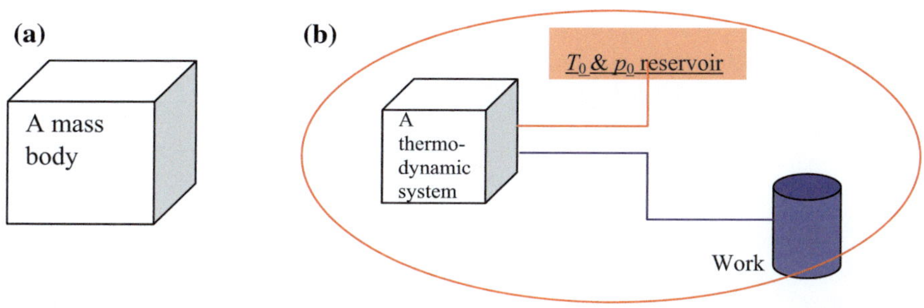

Fig. 1.7 Shown on left is Fig. 1,7a, A mass body which can be studied in itself in terms of local forces. Shown on right is Fig. 1.7b, A thermodynamic system which has to be considered as a part of its surrounding T_0 and p_0 reservoir and in interaction with, or **in absence of**, a work reservoir. Figures 1.7b and 6.6 (see below in Chap. 6) will be the standard schematic when we consider a system throughout the book with the surrounding reservoir and work reservoir explicitly or tacitly present, or explicitly absent

1.10 Quantity of Heat

Temperature is the quantitative measure of the *degree* of heat, not the *quantity* of heat itself. The traditional method of the determination of the quantity of heat is called calorimetry, which is treated in Chap. 2.

According to the *mechanical equivalent of heat* (MEH, to be introduced in Chap. 3), any change in heat is quantitatively linked to a corresponding change in mechanical energy. The method for the determination of the quantity of heat by connecting it to the mechanical energy is treated in Chap. 3. Correspondingly, heat has been considered to be a form of energy—i.e., heat is defined as

Definition of heat : heat is energy in transition; or, heat is a form of energy that moves from a hotter object to a colder one.

This "energetic" definition of heat offers too narrow an understanding of heat, one associated with the widely accepted but erroneous interpretation of MEH as universal interconvertibility (see Sect. 8.6.2), which gave rise to the notion of "consumption of heat" as the cause of producing work.

The true meaning of MEH will be critically studied in Chaps. 3–8, in which a case will be made that the interpretation of MEH that is associated with this definition of heat amounts to accord the first law of thermodynamics preeminence over the second law as it is presently formulated (as it will be discussed in Chap. 8). Without a complete second law, it is not possible to comprehend heat adequately. Correspondingly, this disquisition takes the position that the aforementioned definition of heat merely captures one aspect of heat, denoted as Q (see Chaps. 2 and 3), and proposes the definition of heat (as it was recorded at the beginning of this chapter) in Sect. 5.6, with further elaboration for the meaning of heat when a more precise interpretation of MEH is given in Chap. 8.

Problems

1.1 Calculate the work (in unit J) performed by a body expanding* from an initial volume of 3 L (1 L = 10^{-3} m^3) to a final volume of 4 L at the pressure of 2.5 atm (1 atm = 101, 325 Pa).

253.3 J

1.2 Calculate the pressure of 0.03 kilograms of hydrogen (hydrogen molar mass $M_{hydrogen}$ = 2.016 kg/kmole) inside a container of 1 m^3 at the temperature of 18 °C.

36.02 kPa

1.3 Calculate the density and specific volume of nitrogen ($M_{nitrogen}$ = 28.013) at the temperature of 0 °C and the pressure of 1 atm.

1.25 kg/m^3 (0.0446 kmol/m^3; 22.4 m^3/kmole).

1.4 Calculate the work performed by 0.01 kg of oxygen ($M_{oxygen} = 31.999$) expanding isothermally* at 20 °C from 1 to 0.3 atm of pressure.

0.917 kJ

1.5 Everything we experience is at disequilibrium (the universe is expanding and we're on a warm planet in cold space but exposed to the sun's hot rays). How come, or why, we develop a physics based on equilibrium states?

*All these processes are internally reversible and quasi-static (see Chap. 6).

References

1. Henderson LJ (1917) The Order of Nature. Harvard University Press, Cambridge
2. Schumacher EF (1973) Small is Beautiful. Harper and Row, New York (p. 47)
3. Planck M (1969) Treatise on Thermodynamics, 3rd edition. Dover, New York
4. Callen HB (1st edition, 1960; 2nd edition, 1985) Thermodynamics and an Introduction to Thermostatistics. Wiley, New York
5. Kondepudi D, Prigogine I (1998) Modern Thermodynamics: From Heat Engines to Dissipative Structures. Wiley, New York
6. Zemansky MK (1943) Heat and Thermodynamics, 2nd edition. McGraw-Hill, New York

Calorimetry and the Caloric Theory of Heat, the Measurement of Heat

2

Abstract
Joseph Black distinguished between the quantity of heat in a body and its intensity, or temperature, which was introduced in Chap. 1. This chapter introduces calorimetry, the measurement of the quantity of heat, and the contemporary theory of heat, the caloric theory of heat. Development of calorimetry and the caloric theory was the beginning of the quantitative study of heat, and the important stepping stone toward the mechanical theory of heat, the idea that heat is a form of energy.

Keywords
Calorimetry · Caloric theory of heat · Sensible heat · Latent heat · Calorie · Heat capacity at constant volume · Heat capacity at constant pressure · Adiabatic heating

2.1 Theories of Heat

"Joseph Black died quietly in his chair on November 26, 1799, with a cup of tea balanced in his lap. Five years earlier, Antoine Lavoisier's head had been publicly removed by the French Revolutionary Tribunal. These two men defined our understanding of heat phenomena on the eve of the 19th century…" noted the mechanical engineer Lienhard [1].

In 1789, Lavoisier published a treatise on chemistry which, among other things, recognized that the newly discovered chemical element of oxygen was involved in combustion processes rather than the hypothetical *phlogiston* (one issue of the phlogiston theory was phlogiston's weight: the necessity of entertaining phlogiston's negative weight). And, proposed instead the hypothesis of heat (the *caloric theory of heat*)—an invisible, weightless, tasteless, odorless, fluid-like

L.-S. Wang, *A Treatise of Heat and Energy*, Mechanical Engineering Series,
https://doi.org/10.1007/978-3-030-05746-6_2

substance/material that is distinctive from matter substances. It became known as *calorific fluid* or *caloric*. The caloric theory considered "heat energy" and "heat flow" synonymous terms, i.e., caloric. A heat flow process was simply the flow of the caloric fluid, which was measured by its own unit, the *calorie*—one calorie is defined as that amount of caloric fluid when it is absorbed by one gram of water at constant atmospheric pressure leads to a temperature rise from 14.5 to 15.5 °C. When caloric flows into a body, it results in a gain in the body of the same caloric.

It is useful to view caloric theory of heat in terms of what is heat, the process of heat flow, and how is it measured: the materiality of heat hypothesizes what heat is, the Fourier law and the Clausius Statement (see Sect. 4.1) describe/observe how heat flows naturally from hot to cold, and calorimetry provides an operational definition of heat, thus, how heat is measured quantitatively. It turns out that, of the three ways of understanding heat in accordance with the caloric theory, only the notion of heat as "material-like" caloric was an illusion and, in any case, a notion that we can do without, while the latter two are still valid today: we still study how heat flows down the temperature gradients and we still measure the heat value of foods in terms of their calorie. This chapter provides a summary of the operational definition of heat or heating according to the caloric theory of heat.

A rival hypothesis, the mechanical theory of heat (or, the dynamical theory of heat)—heat is vis viva, the result of the internal random motions of the molecules of bodies—was less developed than the caloric theory at that time. But, it could also explain the transmission of heat and the approach to thermal equilibrium between unequally heated bodies put in contact. Proponents of the caloric theory, such as Lavoisier and Laplace, considered the dynamical theory as a serious but less successful competitor of the caloric theory. In fact, they took a suspension-of-judgment position with regards to the ontological nature of heat as shown in this statement from *Memoir on* Heat [2].

> We shall not decide between these two hypotheses. Several phenomena seem favorable to the latter [the dynamical theory hypothesis], such as, for example, the heat produced by the friction of two solid bodies. But there are other phenomena that are explained more simply by the first hypothesis [the caloric theory]. Perhaps both occur simultaneously. Whatever may be the case, since only these two hypotheses about the nature of heat can be devised, we should accept the principles common to both…*in a simple mixture of bodies the quantity of free heat remains always the same.* This is evident, if heat is a fluid that tends to reach equilibrium and also if it is only the *vis viva* which results from the internal motion of matter, where the principle in question follows from the conservation of *vires vivae.* The conservation of free heat, in the simple mixture of bodies, is thus independent of any hypothesis about the nature of heat.

It is certainly true that, in the simple problems of thermal mixing of bodies, heat is conserved since the totality of all interacting bodies is a thermally isolated system (see Sect. 2.2). For such problems, the conservation of heat is simply a special case of the conservation of energy.

The more intriguing point is that caloricists such as Lavoisier, Laplace, Poisson, and Clapeyron were able to advance the science of heat by making use of the caloric theory formalism to problems that are not thermally isolated, either locally or globally without being straitjacketed to the (incorrect) caloric theory doctrine that heat is a material substance. This point was made nicely by the philosopher Psillos [3]. One example of locally non-isolated thermal system is the problem of adiabatic compression/expansion associated with sound propagation in gases as developed by Laplace and Poisson, which will be considered in Sect. 2.4; Sect. 6.6 will provide a reason why the Laplace–Poisson solution works. The example of a globally non-isolated thermal system is the problem of transformation of heat into work, which is the central problem of the book, and the "caloric formalism" approach to the problem taken by Carnot–Clapeyron is covered in Chap. 4; in Chap. 8 an interpretation will be given as to what Carnot's "caloric formalism" really means.

2.2 Direct Heating: Sensible Heat and Latent Heat

Before Lavoisier and Laplace formulated the caloric theory of heat, Black had made the first advance in the quantitative understanding of heat. He was the first to recognize the importance of making a distinction between temperature and *quantity of heat*, between what we describe as the *intensive* and *extensive* measures of heat—and he together with Berthelot, Lavoisier, and Laplace were the founders of the science of calorimetry. At the time, it had been known how to read temperature directly from the thermometer. But, how could one measure the quantity of heat?

In the earlier experiments of Black on the latent heat (see below), which required melting ice and boiling water, the heat was applied by means of a flame. And, as the rate of heat produced by flame was assumed to be uniform, the quantities of heat supplied were inferred to be proportional to the time during which the supply continued. A method of this kind was obviously very imprecise, and in order to make it at all accurate, it would need numerous precautions and auxiliary investigations with respect to the laws of the production of heat by the flame and its heat transmission to the body which was heated.

Instead of relying on the thermal interaction of flame and a body, Black came to the conclusion that a more dependable research method relies on the interaction of two thermal bodies. Since when heat is applied to or removed from a body it produces changes of various kinds—raising its temperature, altering its volume or its pressure, and in certain cases it changes the state of the body from solid to liquid or from liquid to gaseous—reading of such changes provides means for inferring the quantities of heat.

$$Q_{net} = Q_1 + Q_2 = 0$$

The principle of calorimetry is based on combining this idea with the supposition, the conservation of heat (Fig. 2.1): if two bodies of different temperature are in thermal contact and if no external heat exchange is allowed to go out or enter into the system of two bodies and if no chemical reactions take place in between the two bodies then heat lost by the hotter body will be equal to the heat gained by the

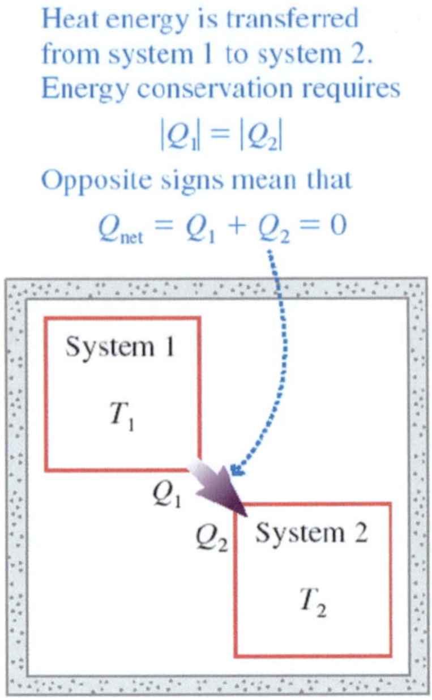

Fig. 2.1 When two or more bodies interact thermally, they come to a common final temperature determined by conservation of heat

colder body, i.e., heat lost = heat gained. Furthermore, body *1* is chosen as a *normal* substance (e.g., water) of certain mass and that *1* is brought in contact with body *2* and body *3* in two separate events, Event *A* and Event *B*. One can surmise again that in both events heat lost = heat gained (for sake of discussion assume *1* gains heat and correspondingly both body *2* and body *3* lose heat [*Q* in reversed direction as shown in Fig. 2.1]). It follows

$$(Q_1)^A + (Q_2)^A = (Q_1)^A - |(Q_2)^A| = 0$$

$$(Q_1)^B + (Q_3)^B = (Q_1)^B - |(Q_3)^B| = 0$$

Then, the question of whether *2* or *3* loses more heat can be determined by the changes to *1* (e.g., temperature changes to *1*). For instance, a greater temperature change in the former case, i.e., Event *A*

$$(Q_1)^A = (C_p)_1(\Delta T_1)^A > (Q_1)^B = (C_p)_1(\Delta T_1)^B$$

will correspond to greater heat loss in body *2*.

Alternatively, one can adjust the masses of substance *1*—designated as $(m^A)_1$ and $(m^B)_1$ for event *A* and event *B*, respectively, so that temperature changes to substance *1* (the normal substance), for instance, for both events are kept the same, and thereby the greater mass of *1* in Event *A* will correspond to greater heat loss in *2*

$$(Q_1)^A = \left(m^A c_p\right)_1 \Delta T_1 > (Q_1)^B = \left(m^B c_p\right)_1 \Delta T_1$$

The temperature change, or the mass of the normal substance, therefore serves to be the quantitative measure of heat by convention. It is customary to take the quantity that, when it is added to one gram of water, raises the temperature of water from 0 to 1 °C as the definition of heat unit calorie or "zero" *calorie*. This was thought to be almost equal to the quantity of heat which will raise 1 g of water 1 °C at any temperature. The refinement of calorimetric measurement has since made it necessary to take account of the initial temperature of the water, and a common initial temperature reference is 14.5 °C, i.e., a **calorie** is defined

$$1[calorie] = 1[gm] \times \left(c_p\right)_{water} \times (15.5 - 14.5)[^\circ C] \tag{10}$$

where $\left(c_p\right)_{water} = 1[calories/\mathrm{gm} \cdot \mathrm{C}]$. That is, the selection of water as the normal substance of calorimeter is equivalent to the specific heat of water being unity at 15 °C.

Consider a change in a body that corresponds to a heat loss of a given amount Q. The calorimetric definition of Q is defined as follows: The body is brought into contact with a calorimeter initially at 14.5 °C while the body undergoes the same change. The water mass of the calorimeter, m_{water}, is adjusted so that heat exchange during body-calorimeter thermal contact causes temperature rise in calorimeter water from 14.5 to 15.5 °C. Correspondingly,

$$Q[calorie] = -m_{water}[\mathrm{gm}]$$

in which the negative sign corresponds to the fact that the body loses heat. Alternatively, if the specific heat of water is taken to be approximately independent of temperature, thereby the mass of water in the calorimeter can be arbitrary allowing the water temperature rise ΔT_{water} to be unrestricted, the quantity of heat can be determined as

$$Q[calorie] = -m_{water}[gm] \times 1[calories/gm \cdot C] \times \Delta T_{water}[C]$$

That is, heat loss is equal to the mass of water in the calorimeter times the change in the water temperature.

The aforementioned thermal event involves temperature change in water. Now, if the temperature of the body also changes as the result of thermal contact, both the heat gained by water and heat lost by the body are called *sensible heat*—which is *heat* exchanged (gained or lost) by a body that changes the body temperature under specified conditions such as constant pressure or constant volume. Let the temperature change of the body be ΔT_{Body}. The ratio of the quantity of heat Q released (or absorbed) by the body to the corresponding decrease (or increase) in the body temperature ΔT_{Body} is called the mean *heat capacity* of the body:

$$C = \frac{Q}{\Delta T_{Body}} \tag{11}$$

Which may be written as

$$Q[calorie] = C\Delta T_{Body} = m_{Body}[gm] \times c_m[calories/gm \cdot C] \times \Delta T_{Body}[C] \tag{11A}$$

where c_m is the mass-based *specific heat* of the body. Correspondingly, the molar specific heat c is given as

$$Q[calorie] = N_{Body}[gmol] \times c[calories/gmol \cdot C] \times \Delta T_{Body}[C] \tag{11B}$$

At specific temperature and specific pressure, the thermal event of direct heating does not produce a change of temperature in the substance and such an event is called *phase change*. For instance, during the evaporation of water, the temperature of the liquid–water/vapor system is constant and will rise only after *all* liquid in the system has evaporated. In 1761, Black deduced that the application of heat to ice at its melting point does not cause a rise in temperature of the ice/water mixture, but rather an increase in the amount of water in the mixture. Additionally, Black observed that the application of heat to water at its boiling point does not result in a rise in temperature of a water/steam mixture, but rather an increase in the amount of steam (vapor). From these observations, he concluded that the heat applied must have combined with the ice particles and boiling water and become latent—the heat in both cases is called *latent heat*. The introduction of heat capacity and the theory of latent heat marked the beginning of the science of heat. The following table shows the latent heats and the phase transition temperatures of some common substances (Table 2.1).

Note the high latent heat of the vaporization of water. The theory ultimately proved important not only in the development of abstract science but in the development of the steam engine. The latent heat of water is large compared with many other liquids, so giving impetus to James Watt's attempts to improve the efficiency of the steam engine invented by Thomas Newcomen. Black and Watt became friends after meeting around 1757 while both were at Glasgow. Black provided significant financing and other support for Watt's early research in steam power.

Example. MIXING

A calorimeter contains 100 gm of water at 30 °C. A 10 gm block of copper heated to 60 °C is added to be submerged in the water. What is the final temperature of the mixture?

Principle:

Heat released by the copper = Heat absorbed by the water

That is,

$$m_{copper}c_{copper}\left(T_1 - T_{final}\right)_{copper} = m_{water}c_{water}\left(T_{final} - T_1\right)_{water}$$

This may be alternatively expressed in terms of the conservation of "total heat",

$$m_{cop}c_{cop}(T_1)_{cop} + m_{wat}c_{wat}(T_1)_{wat} = \left[m_{cop}c_{cop} + m_{wat}c_{wat}\right]T_{final}$$

The final temperature is, thus,

$$\begin{aligned} T_{final} &= \frac{m_{cop}c_{cop}(T_1)_{cop} + m_{wat}c_{wat}(T_1)_{wat}}{m_{cop}c_{cop} + m_{wat}c_{wat}} \\ &= \frac{10[\text{gm}] \times 0.0921[\text{cal/gm}\cdot\text{C}] \times 60[\text{C}] + 100 \times 1 \times 30}{10 \times 0.0921 + 100 \times 1} \\ &= 30.274^\circ\text{C} \end{aligned}$$

Table 2.1 Latent heat for common substances

Substance	Latent heat of fusion (cal/g)	Melting point (°C)	Latent heat of vaporization (cal/g)	Boiling point (°C)
Alcohol, ethyl	25.8	−114	204.3	78.3
Ammonia	79.36	−77.74	327.1	−33.34
Carbon dioxide	43.96	−78	137.14	−57
Helium			5.02	−268.93
Hydrogen	13.86	−259	108.7	−253
Lead	5.5	327.5	208.1	1750
Nitrogen	6.14	−210	47.78	−196
Oxygen	3.32	−219	50.89	−183
Refrigerant *R134a*		−101	51.58	−26.6
Refrigerant *R152a*		−116	78.01	−25
Toluene	17.23	−93	83.86	110.6
Turpentine			70	
Water	79.8	0	539.96	100

2.3 The Doctrine of Latent and Sensible Heats in an Internally Reversible Medium

When heat is applied to a body, it may raise its temperature or change its state without changing its temperature. Truesdell noted "All the pioneers of thermodynamics assumed that in every process the heating Q would equal a linear function of the rates of increase of volume and temperature, with coefficients which were functions of volume and temperature only and hence independent of the process. That is, at all times when dV and dT exist," [4]

$$\delta Q = C_T^{(V)}(V,T)dV + C_V^{(T)}(V,T)dT \tag{12}$$

In the above Eq. (12), instead of Q, the symbol δQ is used to indicate that heat exchange is not an exact differential whereas the use of Q or dQ in the equation might lead to the impression that $Q(V,T)$ is a function (see Sect. 2.4), which would be incorrect. In Eq. (12), $C_T^{(V)}(V,T)$ denotes the latent heat *with respect to volume* of the calorimetric material at the constant controlled temperature, and $C_V^{(T)}(V,T)$ or $C_V(V,T)$ the *heat capacity* of the calorimetric material at constant volume. $C_T^{(V)}(V,T)$ is also called "latent heat of expansion" as noted by Maxwell [5]:

> We here recognize the fact that heat when applied to a body may act in two ways—by changing its state, or by raising its temperature—and that in certain cases it may act by changing the state without increasing the temperature.
>
> The most important cases in which heat is thus employed are
>
> 1. The conversion of solids into liquids. This is called melting or fusion. In the reverse process of freezing or solidification, heat must be allowed to escape from the body to an equal amount.
> 2. The conversion of liquids [or solids] into the gaseous state. This is called evaporation [or sublimation] and its reverse condensation.
> 3. When a gas expands, in order to maintain the temperature constant, heat must be communicated to it, and this, when properly defined, may be called the latent heat of expansion.

The above rule of calculation of heat with respect to volume suggests an alternative expression for δQ with respect to the rate of increase of pressure. In a process described in terms of increments of pressure and temperature, the heat gained by the body of calorimetric material is given by

$$\delta Q = C_T^{(p)}(p,T)dp + C_p^{(T)}(p,T)dT \tag{13}$$

where $C_T^{(p)}(p,T)$ denotes the *latent* heat *with respect to pressure* of the calorimetric material at a constant temperature, and $C_p^{(T)}(p,T)$ or C_p the *heat capacity* of the calorimetric material at the constant pressure. Heat capacity introduced in Sect. 2.2

in reference to water as the calorimetric substance was constant pressure heat capacity. Here a distinction is made between constant volume heat capacity and constant pressure heat capacity.

We can obtain the relations between the two heat capacities: Combine Eqs. (12) and (13) yielding

$$C_T^{(V)}(V,T)dV + C_V^{(T)}(V,T)dT = C_T^{(p)}(p,T)dp + C_p^{(T)}(p,T)dT$$

Taking differentiation of the equation with respect to p under the constraint of constant T

$$C_T^{(p)}(p,T) = C_T^{(V)}(V,T)\left(\frac{\partial V}{\partial p}\right)_T$$

Taking differentiation of the equation with respect to T under the constraint of constant p

$$C_p(p,T) - C_V(V,T) = C_T^{(V)}(V,T)\left(\frac{\partial V}{\partial T}\right)_p \tag{14}$$

It is common to refer to the ratio of heat capacities as

$$k = \frac{C_p}{C_V}$$

Equations (12) and (13) are valid for representing heat exchange in material media, despite that Q itself is not a state function, under the condition that the material media are internally reversible, a notion that will be discussed in Sect. 6.6.

2.4 Adiabatic Heating

Specific heat capacity is the amount of heat needed to cause a one-degree rise in temperature for a unit mass of the substance. This refers to heat *directly applied* to the substance, for example, a calorimetric substance of water being heated from a flame or a hot block of copper. In the case of gaseous substance, a temperature rise can also happen in a completely different way, by sudden compression of the gas. The suddenness is necessary: Rise in temperature would be suppressed in a slow compression by the heat transfer process resulting from dissipating temperature gradient. Heat transfer, however, takes time, thus, when compression is sufficiently rapid the compression is approximately adiabatic. This is then the second kind of heating, *adiabatic heating*.

There was a historically important connection in the study of adiabatic heating to the study of two problems: the problem of the velocity of sound and investigations that gave rise to the idea of energy conservation. Detailed historical account for these developments can be found in these references [3, 4, 6–8]. The idea of energy conservation will be treated in Chap. 3.

This section gives a brief account of how adiabatic heating of gases was formulated in association with the study of the velocity of sound: Newton assumed that isothermal conditions were maintained during the passage of a sound wave and deduced that the velocity of sound in air was

$$c = \sqrt{\left(\frac{\partial p}{\partial \rho}\right)_T} = \sqrt{\frac{p}{\rho}}$$

where p and ρ are related as $p = \rho RT$. It had been well known that a discrepancy existed between Newton's formula and experimental data.

Laplace made the suggestion that heating by compression and cooling by expansion might plausibly account for the error in the theoretical value based on isothermal conditions, which ignored heating/cooling. Whereas p is proportional to ρ under isothermal conditions, under adiabatic conditions $f(p, \rho)$ assumes a different form, which depends on C_p and C_V values. Laplace sought both theoretical answers and experimental data to this $C_p - C_V$ question—in a convoluted procedure involving the use of particles of the caloric fluid [4].

As recounted by Kuhn, "In the same year [1823, which was just one year before the publication of Carnot's *Réflections*] that Laplace first reissued his theory in collected form, Poisson published an incisive paper showing that the same [theoretical] results and others besides could be derived from far more restricted caloric hypotheses. He began by taking from the caloric theory only the hypothesis that the heat content of a gas is a state function; that is, heat content, Q, depends exclusively on pressure and density. From this single caloric premise [without having to assume 'particles of caloric fluid' as Laplace did, thus 'free of its more suspect elements' [7]], he was able to derive Laplace's value for the speed of sound, as well as… relations governing pressure, volume, and temperature during adiabatic change" [6].

Poisson did explicitly made use of the assumption of Q being a state function, $Q(p, \rho)$ or $Q(p, V)$, which would have cast doubt on the validity of the results. In fact, however, this more restricted premise was not required. Poisson's results depended on the less restricted starting point of (12) and (13).

As it was aforementioned at the end of Sect. 2.3, use of Eqs. (12) and (13) in the treatment of δQ as well as the treatment of δQ according to the following is valid:

$$\delta Q = \left(\frac{\delta Q}{\partial p}\right)_V dp + \left(\frac{\delta Q}{\partial V}\right)_p dV = \left(\frac{\delta Q}{\partial T}\right)_V \left(\frac{\partial T}{\partial p}\right)_V dp + \left(\frac{\delta Q}{\partial T}\right)_p \left(\frac{\partial T}{\partial V}\right)_p dV$$

That is,

$$\delta Q = C_V \left(\frac{\partial T}{\partial p} \right)_V dp + C_p \left(\frac{\partial T}{\partial V} \right)_p dV \tag{15}$$

For the adiabatic process in gaseous media,

$$0 = \delta Q = C_V \left(\frac{\partial T}{\partial p} \right)_V dp + C_p \left(\frac{\partial T}{\partial V} \right)_p dV = c_V \cdot \frac{V}{R} dp + c_p \cdot \frac{p}{R} dV$$

Divide the equation by pV

$$0 = \frac{c_V}{R} \cdot \frac{dp}{p} + \frac{c_p}{R} \cdot \frac{dV}{V} = \frac{c_V}{R} [dlnp + \gamma dlnV]$$

Integration of which yields

$$pV^k = constant$$

or for adiabatic processes

$$p = constant \cdot \rho^k$$

One finds, thus,

$$c = \sqrt{\left(\frac{\partial p}{\partial \rho} \right)_{Q=0}} = \sqrt{k \frac{p}{\rho}} \tag{16}$$

which is in perfect agreement with the speed of sound in gases.

Again, it is noted that the δQ equations, Eqs. (12), (13), and (15), are valid under the condition that the material media are internally reversible, a notion that will be discussed in Chap. 6. This does not infer that $Q(T, V)$ itself is a state function: while the condition that $Q(T, V)$ is a state function infers that $\left(\frac{\delta Q}{\partial T}\right)_V$ and $\left(\frac{\delta Q}{\partial T}\right)_p$ are state functions, the opposite inference—that $\left(\frac{\delta Q}{\partial T}\right)_V$ and $\left(\frac{\delta Q}{\partial T}\right)_p$ are state functions infers that $Q(T, V)$ is a state function—is not true.

Still, we have the intriguing possibility that the fact that validity of results derived from treating $\left(\frac{\delta Q}{\partial T}\right)_V$ and $\left(\frac{\delta Q}{\partial T}\right)_p$ as state functions might have led caloricists to their erroneous belief that the heat content, Q, of a substance is a state function too. That mistaking inference might, in turn, strengthen their belief in the ontological status of caloric as matter-like. As a result of the mechanical equivalent of heat (MEH), as it will be discussed in Chap. 3, heat content is not a state function. The notion that heat content is a state function and the materiality of caloric are the two principal errors of the caloric theory. Both turn out to be inessential in the application of the theory.

That the first error is not crucial in treating the problem of adiabatic heating is shown in Sect. 6.6 by the demonstration that the treatment of δQ in Eqs. (12), (13), and (15) is valid does not require that heat content Q itself is a state function. The argument that the materiality of caloric as treated by Carnot is *not* a crucial requirement of Carnot's theory is the main discovery of this disquisition: In the first place, it will be noted that the caloric theory is a *one-place relation* theory dealing with heat and its exchange. And, it will be argued that whereas caloric theory is a one-place relation theory, Carnot's theory is fundamentally different kind of relational theory and, instead of being committed to a rigid ontology of heat doctrine, it has in fact flexibility for adopting either the caloric theory's or a rival theory's ontology of heat.

Problems

2.1 A calorimeter contains 2000 g of water at 20 °C. A 500 g block of copper heated to 90 °C is added to be submerged in the water. What is the final temperature of the mixture?

2.2 We have one 3 kg block of ice at −40 °C. Determine how much heat has to be added to bring the block of ice all the way to superheated steam vapor at 160 °C under the constant pressure of 1 atm. Note: Specific heat of ice is 0.499 cal/gK; latent heat of fusion for water is 79.56 cal/gm K; latent heat of vaporization is 539.96 cal/gK; and specific heat of water vapor is 0.48 cal/gK. Make a temperature versus heat-input plot for the process.

2.3 The speed of sound in non-humid air at 20 °C is 343 m/s. Given Eq. (12), can the speed of sound be changed by adjusting the speaker's frequency or the speaker's amplitude? What will be the speed of sound in a non-humid air at 30 °C?

References

1. Lienhard JH (1983). Notes on the origins and evolution of the subject of heat transfer. Mechanical Engineering June 1983:20–27
2. Lavoisier AL, Laplace PS (1783) Memoir on Heat (Reprint of translated version in Obesity Research 2(No. 2), March 1994:189–202)
3. Psillos S (1994) A philosophical study of the transition from the caloric theory of heat to thermodynamics: Resisting the pessimistic meta-induction. Stud Hist Phil Sci 25(No. 2):159–190
4. Truesdell C (1980) The Tragicomical History of Thermodynamics: 1822–1854. Springer, New York
5. Maxwell JC (1888) Theory of Heat. Dover, 2001 (p. 73)
6. Kuhn TS (1958) The caloric theory of adiabatic compression. Isis 49(No.2):132–140
7. Mendoza E (1961) A sketch for a history of early thermodynamics. Physics Today 14 (No. 2):32–42
8. Fox R (1971) The Caloric Theory of Gases. Oxford Univ. Press

The First Law: The Production of Heat and the Principle of Conservation of Energy

3

Abstract

Pedagogically speaking, the formulation of the first law of thermodynamics is the real beginning of the study of heat and energy. Rejecting caloric's materiality, the mechanical equivalent of heat proved that heat can be measured in terms of mechanical energy and the heat—mechanical energy equivalence led to the first law (the energy conservation principle) and that heat is a form of energy, the lowest-grade form of energy. The conservation principle infers that energy can be neither created nor destroyed, thus, only its form can be transformed. This conceptual understanding and the application of $dU = \delta Q - pdV$ for thermodynamic processes are two key takeaways of this chapter. The nature of the transformation of energy forms is the outstanding question remaining for further investigation.

Keywords

Adiabatic work · Internal energy · Mechanical equivalent of heat (MEH) · The MEH constant (*J*) · Conceptual differentiation of caloric · The first law of thermodynamics · Heat versus *heat* · Enthalpy · Heat capacities · Caloric equation of state for ideal gases · Polytropic processes of ideal gases

3.1 Introduction

With the successful determination of $c_p - c_V$ and the success of the theory of adiabatic heating, the caloric theory was showing itself to be a powerful instrument indeed! Just at this point, the scientific tide began to turn against the Laplacian orthodoxy. The dynamical theory of heat, an older but at that point a less developed theory of heat, was making a comeback to challenge the caloric theory as part of the anti-Laplacian movement.

L.-S. Wang, *A Treatise of Heat and Energy*, Mechanical Engineering Series,
https://doi.org/10.1007/978-3-030-05746-6_3

By the middle of the nineteenth century, the idea of vis viva was refined into the *mechanical equivalent of heat* (MEH, Joule [1843–1850]), which transformed the notion of vis viva into a quantitative proposition by demonstrating that the same effect of heating through heat exchange can be obtained by dissipative heating of mechanical work. Thereby, the unit of heat, the calorie, can be converted into an equivalent unit of mechanical work/energy.[1] This demonstration, of course, showed the idea of materiality of heat to be categorically false. An extension of MEH is the more comprehensive "*mechanical theory of heat*," which sought to understand the nature of heat: What is heat, if it is not a material caloric?

In other words, the program of the mechanical theory, as it will be discussed in Sect. 4.8, sought to place heat in its *ontological category*. While the placing of heat correctly in its ontological category is of foremost importance philosophically, it does not provide the best way to understand heat or to define heat (certainly, it alone is not sufficient). Instead, understanding heat can be sought indirectly in terms of the two laws of thermodynamics (Sect. 1.1) as equal partners. One critical part of the indirect answer is the first law of thermodynamics.

A key in the formulation of the first law is the concept of internal energy U. With the introduction of U, the concept of energy became an extraordinarily useful concept in unifying many branches of knowledge. The present chapter focuses on the energetic issues of internal energy (U) and heat exchange (Q).

Defining heat requires both energetic and entropic understanding of heat and will be given in Sect. 5.6.

3.2 Adiabatic Work and Internal Energy

Joule, the unit of energy *joule* is named in honor of him

James Joule (1818–1889)

[1]Mechanical energy in the strict sense is kinetic energy and potential energy. The term can also be defined broadly as *the form of energy that can be converted to mechanical work completely and directly by an ideal device* such as an ideal electric motor. Mechanical energy in this more general sense, therefore, includes electric energy and other forms of energy that can be converted *completely* into mechanical work.

The crucial experiment that produced the conclusive evidence of conversion of mechanical work into heat energy was a series of experiments performed by Joule in the period from 1840 to 1849, which showed that the same effect of calorimetric heating could be obtained by mechanical work or the input of the mechanical energy (MEH). It was Rudolf Clausius, however, who first introduced the concept of a new state function, internal energy, and thus made the first explicit statement of the first law of thermodynamics [1].[2]

Joule's original demonstration of the MEH made no use of the concept of energy; in his 1850 paper he summarized MEH this way,

> The quantity of heat produced by the friction of bodies, whether solid or liquid, is always proportional to the quantity of force [mechanical energy] expended.

Clausius, who introduced both concepts of *internal energy* and *entropy* (see Chapter 5)

Rudolf Clausius (1822–1888)

He then asserted that this proportionality constant is a universal constant, i.e., in that sense the MEH is the assertion of equality between expended mechanical energy and produced heat or heat energy (see Eqs. (21) and (24)).

The treatment here, departing from the historical chronology of establishing first the MEH then the first law of thermodynamics, follows instead the modern heuristic treatment of defining the state function of internal energy in terms of *adiabatic work* first, then formulating the heat exchange (see Footnote 6, Chap. 4) in terms of internal energy and work in accordance with the supposition of the conservation of energy. The MEH is then shown to be the empirical confirmation of the supposition.

Imagine that work is performed on a system with a "restricted adiabatic envelope" (see Sect. 1.3) in an "adiabatic work" experiment to bring the system from an initial end state to a final end state. The "adiabatic work" can be one of the

[2]The first law of thermodynamics is the principle that energy is conserved. Although the idea of energy conservation is generally credited to Rumford, Mayer and Joule, the introduction of the internal energy, which was critical in the formulation of the first law, was given by Clausius in 1850. Historically, the first and the second laws were simultaneously formulated by Kelvin and Clausius (this chapter and Chap. 4). For heuristic purpose, the first law is presented first in this chapter.

possible combinations of several elementary steps of work[3] done on the system or, in some cases, elementary steps of adiabatic work performed by the system, or a mixture of both kinds of step(s). Suppose that these combinations are made to bring about the same change of a system's state from an initial state to a final state. Carrying out various combinations of adiabatic work setups, our imaginary experimentalist is able to draw the generalization that "if a system is caused to change from an initial state to a final state by only adiabatic work, *the* work *done on the system for all adiabatic paths connecting the two states is the same.*"

This generalization leads to an important conclusion: there exists a *function* of the thermodynamic coordinates of the system, denoted as U, whose value at the final state minus its value at the initial state equals to the adiabatic work bringing about the change,

$$-W_{i \to f}(adiabatic) = U_f - U_i \tag{17}$$

where the sign convention of $W_{(adiabatic)}$ is defined as positive for work performed by a system so that a $-W$ represents adiabatic work done on the system corresponding to a positive $(U_f - U_i)$. This function U is known as internal energy.

We consider an arbitrarily chosen state of the system as the *standard state* 0 (usually specified in terms of "standard temperature and pressure", T_0 and p_0). The internal energy at state 0 is assigned to be zero (i.e., $U_0 = 0$). The internal energy function of the system at any state A is

$$U_A = U(V_A, T_A) = U_A - U_0 = -W_{0 \to A}(adiabatic) \tag{17A}$$

where U is shown to be a function of V and T as an example—it can just as well be a function of p and T, for instance.

Note that the definition, Eq. (17A), of the internal energy U involves an element of arbitrariness since U_A, though independent of the particular path connecting 0 and A, does depend on the particular choice of the standard state 0. Energy, therefore, has no quantitative absolute value and the value of U function depends on the standard state "convention", adopted by the body of the thermodynamic literature. The arbitrariness presents no difficulty, however, as only energy differences between states, e.g., state A and state B, are considered in thermodynamic energy analysis. As Eq. (17A) applies to all possible states including state B

$$U_B = U(p_B, T_B) = -W_{0 \to B}(adiabatic) \tag{17B}$$

[3]It should be noted that the adiabatic work considered here includes mechanical work (e.g., stirring paddle motion) or electromechanical work leading to gain in the potential energy or/and kinetic energy of the system. The potential and kinetic energy will then be transformed into the internal energy as shown in (22B). Principally, the types of internal energy considered in this book are thermal energy, chemical energy (internal energy in the form of chemical bond), and nuclear energy. Thermal energy is a subset of internal energy. The term heat energy is synonymous with thermal energy.

Let us imagine that the collected data resulting from a systematic experimental program are compiled into a database. Once the compilation of U as a function of p and T is completed, the internal energy change $\Delta_{A\to B}U$ can be readily determined by looking up the values of U_A and U_B from the property database

$$\Delta_{A\to B}U = U_B - U_A = -W_{0\to B}(adiabatic) + W_{0\to A}(adiabatic)$$

Or

$$= [U_0 - W_{0\to B}(adiabatic)] - [U_0 - W_{0\to A}(adiabatic)]$$

Note that the result of any energy analysis involving change in energy between states is unique; the analysis result is not dependent on the choice of U_0 at the standard state 0. Both the adiabatic work and the internal energy have the same dimension and their SI unit is the *joule, J*, in honor of Joule.

As shown in Eqs. (17A) and (17B), the internal energy of simple systems may be considered to be a function of two (any two) of the thermodynamic coordinates. If the coordinates characterizing the two states A and B differ from each other only infinitesimally, the change in internal energy, dU, may be expressed in terms of dV and dT, or dp and dT

$$dU = \left(\frac{\partial U}{\partial V}\right)_T dV + \left(\frac{\partial U}{\partial T}\right)_V dT \tag{18A}$$

$$dU = \left(\frac{\partial U}{\partial p}\right)_T dp + \left(\frac{\partial U}{\partial T}\right)_p dT \tag{18B}$$

It should be noted that the two partial derivatives $(\partial U/\partial T)_V$ and $(\partial U/\partial T)_p$ are not equal. The first is a function of V and T, and the second a function of p and T. They are different mathematically and also have different physical meanings (see Chap. 9).

To recapitulate: Heating or gain in internal energy can be a result of adiabatic work, instead of a result of heat flow or calorimetric heating as it had been known previously (Chap. 2). It is clear, therefore, that "heat energy" and "heat flow" are not synonymous terms since heat energy can be gained as a result of adiabatic work. The concept of caloric must be conceptually differentiated into the concepts of heat (i.e., thermal) energy and heat flow—which will be considered in more detail in the next section.

3.3 Heat Exchange and the First Law of Thermodynamics

A temperature increase from 14.5 to 15.5 °C in one gram of water at constant atmospheric pressure can be a result of the absorption of heat (Chap. 2), or a result of adiabatic work (Sect. 3.2)—or some combination of work and heat.

Consider again the system in Sect. 3.2. Imagine that we already compiled the internal energy database of the system. Now, we perform a different kind of experiment on the same system: In this new experiment, the system is not thermally insulated in its change $i \rightarrow f$, during which the system may interact with its surroundings in both work, W, *and* heat exchange, Q. Suppose that the resulting work (*the nonadiabatic work* $W_{i\rightarrow f}$ which is not necessarily equal to $U_f - U_i$) is determined experimentally. There are two ways to consider the heat exchange, Q.

It can be determined by using an available calorimetric experimental apparatus. The result $(Q_{i\rightarrow f})_{\text{calorie}}$,

$$-m_{water}c_{p,water}\Delta T = \left(Q_{i\rightarrow f}\right)_{calorie} \tag{19}$$

is expressed in the unit of *calorie*, where ΔT is the temperature rise in the calorimetric substance of water, as it was described in Chap. 2, Eq. (11A).

At the same time, the supposition of "the conservation of energy" suggests that heat exchange is quantitatively balanced with the difference between the internal energy change of the system and the (general) work exchange of the system. That is, one can define a theoretical $Q_{i\rightarrow f}$ (based on the conservation supposition) in the unit of *joule* as

$$\left(U_f - U_i\right) - \left(-W_{i\rightarrow f}\right) \equiv Q_{i\rightarrow f} \tag{20}$$

If the supposition is correct, an equivalent relationship must exists between the theoretical $Q_{i\rightarrow f}$ in (20) and the calorimetric $\left(Q_{i\rightarrow f}\right)_{calorie}$.

Note that the theoretical $Q_{i\rightarrow f}$ and the calorimetric $\left(Q_{i\rightarrow f}\right)_{calorie}$ are measured in terms of different units – the unit of *joule* and the unit of calorie, respectively. It is, therefore, necessary to introduce a conversion factor J in the expression of the equivalent relation

$$Q_{i\rightarrow f} = \boldsymbol{J}\left(Q_{i\rightarrow f}\right)_{\text{calorie}}$$

or

$$\left(U_f - U_i\right) + W_{i\rightarrow f} = J\left(Q_{i\rightarrow f}\right)_{\text{calorie}}$$

(The equation is reducible to Eq. (17) in the limit of $Q \rightarrow 0$.) Imagine that experiments of the kind were repeatedly carried out along different paths with different works and heat. Joule and others obtained experimental outcomes showing that in *all* cases the conversion factor J's are found to be the same constant, i.e., a universal constant and thus provided the proof that the supposition is true: the supposition of the conservation of energy became the principle of the conservation of energy.

There are various definitions of the calorie unit; the definition that is based on "a temperature rise from 14.5 to 15.5 °C in one gm of water at constant atmospheric pressure" may be referred to as *calorie*$_{15}$,

$$Q_{i \to f} = J\left(Q_{i \to f}\right)_{calorie15}$$

This J has been found to be 4.1868 J/calorie$_{15}$. Other slightly different values of J also exist in terms of different definitions of calorie. The various definitions of the calorie unit give rise to a degree of ambiguity[4] with regard to the quantitative value of the J conversion factor—though not its meaning. Unlike the triple point of water for temperature fix point *reference* provided satisfactory resolution for temperature measurement, the ambiguity about J values is the reflection of no similarly unique operational definition of calorie for reference yet in existence.

The rational thing to do is to embrace the *mechanical equivalent of heat*, abandon the calorie unit in theoretical physics and express the calorimetric heat measurement in the mechanical energy unit of *joule* as the preferred practice,[5]

$$J = 1\frac{W(joule)}{Q(joule)} \tag{21}$$

The dimensionless conversion factor, J, becomes unity and thus *disappears* from the equivalent relationship equation, which becomes the first law of thermodynamics,

$$U_f - U_i = Q_{i \to f} - W_{i \to f} \tag{22}$$

or

$$dU = \delta Q - \delta W \tag{23}$$

By integrating Eq. (23) over a cyclic process, the first law for a cyclic process becomes

$$\oint \delta Q = \oint \delta W \tag{24}$$

which is a useful equation for the analysis of *closed heat engine cycles*: the cyclic net heat exchange of a closed cycle is equal to the cyclic net work exchange.

[4]The various definitions of the caloric unit depend on the particular property (specific heat capacity) of water at a given state, in which a particular definition is based.

[5]This does not mean that we should abandon calories unit in applied practices, only that it is not required in theoretical physics and engineering fundamentals.

In the general case of system changes that may involve change in the kinetic energy *KE* or the potential energy *PE* (see Footnote 3), we can express the first law in terms of *total internal energy*, E ($\equiv U + KE + PE$),

$$E_f - E_i = Q_{i\to f} - W_{i\to f} \tag{22A}$$

$$dE = \delta Q - \delta W \tag{23A}$$

This may be considered for the effect of heat and work on a system, which is initially of negligible *KE* and *PE* with internal energy U_i,

$$Q_{i\to f} - W_{i\to f} = E_f\left(U_f' + KE' + PE'\right) - U_i = U_f - U_i \tag{22B}$$

That is, the system will go through a transitory state in which its kinetic and potential energy will be transformed into internal energy ending with final internal energy of $U_f = E_f$.

Equation (21) is the mathematical expression of the MEH. Equations (21), (22), (23), (24), and (22B) collectively represent the first law of thermodynamics, which supplants the caloric theory in terms of five closely related steps/ideas as shown in Table 3.1.

Table 3.1 The concept of energy and the first law of thermodynamics

1. The initial conceptual differentiation of caloric into heat energy[a] and heat exchange (or heat flow): the former is a property or state function quantity, whereas the latter is a path-dependent quantity
2. There are two modes of system energy exchanges: *work*, δW, and *heat-exchange*, δQ, and both W and Q are path-dependent quantities (see also the caption of Fig. 1.6)
3. The state function of *internal energy*, U, can be in the forms of thermal, chemical, and nuclear; a change in internal energy can be either the result of heat exchange, or a result of adiabatic work (Sect. 3.2), or some combination of work and heat exchange (Eq. (22))
4. The MEH constant, J, implies the principle of heat and mechanical work equivalence. Heat and mechanical energy, therefore, are of the same ontological category (see Sect. 4.8)
5. The principle of the conservation of energy and its transformation: *energy can be neither created nor destroyed; only the form in which energy exists can be transformed from one form into another*

[a]Heat energy is thermal internal energy, which is a special form of internal energy, #3—the introduction of *internal energy* by Clausius and *energy* (= internal energy + mechanical energy, see Chap. 4), and transformation of energy forms (including MEH) are, therefore, additional steps beyond the initial conceptual differentiation of caloric, #1. The *complete* conceptual differentiation (An additional, related conceptual differentiation will be discussed in Chap. 4. That idea will be connected to the simultaneous formulation of the first law and the second law.) of caloric into heat and energy consists of steps #1, #3, #4 and #5 together. In common practice, the same term, heat, is often used for heat energy and heat exchange and this practice has been accused of committing error of caloric theory thinking. But, this linguistic use of heat does not risk the implication of treating heat as a substance as long as it is used without denying difference between heat as state function quantity and heat as path-dependent quantity and conceptual differentiation in its full sense. The linguistic issue will be further discussed in Sects. 3.5 and 5.6

The first law rendered obsolete the caloric theory of heat. The caloric theory of heat is a *one-place relation* theory, in which heat (or caloric) is conserved in calorimetric processes. With the conceptual differentiation of caloric, we have the foundation of a *two-place relation* "mechanical theory of heat" dealing with heat and work, and their interconversion: the transformations of heat into mechanical work and the opposite transformations of mechanical work into heat. In these transformations, it is energy, rather than heat as in calorimetric processes, which is conserved.

Production of heat, and additional comments on the transformation of heat and work and the transformation of energy

In establishing the MEH, Joule demonstrated equivalence (i.e., equality) between expended mechanical energy and produced heat when mechanical energy is converted into heat. In addition, Joule asserted that equality holds also during the interconversion of heat into mechanical energy, even though he did not demonstrate such equality during heat → work conversion. His bold assertion of equivalence of heat and work turned out to be true. The principle of equivalence of heat and work, #4, Table 3.1, may be stated, as by Clausius, as,

> In all cases where work is produced by heat, a quantity of heat proportional to the work done is expended; and inversely, by the expenditure of a like quantity of work, the same amount of heat may be produced.

This understanding and the corresponding principle of the conservation of energy, #5, Table 3.1, represented one of the greatest achievements in the history of science.

It is important to note, however, that whereas, for the *production of heat*, the MEH teaches us it is always possible to have the *complete* conversion of mechanical energy into useful heat (if all ancillary heat losses are eliminated), the MEH, in asserting that the *production of work* requires the consumption (expenditure) of heat, provides no guidance as to how much of this heat is "consumed" in the production of work. The question of the fraction of heat that may be consumed for the production of work is to be addressed in Chaps. 4, 5, 7, and 8. In fact, the notion that heat is expended or consumed will be called into question as to what is really expended (see Sect. 3.11 and Chap. 8).

It should be noted furthermore that the complete statement of the principle of the conservation of energy, as stated in #5, amounts to the idea of energy conservation and the idea of energy availability which, strictly speaking, is beyond this chapter and belongs to the treatment in Sect. 4.7, Chapter 4. That is, statement #5, which is a powerful statement in its full meaning, is the most consequential statement in thermodynamics that, in fact, captures partially the second law. For that reason, it is the most misunderstood statement and, at the same time, the correct understanding of which will offer true comprehension of heat and energy.

3.4 Energy Conservation in a Reversible Universe

Adequate understanding of heat and energy will have to be obtained after both the first law and the second law are introduced; as it will be seen, the introduction of the second law will require several stages in Chaps. 4, 5, 7, and 8. At this point, we can comment preliminarily, in this section and Sect. 3.5, on heat in terms of a peculiar notion of heat that has been widely promoted in thermodynamics literature:

For a purely mechanical system in a reversible universe, the notion of energy (mechanical energy) is straightforwardly based on energy conservation alone. Equation (22A) reduces to

$$(KE + PE)_i - (KE + PE)_f = W$$

The external work performed by the system equals the decrease in mechanical energy (KE + PE). Therefore, the external work in this case (work in a conservative field) is not path-dependent (in contrast to the general cases as stated above). There is no work reservoir involved either. In fact, no explicit consideration of W itself is necessary as one mass-body's mechanical energy loss is balanced by another mass-body's mechanical energy gain; the total mechanical energy of a Newtonian universe is constant. In a sense, such a reversible universe is a *block universe*,[6] which is deterministic and in which, since its future as well as its past is already determined, "nothing changes."

3.5 Irreversible Universe: Heat versus *Heat*

Mechanical energy is defined in the reversible world of conservative force field in terms of force-driven interactions. In the generalization of the conservative force field to an irreversible one, the concept of energy has to be generalized to include internal energy. In this generalization, internal energy can still be comprehended in terms of microscopic force-driven interactions among molecules and atoms.

[6]"Block universe" was a term introduced by W. James, the pragmatist philosopher. S. Krishnananda in *Studies In Comparative Philosophy* summarized James' philosophy,

> William James, the great teacher of pragmatism in America, repudiates the claims of the logical reason in constructing systems of absolute monism, which, according to him, gives us an unmanageable "block-universe" and set at naught moral responsibility, free will, effort and aspiration, indeterminacy, want, and struggle which are main characteristics and daily occurrences of life. The pragmatism of William James is a theory of the will which looks with disfavor on the intellectual philosophies that make a self-complete Absolute the entire reality.

S. Krishnananda made this comment on James' repudiation of the "absolute monism as the entire reality" with the intention to defend the monism of the Absolute. But, James is correct in repudiating monism, reductionism, and block universe.

Force-driven interactions offer representation of energy in both reversible and irreversible worlds. Heat is a different matter.

Heat phenomena only exists in an irreversible universe. Before the nineteenth century, there was only the intuitive view that heat was something that caused a temperature rise. I shall refer to this as *heat*. It is this *heat*, broadly understood, that was studied by Black, Lavoisier, Carnot, Clapeyron, Mayer, Joule, Helmholtz, Thomson, Rankine, and Clausius. Initially, each of them—like a blind man in the parable of "blind men and the elephant", apprehended one part or parts of its manifold essences. The emergence of the MEH as an undisputed principle made it into the privileged position as the "principal corner stone of classical thermodynamics," [2] on which the substance theory of heat—that heat is a substance—was vanquished. The conception of *heat* underwent a transformation too, to be encapsulated narrowly in terms of heat, Q, as the process of heat exchange.

In association with this transformation, there has been a mistaken attempt to treat heat and work, Q and W, in Eq. (22) in parallel fashion by adopting the official doctrine of forbidding the use of "the heat in a body." It is a doctrine that physics has not been able to enforce. For instance, in a 2004 paper Brookes, et al. [3] reports: "When it comes to language about heat, there is consensus that physicists' language can be misleading, but little agreement about why it is misleading or how it can be corrected." They concluded their study with the astonishing admission that physicists talk about heat predominately as if it were a substance!

This astonishing admission was commented on in Table 3.1 and will be further examined in greater detail in Sect. 5.6, which will conclude that the critique of Brookes et al. was misdirected: when physicists talk about heat in a body, they obviously talk about heat in the broader sense than heat as a process, i.e., heat energy as a state function, with no implication of using the term to mean heat as a substance.

The whole issue was a red herring resulting from the inability of providing a concise definition of *heat*. As it will be presented in Sect. 5.6, a middle ground can be found between the broadest meaning of *heat* and the narrowest interpretation of heat (as heat exchange, Q, alone) by defining heat narrowly as a process *and* as a state function corresponding to natural end states of spontaneous changes. Such a definition serves how we intuitively use the term of heat as manifestation of irreversibility in nature.

Correspondingly, though force-driven interactions offer representation of energy even in an irreversible world, force-driven interactions cannot explain how we use energy because such uses depend on irreversibility in nature, which can be explained only statistically. To put it in another way, energy conservation prevails in both reversible and irreversible worlds but energy availability (see Sect. 4.7) is a concept manifested only in irreversible worlds.

3.6 Enthalpy

Consider some immediate application of the first law. Equation (23) may be written for internally reversible, quasi-static processes as

$$dU = \delta Q - pdV \tag{25}$$

It is useful to introduce an alternative energy function, the enthalpy function denoted by H,[7]

$$H \equiv U + pV \tag{26}$$

It follows that

$$dH(T,p) = dU + d(pV) = \delta Q - pdV + pdV + Vdp = \delta Q + Vdp \tag{27}$$

Taking the differential of the enthalpy function

$$dH = \left(\frac{\partial H}{\partial T}\right)_p dT + \left(\frac{\partial H}{\partial p}\right)_T dp \tag{28}$$

We shall see in the next section that enthalpy is a particularly useful function for constant pressure processes.

3.7 Heat Capacity and Molar Heat Capacity

We now consider an infinitesimal transformation of a system, that is, a quasi-static heating[8] of the system for which the change meets internal reversibility criterion (see Sect. 6.5). Substituting Eq. (18A) into (25),

$$\left(\frac{\partial U}{\partial T}\right)_V dT + \left[\left(\frac{\partial U}{\partial V}\right)_T + p\right] dV = \delta Q \tag{29}$$

[7]For an excellent introduction to the concept of enthalpy the reader is referred to *The Laws of Thermodynamics: A Very Short Introduction* by Peter Atkins [4].

[8]When heat is absorbed by a system, a change of temperature may or may not take place. For instance, the evaporation of water as a result of heating is not accompanied by temperature change (the constant temperature is called saturation temperature); heat, in this case, is referred to as *latent* heat. Heating of water without phase change is accompanied by a temperature rise; heat, in this case, is called *sensible heat*. The processes considered here are heating or cooling processes that produce temperature change in systems.

Substituting Eq. (18B) and $dV = \left(\frac{\partial V}{\partial T}\right)_p dT + \left(\frac{\partial V}{\partial p}\right)_T dp$ into Eq. (25),

$$\left[\left(\frac{\partial U}{\partial T}\right)_p + p\left(\frac{\partial V}{\partial T}\right)_p\right] dT + \left[\left(\frac{\partial U}{\partial p}\right)_T + p\left(\frac{\partial V}{\partial p}\right)_T\right] dp = \delta Q \tag{30}$$

Alternatively, substituting Eq. (28) into (27),

$$\left(\frac{\partial H}{\partial T}\right)_p dT + \left[\left(\frac{\partial H}{\partial p}\right)_T - V\right] dp = \delta Q \tag{31}$$

The *heat capacity* of a body is, by definition, $\delta Q/dT$ (see Eq. (11)), the ratio of the infinitesimal amount of heat δQ absorbed by the body to the infinitesimal increase in temperature dT produced by this heat. In general (one exception is water at 4 °C and 1 atm, see Sect. 9.5.2), the heat capacity of a body will be different according to whether the heating process is taking place at constant volume or at constant pressure. Let C_V and C_p be the heat capacities at constant volume and at constant pressure, respectively.

The expression for C_V can be obtained from Eq. (29),

$$C_V = \left(\frac{\delta Q}{dT}\right)_V = \left(\frac{\partial U}{\partial T}\right)_V \tag{32}$$

It is noted that in the case of isochoric (constant volume) heating, heat exchange Q directly leads to gain in the internal energy U.

Similarly, the expressions for C_p can be obtained from with Eqs. (30) or (31),

$$C_p = \left(\frac{\delta Q}{dT}\right)_p = \left(\frac{\partial H}{\partial T}\right)_p \tag{33}$$

$$C_p = \left(\frac{\delta Q}{dT}\right)_p = \left(\frac{\partial U}{\partial T}\right)_p + p\left(\frac{\partial V}{\partial T}\right)_p \tag{33A}$$

Equation (33A) can also be obtained by substituting Eq. (26) into (33). The second term on the right-hand side of Eq. (33A) represents the effect on the heat capacity of the work performed during the constant pressure heating expansion. This effect is incorporated in the enthalpy function; ∂H in Eq. (33) combines the internal energy change and expansion work.

The internal energy and enthalpy per unit mass m or per unit mole N are examples of *specific properties* of a substance. They are called (mass based) *specific* internal energy and *specific* enthalpy and, in the case of mole-based terms, *molar internal energy* (denoted by u) and *molar enthalpy* (denoted by h). Unless it is specifically pointed out, we shall use mole-based specific properties and adopt the latter names. The corresponding *molar* heat capacities are denoted by c_V and c_p; the formulas of which are

$$c_V = \frac{1}{N}\left(\frac{\delta Q}{dT}\right)_V = \left(\frac{\partial u}{\partial T}\right)_V \tag{32B}$$

$$c_p = \frac{1}{N}\left(\frac{\delta Q}{dT}\right)_p = \left(\frac{\partial h}{\partial T}\right)_p \tag{33B}$$

In general, heat capacities and molar heat capacities are functions of temperature and pressure for a given substance. In application to processes that take place within narrow ranges of temperature and pressure, it is often an acceptable approximation to treat their values as constants.

3.8 Joule's Law (Joule Free Expansion): The Caloric Equation of State for Ideal Gases

We call Eqs. (3), (4), and (5) the *thermal equation of state* for ideal gases. Another important property relation for an ideal gas is Joule's law, which shall be called the *caloric equation of state* for ideal gases. Joule's law states that u is a function of T alone, i.e., the change in internal energy of an ideal gas depends on the temperature change only.

$$\left(\frac{\partial u}{\partial v}\right)_T = 0 \,\text{and}\, u(T, v) = u(T)$$

i.e.,

$$dU = Nc_V dT \tag{34}$$

$$\Delta U = N\int_{T_1}^{T_2} c_V(T)dT = Nc_V\Delta T \tag{34A}$$

$$U = U_0 + Nc_V(T - T_0) \tag{34B}$$

This result was originally discovered through experimental method by Joule. It turns out that Joule's law, Eq. (34), is an inference of Eq. (3), as shown in Sect. 9.5.2, Eq. (175A): any gas that satisfies the thermal equation of state obeys the caloric equation of state.

From the thermal equation of state, we obtain $h = u + RT$. The enthalpy of an ideal gas, thus like the internal energy in accordance with its caloric equation of state, is also a function of temperature alone,

$$\left(\frac{\partial h}{\partial p}\right)_T = 0 \text{ and } dH = Nc_p dT \tag{35}$$

Since both the internal energy and the enthalpy of an ideal gas are functions of temperature alone, the molar heat capacities, c_v and c_p (see Eqs. (32B) and (33B)), are functions of temperature alone as well. From this result and $h = u + RT$, we find the relationship between the two molar heat capacities,

$$c_p(T) = c_v(T) + R \tag{36}$$

Note that in general, both molar heat capacities individually are temperature dependent, but their difference is a constant, the universal gas constant R. Introducing the ratio of heat capacities k by the definition $k \equiv c_p/c_v$, we obtain for ideal gases

$$k(T) = \frac{c_P}{c_V} = \frac{c_V + R}{c_V} = 1 + \frac{R}{c_V(T)} \tag{37}$$

Correspondingly,

$$R = c_p - c_V = (k-1) \cdot c_V = \frac{k-1}{k} c_P \tag{37A}$$

Within the appropriate narrow temperature ranges, c_p, c_v, and k may be treated as constants. It can be shown by an application of kinetic theory of gas that

$c_v = 3/2R$ for a **monatomic gas**, correspondingly
$c_p = 5/2R$,
$k = 5/3$.

$c_v = 5/2R$ for a **diatomic gas**, correspondingly
$c_p = 7/2R$,
$k = 7/5 = 1.4$.

These are the acceptable constant values used in the "room temperature" range. Over a broad temperature ranges, both molar heat capacities monotonically increase with temperature, but their difference is again the universal gas constant over the entire range as long as the ideal gas equation of state applies.

3.9 Quasi-static Heating and the Adiabatic Transformation of a Gas

We consider for the cases of constant molar heat capacities c_p and c_v, some examples of ideal gas processes.

3.9.1 Isochoric processes

For an internally reversible quasi-static *isochoric* (constant volume $V_f = V_i$) process, Eq. (29) becomes

$$\left(\frac{\partial U}{\partial T}\right)_V dT = \delta Q \tag{29A}$$

Integrating Eq. (29A) with the definition of heat capacity, Eq. (32),

$$Q_{i \to f} = Nc_V\left(T_f - T_i\right)$$

From the thermal equation of state for an ideal gas Eqs. (4) or (5).

$$\frac{p}{T} = const. \tag{38}$$

or

$$\frac{p_f}{p_i} = \frac{T_f}{T_i}$$

Heating leads to an increase in gas pressure in a constant volume container.

3.9.2 Isobaric processes

Note that this is a prediction of heating in a *closed* constant volume system. A student may sometimes mistakenly assume that heating associates generally with pressure increase, and apply this relation to an open flow system. In that case, the correct prediction will be that the pressure of an open system is determined by the overall flow configuration (based on Poisson equation for the pressure) and thus remains little changed as a result of heating, effect of which is limited to local region; instead, the specific volume of the fluid increases (corresponding to the result shown below for constant pressure cases) as a result of local change in temperature.

For an internally reversible quasi-static *isobaric* (constant pressure) process, Eq. (31) becomes

$$\left(\frac{\partial H}{\partial T}\right)_p dT = \delta Q \tag{31A}$$

Integrating Eq. (31A) with definition, Eq. (33),

$$Q_{i\to f} = Nc_p\left(T_f - T_i\right)$$

From the thermal equation of state for an ideal gas,

$$\frac{V}{T} = const. \tag{39}$$

or

$$\frac{V_f}{V_i} = \frac{T_f}{T_i}$$

Heating leads to an increase in gas volume, or gas specific volume (also see comment in the above paragraph).

3.9.3 Adiabatic Transformation of an Ideal Gas

Another simple application of the first law is the determination of the relation between state variables (property functions or thermodynamic coordinates) for an internally reversible quasi-static *adiabatic* process. This problem was treated in Chap. 2 based on the caloric theory of heat. Here, we consider the problem again based on the first law. Since $\delta Q = 0$, Eq. (25) for an ideal gas in view of the caloric equation of state becomes

$$C_v dT + pdV = 0$$

Using the thermal equation of state, we can express p in terms of T and V. The above equation becomes

$$C_V dT + \frac{RT}{V} dV = 0$$

or

$$\frac{dT}{T} + \frac{R}{C_V}\frac{dV}{V} = 0$$

Integration yields

$$\ln T + \frac{R}{C_V}\ln V = CONSTANT_a$$

Using Eq. (37A),

$$TV^{R/C_V} = TV^{k-1} = CONSTANT_b \tag{40}$$

Using the thermal equation of state of an ideal gas to eliminate variable T or variable V, respectively, in Eq. (40), we can express Eq. (40) in the following two alternative forms as

$$pV^k = CONSTANT_c \tag{41}$$

$$Tp^{-\left(\frac{k-1}{k}\right)} = CONSTANT_d \tag{42}$$

Equation (41) is to be compared with the hyperbola,

$$pV = CONSTANT_e \tag{43}$$

of the quasi-static isothermal process of an ideal gas (recall the example in Sect. 1.8).

On a (p, V) diagram, the isotherms are a family of equilateral hyperbola; the adiabatic lines represented by Eq. (41) are qualitatively similar to hyperbola, but their (negative) slopes are steeper because $k > 1$. Schematic curves in (p, V) diagram of isobaric line, isothermal hyperbola, adiabatic curve, Eq. (41), and isochoric line are shown in Fig. 3.1.

Consider the example of an internally reversible, adiabatic compression of an ideal gas. The "quasi-static" work is

$$W = \int_{V_i}^{V_f} pdV$$

The pressure p of an ideal gas is related to its volume V through the isentropic p-V relation,

$$pV^k = p_1V_1^k (= \text{constant}) = p_2V_2^k$$

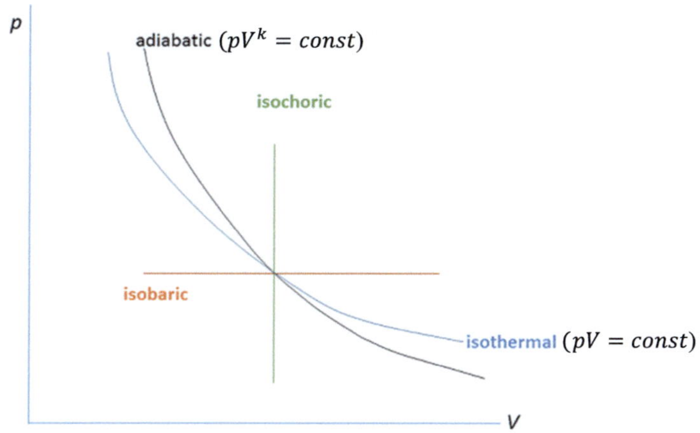

Fig. 3.1 Examples of polytropic processes, $pV^n = constant$: *n = 0 (isobaric). n = 1 (isothermal). n = k (adiabatic). n → ∞ (isochoric)*

Substituting p in terms of $p_1V_1^k$ and V^k, we get

$$W = p_1V_1^k \int_1^2 V^{-k}dV = p_1V_1^k \cdot \frac{1}{-k+1} \cdot \left[\left(V^{-k+1}\right)_2 - \left(V^{-k+1}\right)_1\right]$$
$$= \frac{p_1V_2 \cdot (V_1/V_2)^k - p_1V_1}{-(k-1)} = \frac{p_2V_2 - p_1V_1}{-(k-1)} = \frac{NRT_2 - NRT_1}{-(R/c_V)}$$
$$= Nc_V(T_1 - T_2)$$

That is, the work done by a system is derived from decrease of the system's internal energy; conversely, a work done to the system will increase the system's internal energy.

If there is 0.1 kmol of air, which is initially kept at a temperature of 0 °C, or 273.15 K, and the gas is compressed from a volume of 4–1 m^3, determine the compression *W*.

We can determine first the initial pressure

$$p_1 = \frac{0.1 \times 8.3143 \times 273.15}{4} = 56.78kPa$$

The final pressure is (from Table A-2, **Cengel/Boles Property Tables Booklet/thermodynamics**, $k = 1.4$)

$$p_2 = p_1 \times (V_1/V_2)^k = 56.78 \times (4)^{1.4} = 395.41kPa$$

Compression work is, therefore,

$$W = \frac{p_2V_2 - p_1V_1}{-(k-1)} = \frac{395.41 - 56.78 \times 4}{-0.4} = -420.77kJ$$

Alternatively,

$$T_2 = T_1 \times (p_2V_2/p_1V_1) = 475.58K$$

From Tables A-1 and A-2, we find
$c_V = 20.80kJ/kmol - K$. It follows
$W = 0.1 \times 20.80 \times (273.15 - 475.58) = -421kJ$, which agrees with the above.
(The minus sign indicates that work is done on the gas.)

3.10 Energy Analyses of Processes in Open Systems

The first law energy analysis in the above is applied to closed systems. Engineering devices are often open systems with mass flow (most of the time with energy flow [energy in transit] as well) into and out of the system. The first law equations, Eqs. (22) and (22A), can be formulated for open systems. A concise derivation of the first law equation (as well as the exergy equation) for open systems is given in Chaps. 7 and 10, and more detailed treatment of such problems can be found in most engineering thermodynamics books.

3.11 The Story of Heat

Up to this point, the question of *useful work* has not been addressed. This chapter presented the first phase of the story of heat: our understanding of heat and work (in general terms) as the matter stood in 1850, the year Joule published his definitive paper on the MEH and Clausius formulated the first law by introducing the internal energy incorporating the MEH. At this point of time, there were two contradictory views of the conversion of *heat* (or heat) into work: Carnot's principle, which viewed work production to be resulting from the *transfer* of heat (see Chap. 4) and the view of Rumford–Mayer–Joule, i.e., the MEH, which viewed work production to result from the *consumption* of heat (this chapter, Eq. (24)) [5].

Historically, Carnot's contribution preceded Mayer–Joule's contribution. It turned out that Carnot's "heat" and Joule's "heat" were not the same entity, i.e., different versions of *heat*. The next two chapters address the question of useful work production and present the second phase of the story in which Kelvin (1848–1854) and Clausius (1850, 1854, 1865) succeeded in incorporating both energy conservation and Carnot's principle by introducing the conceptual differentiation of *heat* into heat flow and entropy flow. Correspondingly, heat phenomena are to be characterized by two fundamental principles: the principle of energy conservation (this chapter) and the principle of unidirectionality (next two chapters).

It is significant to note that why the production of useful work cannot be answered as an issue solely of the consumption of heat is because, in this understanding of heat according to the MEH, the nature of "consumption," i.e., the cause, is not addressed explicitly. Instead, the cause is to be identified with Carnot's *transfer* of heat. That is, the cause of the process is the purview of the second law of thermodynamics (the principle of unidirectionality) which will be treated in Chaps. 4, 5, 7, and 8.

The first law is sometimes called the principle of conservation of energy. This principle is one of the few laws that we call them principles, including the entropy principle as it will be treated in Chap. 5. Holton and Brush commented, "So simple, general, and powerful are they that by the very extensiveness of their application the conservation laws unify the various physical sciences within themselves and with one another. Indeed, they have come to be called principles rather than laws, a terminology that betrays that they are no longer merely summaries of experimental facts but instead have become the starting points of scientific understanding itself." [6].

Problems

3.1 Show the relation, $c_p = c_V + R$ between the isobaric and the isochoric molar heat capacities, holds for an ideal gas based on the definitions of c_p and c_v, the definition of enthalpy, and the Joule's law for ideal gases.

3.2 Calculate the internal energy change of a system which performs 34 *J* of work and absorbs 32 *calories* of heat.

99.98 J

3.3 How many joule (J) of heat are absorbed by 3 kmols of an ideal gas expanding isothermally* from the initial pressure of 5 atmospheres to the final pressure of 3 atmospheres, at the temperature of 0 °C?

3480 kJ

3.4 A piston-cylinder device contains 0.95 kg of oxygen initially at a temperature of 27 °C and a pressure, due to the ambient atmospheric pressure and a weight on the top of the piston, of 150 kPa. Heat is added to the gas until it reaches a temperature of 627 °C. Determine the amount of heat added to the gas during the process. How much of this heat in consumed (1) to increase the internal energy of gas, (2) to perform work of compressing the ambient atmosphere (assuming p_0= 101.3 kPa), and (3) to increase the potential energy of the weight. Also by making necessary assumption of the geometry of the piston cylinder determines the mass of the weight and the height increase of the piston.

$$Q = 518.4\,\text{kJ}, \Delta U = 370.3\,\text{kJ}$$
$$W_{atm} = 100\,\text{kJ}, \Delta PE_{weight} = 48.1\,\text{kJ}$$

3.5 One gmol (i.e., mole) of a diatomic ideal gas performs a transformation from an initial state for which temperature and volume are, respectively, 291 K and 21 L to a final state in which temperature and volume are 305 K and 12.7 L. The transformation is represented on the (p, V) diagram by a straight line*. Find the work performed and heat absorbed by the system.

$$W = -1.3067\,\text{kJ}; Q = -1.0157\,\text{kJ}$$

(work is done to the system and heat is rejected by the system)

3.6 Consider the same one gmol ideal gas at the same initial state of 291 K and 21 L, which is compressed to a final state volume of 12.7 L. What is the initial pressure? What is the pressure and temperature at the final state if the compression is adiabatic (i.e., isentropic—see Chap. 4) and the work performed? What is the pressure of the final state if the compression is isothermal and the work performed and the heat absorbed by the system?

$$232.96\text{kPa}, 355.8\text{K}, -1.347.9\text{kJ}; 190.5\text{kPa}, -1.2168\text{kJ}, -1.2168\text{kJ}$$

3.7 A diatomic ideal gas expands adiabatically* to a volume 1.35 times larger than the initial volume. The initial temperature is 18 °C. Find the final temperature.

$$258.2\text{K}(-14.9\,^{\circ}\text{C})$$

3.8 Two blocks A and B are initially at 100 and 500 °C, respectively. They are brought together and isolated from the surroundings. Determine the final equilibrium temperature of the blocks. Block A is aluminum $\left[c_p = 0.900 kJ/kg \cdot K\right]$ with m_A= 0.5 kg and block B is copper $\left[c_p = 0.386 kJ/kg \cdot K\right]$ with m_A= 1.0 kg. Note: for incompressible solids (liquids too—see Sect. 9.5.2, Eq. (176B)), $du \approx c_p dT$.

$$284.7\,^{\circ}\text{C}(557.85\text{K})$$

3.9 In the adiabatic free expansion of a gas, how is it that the temperature doesn't drop, given that it increases when the gas is compressed back to its original volume?

3.10 An ideal gas of 0.1 kmol at the initial state of 298.15 K and 303.9 kPa occupies one chamber of an insulated composite system. The other chamber (of double volume of the first) contains a vacuum. Determine the volume of the first chamber V_I. After the removal of the partition between the two chambers, the ideal gas undergoes an adiabatic free expansion from its initial volume V_I to its final volume 3 V_I. Explain why the gas remains at the same temperature at its final state (hint: note the values of heat exchange Q and work W in this case as it was considered in Prob. 3–9; give **precise** reason for the W value you assume).

$$0.816\text{m}^3$$

- All these processes are internally reversible and quasi-static.

References

1. Clausius R (1851) On the moving force of heat, and the laws regarding the nature of heat itself which are deducible therefrom. Philosophical Magazine 2(No. 4):1–21, 102–119 (Translated by J. Tyndall)
2. Mares JJ et al. (2008) Phenomenological approach to the caloric theory of heat. Thermochimica Acta 474:16–24
3. Brookes D, Horton G, Van Heuvelen A, Etkina E (2005) Concerning scientific discourse about heat. AIP Conference Proceedings 790:149–152
4. Atkins P (2010) The Laws of Thermodynamics: A Very Short Introduction. Oxford Univ Press
5. Coppersmith J (2010) Energy, the Subtle Concept. Oxford Univ Press
6. Holton G, Brush SG (2001) Physics, the Human Adventure. Rutgers Univ Press (p. 202)

4 Carnot's Theory of Heat, and Kelvin's Adoption of Which in Terms of Energy

Abstract

MEH surmised work production to be resulting from the consumption of heat. Carnot demonstrated that work production is derived from the transfer of heat. Kelvin and Clausius were able to synthesize the two competing theories into one theory by incorporating Carnot's theory to accord for what was meant by "consumption". This chapter covers Carnot's theory of heat and how Kelvin incorporated Carnot's theory into the MTH in terms of absolute thermodynamic temperature, the Carnot–Kelvin formula, and the concept of available energy and the energy principle, which is Kelvin's version of the second law of thermodynamics. That is, Kelvin formulated a thermodynamics theory in terms of energy: in terms of energy conservation and energy availability, as well as the universality in the direction of energy form transformation.

Keywords

The Clausius statement · The Kelvin-Planck statement · Carnot's principle · Absolute temperature · Carnot's function · The Carnot/Kelvin formula · Carnot gas cycle · The energy principle (constancy and availability of energy) · The nature of heat (the doctrine of heat as energy) · Interconvertibility of heat and work · Mechanical theory of heat (MTH)

4.1 Unidirectional Nature of Processes and the Production of Work

A REVERSIBLE UNIVERSE can be defined as a universe that is *governed* by time-reversal-invariant (TRI) dynamical laws of nature, i.e., a universe that is described in terms of force-driven interactions alone (see Sects. 3.4 and 3.5). In contrast, an irreversible universe cannot be accounted for by TRI laws alone.

L.-S. Wang, *A Treatise of Heat and Energy*, Mechanical Engineering Series,
https://doi.org/10.1007/978-3-030-05746-6_4

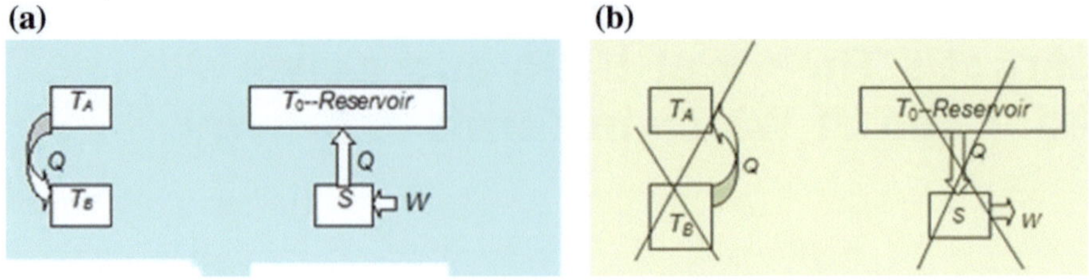

Fig. 4.1 **1a** and **2a** represent spontaneous changes: **1a**, dispersal processes and **2a**, frictional processes. **1b** and **2b** represent "the impossibility statements": **1b**, the Clausius statement and 2**b**, the Kelvin–Planck statement

Energy can be represented in terms of force-driven interactions in a reversible universe as well as an irreversible universe. (Though such representation is sterile, and the meaningful understanding of energy and its definition including the notion of energy's availability require consideration of what makes a universe irreversible.) Heat, i.e., heat phenomena or heat processes, on other hand, can only have meaning in an irreversible universe—one that is far from equilibrium (or, at least, away from equilibrium) abounding with unidirectional processes.[1] Ours is an irreversible universe.

In an irreversible universe, hot lava cools; bouncing balls come to rest; sugar dissolves; batteries discharge; orderly things fall apart; as Thomson (later, Lord Kelvin) put it: mechanical energy dissipates into heat. Thermodynamics began with these simple observations that nature abounds with spontaneous natural (or, *spontaneous* for short) processes. They can be identified into two kinds (archetypes): dispersive kinds such as hot lava cools, sugar dissolves, compost decays and generates heat; dissipative kinds or frictional kinds[2] such as bouncing balls come to rest, wind flows and ebbs, spinning tops wind down. The dispersive ones, unless maintained by "driving forces" (such as temperature gradient, concentration gradient, and chemical affinity), lead to the disappearance of gradients—and the frictional ones convert work, or mechanical and electrical energy, into heat energy. Both kinds of processes occur in their "preferred directions": *spontaneous* changes in the opposite directions are impossible.

With heat transfer as the paradigmatic example of the dispersive process (Fig. 4.1, 1a), the impossibility of a spontaneous change in the opposite direction is formally stated as the

[1]It has been mentioned in Sect. 1.3 and will be suggested repeatedly that irreversibility and "away from equilibrium existence" are synonymous: there is no irreversibility if the world is at equilibrium; it is also impossible to be away from equilibrium without any manifestation of irreversibility.

[2]Frictional is a better term here than dissipative as Kelvin used "dissipative" to represent both dispersal kinds and frictional kinds. In the following, I'll follow Kelvin using dissipative processes as *general* irreversible processes: dispersal and frictional processes as two specific kinds of irreversibility.

Clausius Statement of the second law: No process is possible in which the *sole* result is the transfer of heat from a cooler to a hotter body (Fig. 4.1, 1b).

The frictional kind of spontaneous processes is schematically represented in Fig. 4.1, 2a. The impossibility of a spontaneous change in the opposite direction, i.e., heat spontaneously transformed into work, is formally stated as the

Kelvin-Planck Statement of the second law: No process is possible in which the *sole* result is the absorption of heat from a reservoir and its complete conversion into work *W* (Fig. 4.1, 2b).

In the above Kelvin–Planck Statement, Kelvin hypothesized that, while it is common knowledge that work can be converted to heat without limit, heat may not be completely converted into work (see Sect. 4.6). Thus, heat and work do not have a reciprocal relation; but Kelvin was not consistent on this important point in his writing and famously he experienced a change of mind as reflected from his pre-1851 writings to his writings afterward. In this inconsistency it rests what ills the MEH-based mechanical theory of heat or the energetic theory of heat. This disquisition aims to address this incoherency of the energetic theory, beginning with Sects. 4.7 and 4.8, and the full resolution of the incoherency will be given in Chap. 8. It suffices, for now, to say that MEH, or the first law, does not deal with the question of the amount of the driving force that is expended in the driving of a process. The cause, i.e., the driving force that is expended, of the processes is the purview of the second law of thermodynamics. The energetic bias of the MEH-based mechanical theory of heat prevents us from appreciating this point and, in fact, prevents us from understanding, correctly, heat and energy themselves. The development toward a full meaning of the second law and that understanding will be treated in Chaps. 5, 7, and 8.

Sadi Carnot, the founder of the science of heat, presented the first challenge to the Newtonian worldview that all phenomena can be explained by physical forces.

Sadi Carnot (1796-1832)

4.2 The Carnot Cycle and Carnot's Principle

Before Kelvin formulated the K-P Statement, Sadi Carnot had already realized that heat from a single body cannot be transformed into work. Carnot's study of heat was nominally considered to be based on the caloric theory of heat, even though a case will be made in Chap. 8 that Carnot's theory of heat is not a caloric theory of heat. According to the caloric theory, heat or *caloric* was thought to be an invisible, weightless fluid that could be neither created nor destroyed. Therefore, there can be no consumption of caloric; instead, "the production of motive power is then due in steam engines not to an actual consumption of the caloric, but to its transportation from a warm body to a cold body" [1:7]. Again, in his own words

> The motive power of a waterfall depends on its height and on the quantity of the liquid; the motive power of heat depends also on the quantity of caloric used and on what may be termed the height of its fall, that is to say, the difference of temperature of the bodies between which the exchange of caloric is made. In the waterfall the motive power is exactly proportional to the difference in level between the high and low reservoirs. In the fall of the caloric the motive power undoubtedly increases with difference in temperature between the warm and cold bodies [1:15]. [See Eq. (49) below.]

The motive power of a waterfall is the potential energy (and possibly some kinetic energy also) of the waterfall. In accordance with the modern understanding, the motive power of heat is the "free energy" (see its definition in Chap. 7) of the thermal energy. However, Carnot in his inquiry into the motive power of heat had neither this terminology of free energy nor the concept of energy.

Carnot, as we shall see, was nonetheless able to conduct his inquiry by the method of *thought experiment*[3]: Imagine air in a cylindrical vessel (Fig. 4.2) provided with a movable piston at an initial position *cd*. The cylindrical vessel will be in contact with one of the two bodies, a constant temperature heat source body *A* and a constant temperature heat sink body *B*, at certain points during the course of operations. He *prescribed* the series of operations as follows [1:17–18]:

1. Bring the vessel with the air enclosed in the space defined by the piston at *cd* in contact with the heat source *A*.
2. The piston rises gradually from the position *cd* to a position *ef* (end position of isothermal expansion step) while the vessel is at all time in contact with heat source *A*, which furnishes heat to vessel to keep the air temperature constant.
3. *A* is removed and the air vessel continues to expand without heating with the piston rising from *ef* to *gh* (end of **isentropic expansion** [i.e., reversible adiabatic] step). This "expansion cooling" step lowers the temperature of the air in the vessel to the temperature of *B*.

[3]*Thought experiments* (theoretically constructed experiments) are devices of the imagination used to investigate the nature of things. We need only to list a few of the well-known thought experiments to be reminded of their enormous influence and importance in the sciences: Galileo's tower of Pisa, Newton's bucket, Maxwell's demon, Einstein's elevator, Heisenberg's gamma-ray microscope, Schrödinger's cat and, of course, Carnot's cycle.

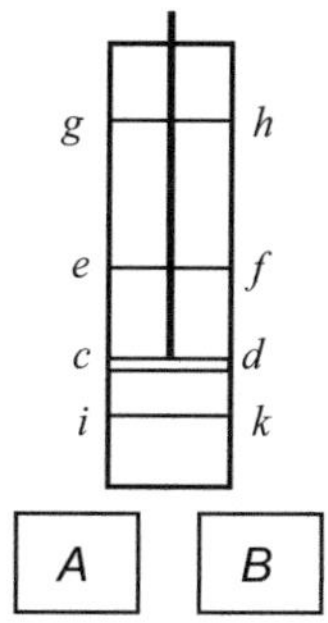

Fig. 4.2 Carnot's thought experiment

4. The air is placed in contact with heat sink B while it is compressed by the return of the piston from gh to cd (end of **isothermal compression** step). Air is kept isothermal by rejecting heat to B.
5. B is removed and the compression of air continues adiabatically until air temperature reaches the temperature of A. The piston correspondingly moves from cd to ik (end of **isentropic compression**).
6. The air is again placed in contact with A undergoing **isothermal expansion** with the piston moving from ik to ef.
7. The step 3 is renewed, which is followed by steps 4 and 5, and again 6, 3, 4, and 5—forming a four-step ($6 \rightarrow 3 \rightarrow 4 \rightarrow 5$) isothermal expansion at $t_A \rightarrow$ isentropic expansion $\rightarrow$ isothermal compression at $t_B \rightarrow$ isentropic compression cyclic process.

Prescribing the four-step cyclic process, Carnot transformed the caloric theory by demonstrating the reversible "fall of the caloric" (see Sect. 4.5 for the possible meaning of "caloric flow") to be the cause of producing the motive work. In fact, the Carnot cycle, Carnot's principle (below), and his conclusion that "Heat alone is not sufficient to give birth to the impelling power: it is necessary that there should also be cold; without it, the heat would be useless", transformed the caloric theory of heat into a new theory of heat—we shall call it Carnot's theory of heat. In Sect. 8.6.2, I shall give the reason why it should be considered a new, different theory of heat rather than a special version of the caloric theory of heat.

Reading his memoir [1], one is struck by the originality Carnot exhibited in its pages: his use of thought experiments, his realization that air is potentially a better working fluid than steam, his invention of using the cyclic process as an analytical tool, and the concept of maximum reversible work. He used expansion cooling making it the "instrument" of bringing about the fall of the caloric "reversibly" and used compression heating to return the working fluid to the same starting point so that the cyclic process can be repeated indefinitely.

The most stunning of all is the concept of *thermodynamic reversibility*—a theoretical construct via thought experiments. A heat engine operates with the outcome of converting heat into work, but a reversible engine can operate as a heat engine as well as, in reverse, as a heat pump. From which Carnot formulated *Carnot's Principle*:

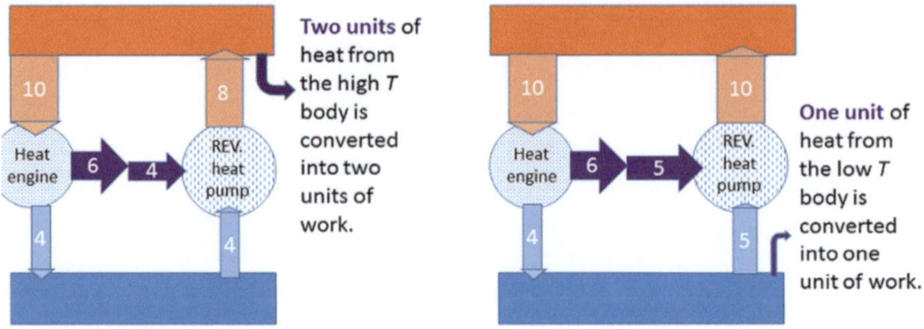

Fig. 4.3 Demonstration of Carnot's principle by supposing, contrary to the principle, that a heat engine has a higher efficiency than a reversible engine. The reversible engine can operate as a heat pump. Two possible arrangements are shown: On left, an arrangement is shown resulting in the extraction of heat from high T reservoir converting it into work. On right, it is shown the extraction of heat from low T reservoir into work. Both are perpetual machines of the second kind

No heat engine can be more efficient than a reversible one operating between the same temperatures.

For suppose (see Fig. 4.3) that a heat engine, contrary to the Principle, can be more efficient than a reversible one, the reversible one can be operated in reverse as a heat pump and the combined outcome of two engines connected mechanically and operating between two reservoirs, the former as a heat engine and the latter in reversed heat pump operation calibrated so that no *net* heat gain/loss to one of the reservoirs. Such operation results in the extraction of heat from the other reservoir and the conversion of this heat into work. That is, the combined set of two engines is a perpetual machine of the second kind, which violates the Kelvin–Planck Statement. The supposition thus is false, which proves Carnot's Principle.

It is important to note that whereas the caloric theory of heat is discredited, *Carnot's theory of* heat is not, because it is not (see Chap. 8) a caloric theory even though it made use of the concept of 'caloric': "To discredit the caloric theory is not to discredit Carnot" as noted Kuhn [2]. Jaynes in a 1984 article *The evolution of* Carnot's principle [3] made the comment, "Carnot's reasoning is outstandingly beautiful, because it deduces so much from so little…but at the same time with such a compelling logical force." It remains compelling today for us to ask what is the meaning of "caloric" as used by Carnot and how we should understand Carnot's theory in view of the MEH, both questions defining in large measure our inquiry beginning with Sects. 4.3 and 4.4.

4.3 The Absolute Thermodynamic Temperature

A corollary of *Carnot's Principle* is as follows:

> The motive power of heat is independent of the **agents** employed to realize it; its quantity is fixed solely by the temperatures of the bodies between which is effected, finally, the transfer of the caloric [1:20].

Expressing the corollary mathematically, it is

$$W = W(Q_A, t_A, t_B) \tag{44}$$

The first decisive step toward the formulation of the second law was taken by Kelvin [4:100–106 (1848)] by introducing the concept of the absolute temperature. What follows is the modern treatment of the topic (e.g., Fermi [5]), which starts with the premise of the MEH firmly in place while, originally in his 1848 treatment, Kelvin had not yet accepted the premise.[4]

Consider a reversible Carnot cycle engine receives Q_A from reservoir at t_A operating in a cyclic process. Work produced in the cyclic process equals to the net heat or $Q_A - Q_B$ (see Eq. [24] in Chap. 3), where Q_B is heat rejected by the engine to reservoir at t_B. Introduce the *efficiency* of the Carnot cycle to be the ratio of the work performed by the cycle to the heat absorbed from the heat source at temperature t_A.

$$\eta \equiv \frac{W}{Q_A} = \frac{Q_A - Q_B}{Q_A} = 1 - \frac{Q_B}{Q_A} \tag{45}$$

Accordingly, the corollary may be restated

> The efficiency of the reversible work derived from a heat source depends solely on the temperature of the heat source, t_A, and the temperature of the heat sink, t_B, independent of working fluids used in the production of work.

That is, the mathematical expression of Carnot's principle becomes

$$W = W(Q_A, t_A, t_B) = Q_A \eta(t_A, t_B) \tag{44A}$$

We shall prove that the function $\eta(t_A, t_B)$ has the following property. Imagine a third body of heat reservoir at a temperature t_C ($t_A > t_B > t_C$). We now consider two reversible Carnot heat engines in series: the first, engine *A*, operates between reservoirs *A* and *B* and a second engine, engine *B*, operates between reservoirs *B* and *C*. We assume, for sake of simplicity, that the two engines operate in such manner that the amount of heat, Q' (at the temperature t_B) rejected by engine *A* to

[4] In the 1848 paper, he had this to say: "the conversion of heat (or *caloric*) into mechanical effect is probably impossible*, certainly undiscovered." In the *footnote, however, Kelvin acknowledged the contrary opinion advocated by Joule, and signaled the move to accept the MEH that he would take in a very short time (1851).

the t_B reservoir equals the amount of heat received by engine B from the t_B reservoir. The t_B reservoir thus experiences no net heat exchange. By a simple mechanical linkage, A and B can be combined into a single reversible engine to be called composite reversible engine, which receives heat from t_A reservoir and rejects heat to t_C reservoir.

Let W_A be the work output of the first engine A, W_B the work output of the second engine B, and W_{COM} the work output of the composite engine, i.e., $W_{COM} = W_A + W_B$. We have then the following relations in accordance with Eq. (44A):

$$\eta_A \equiv \eta_{AB}(t_A, t_B) = \frac{W_A}{Q_A}\left(= \frac{Q_A - Q'}{Q_A}\right)$$

$$\eta_B \equiv \eta_{BC}(t_B, t_C) = \frac{W_B}{Q'}$$

$$\eta_{COM} \equiv \eta_{AC}(t_A, t_C) = \frac{W_A + W_B}{Q_A}$$

From these three relations,

$$\begin{aligned}\eta_{AC} &= \frac{W_A + W_B}{Q_A} = \frac{W_A}{Q_A} + \frac{W_B}{Q'}\left(1 - \left[1 - \frac{Q'}{Q_A}\right]\right)\\ &= \frac{W_A}{Q_A} + \frac{W_B}{Q'}\left(1 - \frac{W_A}{Q_A}\right) = \frac{W_A}{Q_A} + \frac{W_B}{Q'} - \frac{W_A}{Q_A} \times \frac{W_B}{Q'}\end{aligned}$$

That is,

$$\eta_{AC} = \eta_{AB} + \eta_{BC} - \eta_{AB} \times \eta_{BC}$$

Rearrangement yields

$$\eta_{AC} = 1 - (1 - \eta_{AB})(1 - \eta_{BC})$$

That is,

$$1 - \eta_{AC} = (1 - \eta_{AB})(1 - \eta_{BC}), \text{in which } t_A > t_B > t_C$$

This condition can be met if and only if *function* $(t_A, t_B) = 1 - \eta_{AB}$ has the functional form

$$function(t_A, t_B) = f(t_B)/f(t_A)$$

and by Eq. (45)

$$1 - \eta_{AB} = f(t_B)/f(t_A) = Q_B/Q_A$$

Any specific scale of temperature t based on a thermometric substance (like mercury) was an arbitrary one, thus the meaning of t was ambiguous. Turning the argument around, Kelvin perceived that the above functional form derived from Carnot's principle provided the resolution to this ambiguity: Note that the reversible efficiency $(1 - Q_B/Q_A)$, i.e., the ratio of the Q's, Q_B/Q_A, is directly equal to the ratio of functions of the temperatures. Since the ratio of Q's is well defined theoretically according to Carnot's principle, the ratio of functions of the temperatures is thus as well, and we may as well use the function of the temperature f itself as the temperature scale. That is, function f, in effect, *defines* a universal temperature scale that is independent of the properties any particular substance. Writing it instead as θ, the above equation is rewritten in the general form

$$\frac{Q_A}{\theta_A} = \frac{Q_B}{\theta_B} \tag{46}$$

The temperature function just defined is called the *absolute thermodynamic temperature*.

To complete the definition of the Kelvin scale we proceed as in Chap. 1 by assigning the arbitrary value of 273.16 K to be the temperature of the triple point of water (designated as T_3). (The TRIPLE POINT is defined to be the state of a substance all three phases of it exist in phase equilibrium. The triple point of water is 273.16 K and 611.224 Pa.) The temperature of an arbitrary body is theoretically determined by the values of Q and Q_3 of a Carnot engine operating between the body and a body at the water triple-point temperature

$$\theta(K) = 273.16(K) \cdot \frac{Q}{Q_3} \tag{47}$$

This assignment corresponds to the freezing point of water at 273.15 K and 1 atm and the boiling point of water at 373.15 K and 1 atm, that is, a difference of 100 K between the two historically *fixed points* in thermometry.

At first thought, it might seem that the ratio of two Kelvin temperatures would be impossible to measure, since a Carnot engine is an ideal machine, quite impossible to construct. The situation, however, is not as bad as it seems. The ratio of two Kelvin temperatures is the ratio of two heats that are transferred during two isothermal processes bounded by the same two adiabatic. The two adiabatic boundaries may be located experimentally, and the heat exchanges during two "internally reversible" isothermal processes can be measured with considerable precision. As a matter of fact, this is one of the methods used in measuring temperatures below 1 K.

4.3.1 Carnot's Reversible Efficiency

Carnot's reversible efficiency is, in view of Eqs. (45) and (46)

$$\eta = 1 - \frac{Q_B}{Q_A} = 1 - \frac{\theta_B}{\theta_A}$$

We shall revisit this result in the two sections below.

4.4 Carnot's Function and Kelvin's Resolution of the Conflict Between MEH and Carnot's Principle

The modern reconstruction of the thermodynamic concept of temperature in Sect. 4.3 is logical, a poster child of efficiency-in-reasoning with a 20/20 hindsight. But it overlooks an important chapter in the actual evolution of the temperature concept, which was a messy affair (see [6]) but in which a *better* appreciation of Carnot's reasoning can be found.

The centerpiece in this history is the temperature function, which is known as *Carnot's Function*

***Carnot Function, μ*: A relation between the amount of heat given off by a source of heat, and the reversible work that can be derived from it**

$$\delta W = \mu(t)Qdt \tag{48}$$

where W is the reversible work, Q is the caloric (i.e., the amount of heat in modern thermodynamics), t is the temperature of a specific scale. Equation (48) is a special case of Eq. (44), $W = W(Q, t_A, t_B)$.

This history is instructive, especially in how Kelvin resolved the conflict between Carnot's principle and Joule's MEH.

Carnot made several attempts to determine the temperature function and noted, "We do not know what laws it [the isothermal expansion–heat absorption step 6 or the isothermal compression–heat rejection Step 4] follows relative to the variations in volume: it is possible that its quantity changes either with the nature of the gas, its density, or its temperature. Experiment has taught us nothing on this subject" [1:16–17]. Interestingly, it was the absence of the knowledge of the precise amount of caloric that gave him the freedom to assume the μ value in Steps 6 and 4. In one case he concluded that μ is constant so that, since Q is a conserved quantity in the caloric theory, integration of Eq. (48) yields

$$W = \mu Q(t_A - t_B) = Q^*(t_A - t_B) \tag{49}$$

where $Q^* = \mu Q$. That is, "the motive power produced would be found exactly proportional to the fall of caloric." He, however, drew back from endorsing this special case afterward, realizing that he did not have the sound reason for this conclusion.

A detailed account of the intellectual journey began with Eq. (48) that would lead to the birth of thermodynamics can be found in Truesdell's estimable book, *The Tragicomical History of Thermodynamics, 1822–1854* [6]. A scholarly account was given by Cropper [7], which is summarized in the table below.

		Carnot's temperature function	
1824	Carnot	$\mu(t) \equiv \frac{dW}{Qdt}$	Existence of a temperature function as a corollary of Carnot's Principle
1834	Clapeyron	$\frac{1}{\mu} = \frac{1}{R}\left[v\left(\frac{\partial Q}{\partial v}\right)_p - p\left(\frac{\partial Q}{\partial v}\right)_p\right]$	Marred by the use of the *objectionable* heat function
1845	Holtzmann		Marred by using heat function
1847	Helmholtz	$\mu = a/(t + 1/\alpha)$	First identification of μ with the Inverse of absolute temperature—but marred by using heat function
1848	Kelvin	Proposal of thermometry principle	Kelvin's first paper on the absolute temperature
1848	Joule	$\mu = J/(t + 1/\alpha)$	The expression was disclosed in a letter to Kelvin
1850	Clausius	$\mu = J/(t + 1/\alpha)$	This result, obtained Independently by Clausius and Joule, is important because it avoided the use of heat function
1851	Rankine	Treatment of the problem try using microscopic models	
1851	Kelvin	$W = JQ_2\left[1 - exp\left(-\frac{1}{J}\int_{t_1}^{t_2} \mu(t)dt\right)\right]$	Kelvin now fully accepted that heat and work were interconvertible and, at the same time, realized that that acceptance did not have to discard what was essential in Carnot's theory: Carnot's
1854	Kelvin	$\mu = J/\theta$	Instead of the determination off μ as a function of temp., the realization that the Function *determines* the absolute temp θ

As shown in the table, attempts toward the determination of Carnot's function were made by many including Clapeyron, Holtzmann, Helmholtz, Kelvin, Joule, and Clausius, but no resolution was obtained. In 1848, Kelvin was faced with this situation (the greatest unresolved issue in his mind):

1. Carnot's theory of heat expresses work as Eqs. (48) or (49),

$$W = \mu Q(t_A - t_B) = Q^*(t_A - t_B) \tag{49}$$

2. The "Mayer–Joule mechanical equivalent of heat" expresses work

$$W = J(Q_A - Q_B) \tag{24A}$$

That is, the former views work production to be resulting from the *transfer* of caloric, and the latter work production to be resulting from the *consumption* of heat. Both Carnot and Mayer–Joule called Q^* in Eq. (49) and Q in Eq. (24A), respectively, the caloric. Clearly, the two theories were contradictory to each other. There were two possible logical choices: (1) one of the two theories, Carnot's or Mayer–Joule's, was wrong; or, (2) both were right, but Q and Q^* are different entities.

Sometime in 1850 or 1851 K began to realize that the acceptance of the equivalence (i.e., the MEH) of heat and work did not have to discard what was essential in Carnot's theory. He then adopted the view [4:189 and 190] that heat dropped from a high to a low temperature in a reversible heat engine was, instead of being a conserved quantity, being continuously converted to work. If during a microcyclic step of the engine [in Eq. (24A)], an amount of heat $Q + \delta Q$ at $t + dt$ would descend to Q at t, δQ of the heat would be converted to an equivalent amount of work δW. That is,

$$\delta W = J\delta Q \tag{50}$$

The same work was calculated according to (48)

$$\delta W = \mu(t) Q dt \tag{51}$$

Though Eq. (51) had the same mathematical appearance as Eq. (48), there was a difference in their use: the original Eq. (48) was constrained with Q to remaining constant, but Q in Eq. (51) is variable at different levels of t while heat is being converted into work in view of Eq. (50). The combination of Eqs. (50) and (51) resulted in

$$J\delta Q = \mu(t) Q dt$$

which could be rearranged

$$\frac{\delta Q}{Q} = \left(\frac{1}{J}\right)\mu(t)dt$$

Kelvin arrived at, by the integration of which, the result

$$ln\left(\frac{Q_A}{Q_B}\right) = \frac{1}{J}\int_{t_B}^{t_A} \mu(t)dt$$

or

$$Q_B = Q_A exp\left(-\frac{1}{J}\int_{t_B}^{t_A} \mu(t)dt\right)$$

The complete heat engine cycle converted the heat $Q_A - Q_B$ to the amount of work $W = J(Q_A - Q_B)$. Kelvin thus obtained in 1851 the result

$$W = JQ_A\left[1 - exp\left(-\frac{1}{J}\int_{t_B}^{t_A} \mu(t)dt\right)\right] \tag{52}$$

In 1854, Kelvin [4:393 and 394] returned to his 1848 thermometry principle, which redefined the problem of Carnot's function: Instead of the determination of μ as a function of temperature, the realization that the function *determines,* i.e., *defines,* an absolute temperature scale. Representing temperatures on this scale with θ, thermometry principle asserts

$$\theta = J/\mu$$

or,

$$\mu = J/\theta \tag{53}$$

Substitution of which into Eq. (52) yields (because $dt = d\theta$)

$$W = JQ_A\left[1 - exp\left(-\frac{1}{J}\int_{\theta_B}^{\theta_A} \frac{J}{\theta}d\theta\right)\right] = JQ_A\left(1 - \frac{\theta_B}{\theta_A}\right) \tag{54}$$

William Thomson, the future Lord
Lord Kelvin (1824-1907)

Cropper noted, "On the day in 1854 when he safely arrived at Eqs. (34) and (36), i.e., Equations (53) and (54), Thomson (the future Lord Kelvin) must have felt like celebrating. This was the Carnot calculation finally resolved, 30 years after Carnot had originally asked his question about maximum (useful) work output from an ideal heat engine" [7].

The conflict between the two theories of heat was partially resolved by the replacement of μQ (or Q^*) with JQ/θ, which, in view of Eq. (46), is constant. Carnot's Eq. (48) can be integrated to yield Eq. (49) as an exact expression [8] which is a valid conclusion after all even though Carnot did not have the sound reason for reaching this conclusion.

Kelvin achieved the beautiful resolution by making a conceptual differentiation of the caloric flow into heat flow Q and a second flow-entity Q^*, or "JQ/T" (see Sect. 4.5 below). Tentatively, we may identify this flow of JQ/T as a flow of something called caloric (to be renamed "entropy" flow in anticipation of defining entropy properly in Chap. 5).

4.5 Falling of Caloric in Reversible Processes

4.5.1 Absolute Thermodynamic Temperature and the Ideal-Gas Thermometric Temperature

Consider a Carnot cycle performed by an ideal gas (Fig. 4.4) which absorbs heat Q_A from a heat source at T_A (T is an ideal-gas temperature) during the isothermal expansion step 1, $1 \rightarrow 2$, and rejects heat Q_B to a heat sink at T_B during the isothermal compression step 3, $3 \rightarrow 4$. The second step, $2 \rightarrow 3$ is an adiabatic (isentropic) expansion step and the fourth $4 \rightarrow 1$ an adiabatic (isentropic)

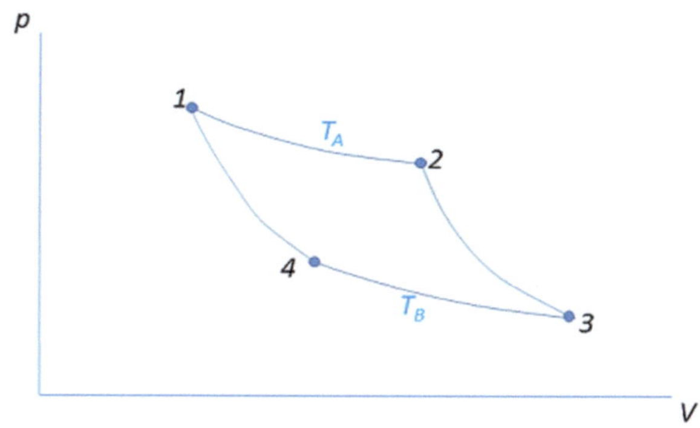

Fig. 4.4 Carnot gas cycle: Step 1 → 2 is isothermal expansion; Step 2 → 3 is reversible adiabatic expansion; Step 3 → 4 is isothermal compression; Step 4 → 1 is reversible adiabatic compression

compression step. The following are the works performed by the ideal gas at each step (a work done by the ideal gas expressed as a positive value and on the ideal gas as a negative value):

Step 1 (see Sect. 1.8)

$$W_{12} = \int_{V_1}^{V_2} pdV = NRT_A \ln\frac{V_2}{V_1}$$

During the isothermal process, there is no change in the internal energy of the ideal gas, it follows from the first law, Eq. (22),

$$Q_A \equiv Q_{12} = W_{12}$$

Step 2

From the first law, Eq. (22), the work performed in an adiabatic step is equal to the change in internal energy (a case that W is directly related to change in property)

$$W_{23} = U_2(T_A) - U_3(T_B)$$

From Eqs. (34) and (37),

$$W_{23} = Nc_V(T_A - T_B) = \frac{NR}{k-1}(T_A - T_B)$$

Step 3

Similar to Step 1, the work is

$$W_{34} = \int_{V_3}^{V_4} pdV = NRT_B \ln \frac{V_4}{V_3}$$

Note, however, using Eq. (40) we have the relation

$$\left(\frac{T_A}{T_B}\right)^{\frac{1}{k-1}} = \frac{V_4}{V_1} = \frac{V_3}{V_2}$$

It follows, therefore,

$$\frac{V_4}{V_3} = \frac{V_1}{V_2}$$

and,

$$W_{34} = -\frac{T_B}{T_A} W_{12} = -\frac{T_B}{T_A} Q_{12}$$

Step 4

It is similarly treated as Step 2 and

$$W_{41} = Nc_v(T_B - T_A) = -W_{23}$$

That is

$$W_{41} + W_{23} = 0$$

The NET Work performed during one cycle by the ideal gas is, thus,

$$W = W_{12} + W_{34} = \left(1 - \frac{T_B}{T_A}\right) Q_{12} = \left(1 - \frac{T_B}{T_A}\right) Q_A$$

A comparison of the expression with Eq. (45) leads to the conclusion that

$$\frac{T_B}{T_A} = \frac{Q_B}{Q_A}$$

or

$$\frac{Q_A}{T_A} = \frac{Q_B}{T_B} (= Q^*) \tag{55}$$

That is, in view of Eqs. (46) and (55), a choice can be made

$$\theta = T \tag{56}$$

The empirically defined ideal-gas thermometric temperature T in Chap. 1 is shown to coincide with the theoretical constructed thermodynamic temperature θ, which is independent of the special properties of any particular thermometric substance. The empirically measurable T is shown to rest on a theoretical basis independent of thermometric assumptions.

Finally, Eq. (54) becomes in view of Eq. (56)

$$W = Q_A\left(1 - \frac{T_B}{T_A}\right) \tag{54A}$$

4.5.2 Falling of Caloric

Carnot's "falling of caloric" is the flow of "entropy"; "entropy" flows as a "conserved" quantity if and only if the process is thermodynamically reversible. Note that work is *not* associated with any entropy flow. In this reversible operation of a Carnot heat engine (Fig. 4.5), both the energy flows (Fig. 4.5a) and the "entropy" flows (Fig. 4.5b) of the Carnot heat engine are balanced: The energy flow balance in Fig. 4.5a is

$$Q_A = Q_B + W_{Carnot\ engine}$$

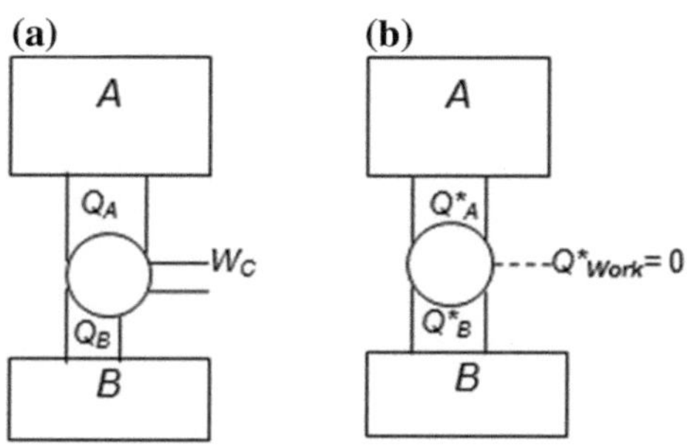

Fig. 4.5 **a** (left diagram) represents the conservation of energy in heat engine; **b** (right diagram) represents the "conservation" of "entropy" in a reversible heat engine

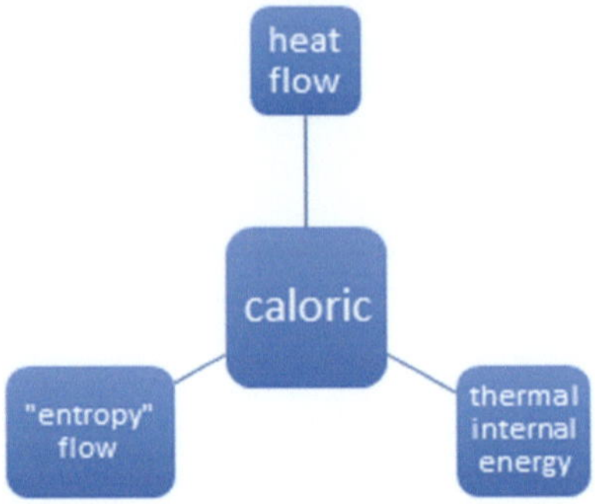

Fig. 4.6 The term **caloric** can be viewed to approximate the idea that I call *heat*

That is,

$$W_{Carnot\,engine} = Q_A - Q_B.$$

The "entropy" flow balance in Fig. 4.5b is

$$Q_A^* = Q_B^* + \left(Q_{\mathrm{Work}}^*\right)_{Carnot\,engine} = Q_B^*$$

where

$$\left(Q_{Work}^*\right)_{Carnot\,engine} = 0$$

since work is not associated with entropy flow.

Figures 4.5a and 4.5b and their interpretation are consistent with the MEH, Eq. (24A), that work results from the consumption of heat and the corollary of Carnot's principle, Eq. (49), that work results from the transfer of heat, respectively.

Note Q's in Eqs. (46) and (55)

$$\frac{Q_A}{T_A} = \frac{Q_B}{T_B} (= Q^*)$$

are all positive terms. By adopting, instead, the sign convention of positive Q for heat received by the engine or the system and negative Q for heat rejected by the engine or the system, the equation takes the form

$$\frac{Q_A}{T_A} = \frac{-Q_B}{T_B} = Q^*$$

That is, Eq. (55) may be written instead as

$$\frac{Q_A}{T_A} + \frac{Q_B}{T_B} = 0 \tag{55A}$$

A generalization of Eq. (55A) may be made by considering a more complicated heat engine that runs cyclically making contact successively with n reservoirs at temperatures $T_1 \ldots T_n$, where the application of the first law results in

$$W = \sum_1^n Q_i.$$

Kelvin in 1854 obtained by the application of Carnot's principle the result of generalized Eq. (55A)

$$\sum_{i[1\to n]} \frac{Q_i}{T_i} = 0 \tag{57}$$

Equations (55A) and (57) are the most important consequence of Carnot's principle that encapsulates the essence of "falling of caloric" in reversible processes. For instance, by making use of which, Carnot's equation, Eq. (49), can be reduced immediately, to Eq. (24A), as suggested in Fig. 4.5

$$W = Q^*(t_A - t_B) = Q^*(T_A - T_B) = \frac{Q_A}{T_A} \times T_A - \frac{-Q_B}{T_B} \times T_B = Q_A + Q_B$$

4.5.3 The Carnot Formula and the Kelvin Formula

Both expressions of Carnot and Kelvin for reversible work are recapitulated here:

$$W_{rev}(Carnot) = Q^*(T_A - T_B) \tag{49}$$

$$W_{rev}(Kelvin) = Q_A\left(1 - \frac{T_B}{T_A}\right) \tag{54A}$$

which will be known as the Carnot formula and the Kelvin formula, respectively. The suggested name of the Kelvin formula, though a departure from the common custom of calling it the Carnot formula, is a proper one in view of Kelvin's pivotal role in its development: the quantitative expression employed by Carnot was Eq. (49), while the expression of Eq. (54A) was first given by Kelvin.

4.5.4 Caloric or *Heat*: Interpreted as Both Heat Flow and "Entropy" Flow

The heat exchange of a system is always balanced with the work exchange regardless of whether the cyclic process is reversible or irreversible. The simple balance of "entropy"-inflow with "entropy"-outflow, i.e., "entropy" conservation, is an entirely different matter; "entropy"-flow is balanced in this simple manner only in this special case of reversible cyclic process. In fact, perfect reversibility, which is a theoretical construct, does not exist in the real physical world. All real processes, natural spontaneous ones as well as artificial ones, are irreversible and, thus, involve entropy production (see Chap. 6). The entropy balance of all real cases must take into consideration this inevitable entropy production in addition to the balance of entropy flow[5] (see Sect. 6.4.1).

The earlier conceptual differentiation of "caloric" into heat flow and heat energy (the MEH differentiation, Sect. 3.3), and the conceptual differentiation here of the "flow of caloric" into heat flow and "entropy" flow (Fig. 4.5a, b) are connected but distinctive moves: That is to say, the two moves together suggest that the single concept of *caloric*, as used differently by Joule (Q), Clausius (Q and U), and Carnot (Q^*), should be conceptually differentiated into the concepts of *heat flow*, *thermal* internal energy, and *"entropy" flow* as shown in Fig. 4.6. With the conceptual differentiation, both Carnot's "falling of caloric" understanding of work production as a result of the transfer of caloric ("entropy") shown in Fig. 4.5b and Joule's understanding of work production as a result of the consumption of heat shown in Fig. 4.5a are consistent with each other. An unsettled question is the meaning of "consumption" used in the MEH.

The aforementioned concept of caloric in Fig. 4.6 is used interpretatively, rather than as a definition. It has the same meaning of the term that I call *heat*; no attempt of giving either term a definition is made here, rather, the use of either term implies their understanding requires both energetic and entropic points of view[6]: The MEH differentiation stresses the quantitatively energetic equivalence of the effects of heat flow and work (flow) on change in system internal energy. The new differentiation here, instead, highlights the qualitatively entropic difference between heat flow and

[5]In the case of a steady-state general system, it may be noted that the entropy flowing out of the system is always larger than the entropy flowing into the system as a result of entropy production. See Chap. 6 for more details.

[6]The treatment of the second law advocated here is closest to that of Planck, which stresses the centrality of the second law and the energetic/entropic understanding of heat. A distinction is made between MEH (in its pure sense, not how Joule and Kelvin interpreted it as discussed in Sect. 4.7) and the "reductive" mechanical theory of heat. In the specific treatment of Q (heat exchange, not heat in the general inclusive sense) in Sects. 3.2 and 3.3 in the above, however, I followed Helmholtz and Born by adopting the "mechanical" definition of Q—which is different from Planck, who considers heat to be a primitive concept and elects to stay away from the reductive mechanical theory of heat approach. For clarity, it is noted again that the use of the "mechanical" definition of Q in Sect. 3.3 does not require the full acceptance of the reductive mechanical theory of heat.

work in terms of whether they are associated with entropy flow or not (see Chap. 5): the answer is heat flow (Yes); work (No).

4.5.5 Equivalence of the Clausius Statement and the Kelvin-Planck Statement

One interesting application of the concept of thermodynamic reversibility is the use of the Carnot cycle to prove the equivalence of the two statements of the second law. Let us assume, in contradiction to the Clausius statement, it is possible to transfer Q_1 from a heat reservoir at T_1 to a heat reservoir at a higher temperature T_2 in such a way that no other change occurs in the state of any system or medium that are involved. Then a Carnot cycle can be used to absorb $Q_1 + Q_2$ from the high-temperature reservoir and reject Q_1 to the T_1-reservoir producing Q_2 amount of work. The combined heat transfer steps result in zero net heat exchange in the T_1-reservoir. On the other hand, the T_2-reservoir gives up Q_2 amount of net heat, which equals the amount of work produced by the Carnot cycle. The sole consequence of the whole operation is the complete transformation of (Q_2)-heat into (Q_2)-work, in contradiction to the Kelvin–Planck statement.

It can be similarly demonstrated that if the Kelvin–Planck statement were not valid, the Clausius statement would not be valid as well. The two statements are thus equivalent.

Carnot's principle and the two Statements, as well as the MEH, were expressed originally in terms of heat (more precisely, heat at high temperature and heat at low temperature) and work. Note that the first law of thermodynamics, while it had had its origin in the MEH in terms of heat and work (mechanical energy), underwent an evolutionary process to becoming expressed in terms of heat, work, and energy. It is important, therefore, to note how the two Statements underwent an idea evolution as well to becoming expressed in terms of energy and its dissipation (Sects. 4.6 and 4.7) and, eventually, in terms of entropy and its growth (Chap. 5).

4.6 Limitation in the Amount of Heat to be Converted into Mechanical Energy

What is the meaning of "consumption of heat" used in the MEH?

The MEH, or equivalence principle, initially meant different things to different investigators. Rumford, Mayer, and Joule committed the MEH to be a principle of conversion between mechanical work and heat. To another group of investigators including Carnot, Clapeyron, Holtzmann, and Thomson (initially before 1850), the concept meant to be an assumption of "a certain association between heat and work, such that the two existed independently of one another but could influence each other" [9]. It meant the coexistence of heat and work: "Sometimes coexistence was hidden behind such words as 'proportionality' or 'equivalence,' which indicated a numerical relation between heat and work and nothing more" [9].

Then, Thomson experienced a decisive change of mind and in his authoritative and influential 1851 paper *On the dynamical theory of heat*, [4:174–200] he opened with these fateful words, "…Considering it as thus established, that heat is not a substance, but a dynamical form of mechanical effect, we perceive that there must be an equivalence between mechanical work and heat, as between cause and effect." In this telling, the equivalence *correlation* was accepted as *causality* between heat and mechanical work, which became known as universal interconvertibility of heat and work [10]. Significantly, Thomson had vacillated about the issue of convertibility and was, famously, the last of the three founders of thermodynamics, Clausius, Rankine, and Thomson, to adopt fully Joule's principle of convertibility. But once converted, he became the most effective champion of Joule's principle.

However, while the spontaneous transformation from work to heat is limitless, *the* convertibility from heat to work is subject to strict limitation because, otherwise, an unrestricted reverse transformation would have contradicted the Kelvin–Planck statement that heat in itself cannot be converted into work. This was where Carnot's theory of heat came into our discussion. For this part of our discussion, we focus on the Carnot heat engine.

There are two ways to consider the Carnot heat engine, the conventional way and a new way of understanding the Carnot heat engine as it will be presented in Chap. 8. In the conventional way, we understand the Carnot heat engine as follows: the process is a reverse energy conversion of heat energy from a T_A heat body to mechanical energy; the three grades of energy involved are mechanical energy, W_{rev}, high temperature heat energy at T_A, Q_A, and low temperature heat energy at T_B; for a given amount of Q_A, the maximal amount of mechanical energy in the reversible Carnot heat engine is

$$W_{rev} = Q_A\left(1 - \frac{T_B}{T_A}\right) \tag{54A}$$

Correspondingly, the conventional takeaway is the following *general statements* (GSs):

> **GS. It is impossible to extract work from a heat source of Q_A amount without at the same time discarding a fraction of the heat—the minimal amount required is $Q_A \cdot (T_B/T_A)$**
>
> **GS-4.b. Heat, therefore, cannot be converted 100% into work.**

This is the first part of the clarification (on the equivalence principle/principle of the conservation of energy) Kelvin and Clausius introduced, between 1850 and 1854, when they attempted to incorporate Carnot's theory into the mechanical theory of heat, which will be referred to as the Kelvin–Clausius synthesis of Carnot's theory and equivalence principle.

4.7 The Energy Principle, A Self-evident Proposition?

The second part of the clarification is Kelvin's energy principle. Useful though the Carnot–Kelvin formula was, the unresolved goal for Kelvin was to formulate an overarching version of the idea of the irreversible world that could be a counterpoint to the principle of conservation of energy, which, as the mature version of MEH, is a quantitative, mathematical equation—unsurpassed in the simplicity of its premises, the disparate things it relates to, and the extent of its area of applicability.

Existing statements, the Kelvin Statement and the Clausius Statement (see Sect. 4.1 above, and Fig. 4.7), fell short of giving the second law statements the same sense of weight and usefulness as the first law Eqs. (22)–(24). Kelvin's own Statement contained the core element of natural tendency in the dissipation of mechanical work into heat. He realized at some point the possibility of interpreting the Clausius Statement in terms of the (same) lens of dissipation. This suggested the transformation of the Kelvin statement and the Clausius statement into the great theme of the universal dissipation of energy. As his biographers Smith and Wise wrote

> Thomson's commitment to a progressionist cosmology and geology has enabled him by 1851 to transform the problem of "loss" of available energy from an engineering issue into a universal cosmological one with clearly defined theological support. This transformation marked the beginning of his quest for a new economy of nature—an irreversible cosmos—as a systematic and coherent vision founded upon the universal dissipation of energy [11].

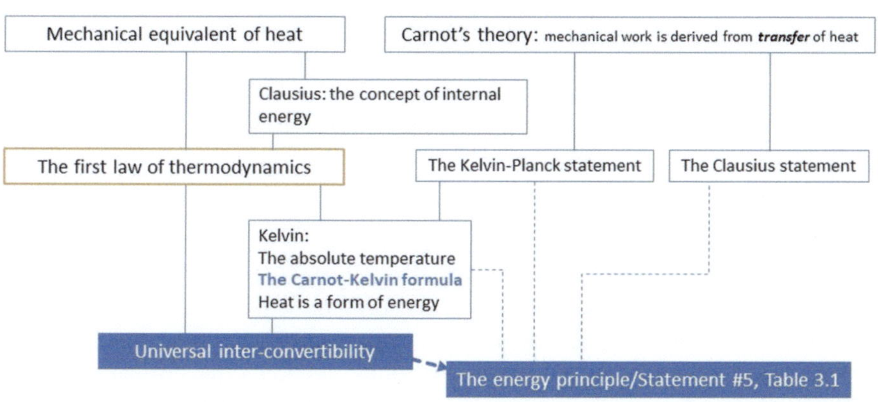

Fig. 4.7 Evolution of thermodynamic thought 1. Carnot's principle and the MEH were the foundations of the theoretical thermodynamic structure. The figure summarizes Kelvin's thermodynamic contributions: the introduction of absolute temperature, mathematical expression of Carnot's principle, Carnot–Kelvin formula, and his formulation of the second law in terms of the energy principle. It is noted that Statement #5, Table 3.1, the verbal statement of the first law which, as it was originally applied in Chap. 3, pointed out that the possibility of energy transformations, now, in view of the energy principle, acquires a new meaning of preferred direction to these energy transformations (see additional discussion in this section)

In his 1852 article [4:511–514], he opened with the statement "The object of the present communication is to call attention to the remarkable consequences which follow from Carnot's proposition, that there is an absolute waste of mechanical energy available to man when heat is allowed to pass from one body to another at a lower temperature, by any means not fulfilling his criterion of a 'perfect thermo-dynamic engine,' established, on a new foundation, in the dynamical theory of heat [referring back to his 1851 article: Ref. 4:174–200]." After a discussion on the possibilities of restoration of mechanical energy spent in processes of heat, he made the same point again as Proposition 3 in the article

> When heat is diffused by conduction, there is a dissipation of mechanical energy, and [with regards to the issue of restoration to the original state] perfect *restoration* is impossible.

Kelvin saw what unified frictional irreversible processes (Fig. 4.1, 2a) and dispersal irreversible processes (Fig. 4.1, 1a) was that all these processes were *dissipative,* giving a cosmic direction to time. It was dissipation that seemed to Thomson to lie behind all the disparate unidirectional phenomena (…bulk energy had been dissipated into increased energy in microscopic motions). The paper ended with a terse declaration of three "general conclusions" [4:511–514]:

GC-1. **"There is at present in the material world a universal tendency to the dissipation of mechanical energy."**

GC-2. **"Any *restoration* of mechanical energy, without more than an equivalent of dissipa tion, is impossible in inanimate material processes, and is probably never effected by means of organized matter, either endowed with vegetable life or subjected to the will of an animated creature."**

GC-3. **"Within a finite period of time past, the earth must have been, and within a finite period of time to come the earth must again be, unfit for the habitation of man as at present constituted, unless operations have been, or are to be performed, which are impossible under the laws to which the known operations going on at present in the material world are subject."**

These general conclusions taken together amounted to a declaration of a universal principle of dissipation of mechanical energy or high-grade energy. I refer to them as the energy principle. Note that the energy principle is *not* the principle of conservation of energy. The principle of conservation of energy in its original narrow sense merely acknowledges the possibility of energy transformation subject to energy constancy during all energy transformations, without any indication of the preferred direction of transformations.

Now, the energy principle gives preferred direction, in addition to energy constancy, of energy transformations, adding new meaning to Statement #5, Table 3.1. Before the 1852 paper, heat and mechanical work and their equivalence were the focus of thermodynamics in the investigation of Carnot, Mayer, Joule, and Clausius (though he also introduced the concept of internal energy). The preferred direction of spontaneous changes was represented by the Clausius statement and Kelvin–Planck statement in terms of heat, heat transfer, and work. With the energy principle, thermodynamics began shifting from heat–work equivalence to the new focus, energy and the transformation of energy (see Fig. 4.7). The process of this change in focus from heat to energy was a long one with detours along the way, the eventual completion of which was the formulation of the theory of exergy. A discussion of that development will be made in Chap. 7.

In a historical study of MEH and the general principle of energy conservation (GPEC), Kipnis made the case, "the 'principle of energy conservation' process did not start with a formulation of a general principle of energy conservation which stimulated the development of particular concepts, such as mechanical equivalent of heat. It will be shown that the opposite happened: it was the development of mechanical equivalent of heat which led to the general principle of energy conservation" [9]. (In Sects. 3.2 and 3.3, the treatment departed from historical development for a pedagogical reason.) In this development from heat to energy and from particular to general, Kelvin and his energy principle played a pivotal role [11].

It is striking in the manner Kelvin presented his principle. Von Baeyer commented on it with these words "The vague, metaphysical character of the principle of dissipation of energy—the natural 'tendency' of energy toward dilution—contrasts curiously with the robust, tangible way in which Thomson described the world [in his other scientific writings]" [12]. Uffink characterized thc *general conclusions* simply as "the un-argued statements of Kelvin" [13]. That is, Kelvin treated the energy principle to be a self-evident proposition.

Nonetheless, von Baeyer had this assessment, "Inasmuch as the second law is one of the pillars of physics, this was Thomson's most significant contribution to the science of thermodynamics, and overshadowed his invention of the absolute scale of temperature, his early recognition of the importance of James Joule's work…" How can an "un-argued statement" have impacts even surpassing the beautifully argued absolute scale of temperature?"

This assessment is correct. The reason for this assessment is that Kelvin extended the characterization of energy from merely in terms of its conservation to its conservation *and* its availability: the former was captured by the principle of conservation of energy and the latter by the energy principle; together, they became Statement #5, Table 3.1. One finds this characterization in the following passage as given in the draft of his 1851 paper [4:174–200]

> The difficulty which weighed principally with me in not accepting the theory so ably supported by Mr. Joule was that the mechanical effect stated in Carnot's Theory to be *absolutely lost* by conduction, is not accounted for in the dynamical theory otherwise than by asserting that *it is not lost* [i.e., the assertion of energy conservation]; and it is not known that it is available to mankind. The fact is, it may I believe be demonstrated that the work is *lost to man* irrecoverably; but [even though energy is] not lost in the material world. Although no destruction of energy can take place in the material world without an act of

power possessed only by the supreme ruler, yet transformations take place which removes irrecoverably from the control of man sources of power which, if the opportunity of turning them to his own account had been made use of, might have been rendered available.[7]

Kelvin, in this remarkable passage, made the fundamental distinction between the destruction of energy and lost energy (or, the term he later introduced in 1852, dissipation of energy): destruction is impossible whereas dissipation happens all the time. He made the connection of "irrecoverable" changes, though involving no destruction of energy, to loss in available energy. Energy transformations among energy forms referred to in Statement #5, Table 3.1, is amended that energy forms can be ordered in accordance to their available energies and spontaneous transformations takes place in the direction of decreasing available energies—transformations in reverse direction are possible but are *made* to happen instead of happening spontaneously.

The technological and social significance of Thomson's contribution here is given by Smith and Wise in *Energy and Empire* [11], and by Smith in the following sentence from one of his articles:

> As every physicist, indeed every schoolgirl and schoolboy know nowadays, energy and its laws are part of a universal language of science if not of the human race. There can be few corners of the planet Earth in which the language of energy is unknown. Yet it is generally much less widely recognized that the terminology of energy had its historical origin in Glasgow, from which it was promoted and propagated in true imperial fashion to the rest of Britain, the British Empire, Europe and the world. The role of William Thomson (later Lord Kelvin) in the origin and promotion of what he himself termed 'the science of energy' is therefore the subject of this paper. [14]

Thomson devoted his life's work to the promotion of industrial use of energy. "Energy" as a word was not new. But the idea of industrial energy, or fungible energy, did not exist before Kelvin. It was through his pivotal role in laying the fundamentals of the science of heat and promoting the industrial use of energy, which coincided with the birth of commercial oil in 1859 and the increasing burgeoning application of coal, that the accelerating worldwide use of fungible energy gave rise to the Second Industrial Revolution, the growth of cities, the Internet Age: our homes, our cars, and the products and essentials we rely on in our daily lives are all operated and produced with fungible energy.

As far as the matter of Kelvin's energy principle as a contribution to the second universal principle (the principle of unidirectionality) is concerned, the assessment is mixed: On the one hand, he did achieve the goal of providing a counterpoint to the principle of conservation of energy by positing the two overarching principles of energy during energy transformations: the conservation (constancy) of energy and the preferred direction (preferred direction of spontaneous changes) of energy transformations. Together, they apply to vastly different kinds of things and phenomena. Specifically, the theory of exergy (Chap. 7) evolved from the two principles; in so far as the understanding of energy is impossible without the concept of exergy, Kelvin's contribution is supremely significant.

[7]This passage is from a draft of Paper [4]. It is taken from a quotation in Ref. [11:329]

On the other hand, a closer examination will be made to the self-evident truth, specifically Kelvin's second *general conclusion* as restated

> Any *restoration* of mechanical energy [of a given amount], without [the expenditure of] more than an equivalent [amount] of [mechanical energy resulting in its] dissipation, is impossible in inanimate material processes…

In Sect. 5.10, this examination will be given with the goal of evaluating its validity. It turned out that science (as we shall go over in Chap. 5) developed, afterward in 1865, the entropy principle as the second universal principle, which is universally accepted by everyone as *the* counterpoint to the principle of conservation of energy. Section 5.10 will conclude that the energy principle is not a universal principle. Whether the entropy principle, itself, is the *true* counterpoint or not will be evaluated in Chap. 8—with a surprising verdict.

4.8 Does the Heat-as-Energy Ontology Infer Equivalence-Convertibility Synonym?

At this point, the first law was encapsulated by the five statements of Table 3.1 and the equation

$$dE = \delta Q - pdV$$

The second law (as captured by Kelvin and Clausius with what they were able to incorporate from Carnot's theory) was encapsulated at this point inclusively by the two claims: the claim that heat cannot be converted 100% into work and the claim of universality in spontaneous changes. At this stage, the theory of heat was very much engineering oriented. This engineering-oriented MTH is based on the MEH, or the ontological revolution against the caloric theory of heat that "heat is not a substance, but a dynamical form of mechanical effect."

But, instead of unqualified acceptance of heat as a form of energy, the MTH accepted, in giving heat its causal agency, a qualification on "heat as energy" by defining heat

Heat is energy in transit

Correspondingly, the energy was standardly defined,

Energy is the capacity for doing work

That is, the capacity of "heat as energy" in doing work lies in the process of heat transfer as described by Carnot heat engine. These definitions of heat and energy, which are still used as standard definitions, are problematic and will be clarified in this discourse below, after the introduction of the entropy principle.

By the beginning of the 1860s, any objection to the new theory of heat came to an end and the MTH was entrenched into science. The MEH was upgraded to, or treated synonymously to be, the interconvertibility principle: *heat and mechanical* work *are universally and uniformly interconvertible in all circumstances* [10].

The crucial matter is that of causality of the heat↔work energy transformations as Kelvin put it "cause and effect." It is clear that mechanical work will spontaneously be degraded into heat so that one can say that work causes heat. How heat causes work is a different matter. As Carnot stated, "Heat alone is not sufficient to give birth to the impelling power: it is necessary that there should also be cold; without it, the heat would be useless." It is in this sense that we say *heat* (see Sect. 3.5 and Figs. 4.6, 4.7) can give birth to work, but heat cannot.

If we do use the same word "cause" for both, the former kind of work → heat causation is *efficient cause*,[8] which is the only acceptable use of cause-and-effect in physics, whereas the latter kind of "cause"-and-effect (*heat* → work as in a steam engine) is a fundamentally different kind (to be called *efficacious cause* [15]), often used in engineering. An engineered process of how *heat* "causes" work is different from the dissipative (Humean) regularity of how work causes heat. All these important issues would have been swept under the rug if one treats MEH and universal interconvertibility to be synonymous. It is a deliberating choice in Fig. 4.7 that a distinction is made between MEH and *universal interconvertibility*. With this distinction, it is possible to embrace the ontological doctrine of heat-as-energy in accordance with MEH, while reject, as we shall in Chapter 8, interconvertibility as an *inference* of the ontological doctrine: that the ontology infers equivalence-convertibility synonym.

The years between 1854 and 1865 marked a maturing stage of thermodynamics from childhood to adolescence in searching of the second universal principle. By 1865, Clausius realized that it was not possible to formulate the second universal principle without moving beyond energy and thereby introduced the new concept of entropy.

Problems

4.1 Show by example such as Fig. 4.3 that a heat engine that is more efficient than a reversible one will lead to the violation of the Clausius statement. Explain why this amounts to, by syllogism,

$$\text{The Clausius statement} \Rightarrow \text{Carnot's principle.}$$

[8]Efficient cause may be characterized by the description,

Physics knows nothing of *causation* except that it is the invariable and unconditional sequence of one event upon another:

July does not *cause* August, though it invariably precedes it.

That is, cause-and-effect in science is characterized by *constant conjunction* (of force and acceleration, e.g.,) and *invariable sequence* (July and August, e.g.,).

4.2 One gmol of a monatomic gas performs a Carnot cycle between the temperatures 400 K and 300 K. On the upper isothermal transformation, the initial volume is 1 L and the final volume 5 L. Find the work performed during a cycle, and the amounts of heat exchanged with the two heat reservoirs.

$$1338\ \text{J}; 5352\ \text{J}; 4014\ \text{J}$$

4.3 Assuming a heat source at (400 + ε) *K* and a heat sink at (300 − ε) *K*, where ε is an infinitesimally small number, being available for the operation of the above Carnot engine in Problem 2, what are the "entropy flows," Q^*, during a cycle into and out of the Carnot engine? Explain why they are equal in value.

$$13.38\ \text{J/K}$$

4.4 What is the maximum efficiency of a heat engine operating between an upper temperature of 400 °C and a lower temperature of 18 °C?

$$0.5675$$

4.5 Find the minimum amount of work needed to extract 4000 J of heat from a body at the temperature of 0 °F, when the temperature of the environment is 100 °F.

$$870.19\ \text{J}$$

4.6 Steam occupies one part of a partitioned insulated chamber of volume V_1. The initial pressure and temperature of the steam are 0.3 MPa and 350 °C. The other part of the chamber contains a vacuum of volume V_2 (=2 V_1). The partition is removed, and the steam disperses and fills the entire volume at the end of the process. Determine the final temperature and pressure of the steam.

$$348.26\ ^\circ\text{C}, 100.08\ \text{kPa}$$

References

1. Carnot S (1824) Reflections on the Motive Power of Fire. [Reprinted from *Reflections on the Motive Power of Fire and Other Papers*, edited by E. Mendoza. Dover Publications, New York (1960)]
2. Kuhn TS (1955) Carnot's version of Carnot's cycle. Am J Phys 23:91–95 (p. 94)
3. Jaynes ET (1984) The evolution of Carnot's principle. EMBO Workshop on Maximum-Entropy Methods (Orsay, France, April 24–28, 1984. Reprinted in Ercksen & Smith [1988], ***1***:267–282)
4. Thomson W (Lord Kelvin) (1911) Mathematical and Physical Papers of William Thomson 1:1–571. Cambridge Univ Press
5. Fermi E (1956) Thermodynamics. Dover (p. 48)

6. Truesdell C (1980) The Tragicomical History of Thermodynamics: 1822–1854. Springer, New York
7. Cropper WH (1986) Carnot's function: Origins of the thermodynamic concept of temperature. Am J Phys 55 (No. 2):120–129
8. La Mer VK (1954) Some current misinterpretations of N.L. Sadi Carnot's memoir and cycle. Am J Phys 22:20–27 (p. 26)
9. Kipnis N (2014) Thermodynamics and mechanical equivalent of heat. Sci and Educ 23:2007–2044
10. Merz JT (1896–1914) A History of European Thought in the 19th Century, Volume 2. (pp. 129–130)
11. Smith C, Wise MN (1989) Energy and Empire: A Biographical Study of Lord Kelvin. Cambridge Univ Press
12. Von Baeyer HC (1998) Maxwell's Demon: Why warmth disperses and time passes. Random House, New York
13. Uffink J (2001) Bluff your way in the Second Law of Thermodynamics. Stud Hist Phil Sci Part B: 2001 Studies in Hist and Philo of Modern Physics 32(No. 2):305–395
14. Smith C (1991) Lord Kelvin: Scientist of energy. Supercond Sci Technol 4:502–506
15. Wang LS (2011) Causal efficacy and the normative notion of sustainability science. SSPP 7 (No. 2):30–40

Entropy and the Entropy Principle

5

Abstract

Clausius generalized the mathematical expression of Carnot's principle obtained by Kelvin, and in 1865, introduced the concept of entropy and formulated the entropy principle. With the concept of entropy and the idea of the entropy principle, a definition of heat can be given, as well as mathematical formalism on equilibrium thermodynamics was formulated for organizing empirical data of thermodynamic properties. The establishment of the entropy principle signaled the beginning of a new era—in which, thermodynamics, which had originated as a branch of engineering knowledge, separated into two distinctive streams. The new science stream was manifested in the mathematical formalism on equilibrium thermodynamics. It is noted that the entropy principle, rather than the energy principle, is the true universal principle.

Keywords

The first Clausius theorem · The second Clausius theorem (Clausius' Inequality) · Entropy · The entropy principle · Gibbs U-V-S surface · Entropy functions for ideal gases · Approximate entropy functions for liquids/solids · Definition of heat · Carnot vapor cycle · Properties of ideal gas mixtures (Gibbs' theorem) · Reversible mixing of ideal gases · Birth of equilibrium thermodynamics

5.1 What Determines the Direction of Natural Processes?

Kelvin made use of the conception of energy to serve the purpose of determining the direction of natural processes. Opinion may still be divided with regards to whether dissipation of energy is universal or not, but the matter was already clear to Planck,

L.-S. Wang, *A Treatise of Heat and Energy*, Mechanical Engineering Series,
https://doi.org/10.1007/978-3-030-05746-6_5

post Clausius' and Boltzmann's contribution, in regard to "What accounts for the direction of natural processes?" or, to use Planck's phrase, what "determine[s] the direction of a thermodynamical process" [1:81]. Planck, being a student of Boltzmann, was in a unique position of possessing the correct answers to that question: Clausius provided the macroscopic inductive account for the unidirectionality, and Boltzmann, the microscopic reductive account. This gave insight to Planck on the limitation in thinking in terms of energy alone, and he had this to say:

> Although these considerations make it evident that the principle of the conservation of energy cannot serve to determine the direction of a thermodynamical process, and therewith the conditions of thermodynamical equilibrium, unceasing attempts have been made to make the principle of the conservation of energy [i.e., the conception of energy] in some way or other serve this purpose…That attempts are still made to represent this law [the second law] as contained in the principle of energy may be seen from the fact that the too restrictive term "Energetics" as sometimes applied to all investigations on these questions. The conception of energy is not sufficient for the second law. [1:81]

Rudolf Clausius (1822–1888)

MaxPlanck(1858–1947)

Clausius was the first to see the necessity of going beyond the conception of energy after his attempts in a series of papers on "proof" of the so-called Clausius Inequality (of different versions), which culminated in his 1865 paper that introduced the concept of entropy.

5.2 A Property of Reversible Cycles, the First Clausius Theorem

Kelvin first obtained and Clausius followed immediately the pivotal expressions for reversible cyclic processes

$$\sum\nolimits_{i[1\to n]} \frac{Q_i}{T_i} = 0 \tag{57}$$

and the corresponding

$$\oint dQ/T = 0 \tag{58}$$

Jaynes noted "It is curious fact, having perceived such an important consequence of Carnot's principle…Kelvin does not seem to have perceived the still more important fact that was now staring him in the face in Eq. (10) [i.e., Eq. (57)]. This was left for Rudolph Clausius…" [2:6]. Jaynes underestimated the degree of difficulty and the required boldness in the conceptual step that Kelvin never took and Clausius did in 1865. The fact remains that Kelvin was committed to the dissipation of energy as the second universal principle throughout his life, and, as evidence, there is no entry of "entropy" in the index to *Energy and Empire: A biographical study of Lord Kelvin* [3]. It is a possibility that there is no single mention of entropy in the 814 pages of the exhaustive biography of Lord Kelvin. It took Clausius, onward from 1854 to 1865, a decade for him to gestate the concept into its final form in 1865, the year Clausius' great paper [4] appeared in print. The following is a reconstruction of the argument following Fermi [5].

This proof is made through another reversible thought experiment on the basis of the *Kelvin–Planck statement of the second law*: "No process is possible in which the sole result is the absorption of heat from a reservoir and its complete conversion into work."

5.2.1 The First Clausius Theorem

Consider a system connected to a heat reservoir at a constant absolute thermodynamic temperature of T_R through a reversible Carnot machine (Fig. 5.1). Suppose that the Carnot machine undergoes through a sequence of microcyclic processes producing work; the system undergoes through a sequence of infinitesimal *reversible* process steps, each step taken by the system corresponds to one micro-Carnot cycle. The system temperature $T(t)$ may change over the course of the thought-experiment operation, where t denotes the time.

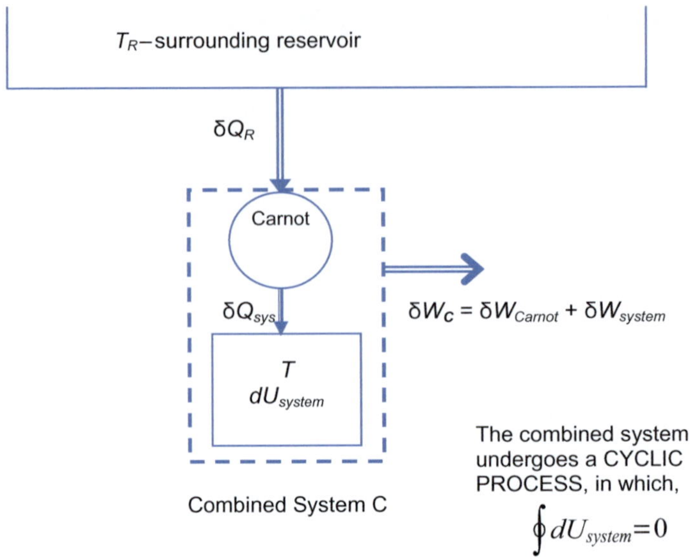

Fig. 5.1 Thought experiment to demonstrate the first Clausius theorem

Imagine one such infinitesimal step taken by the system, during which the Carnot engine goes through one complete microcycle and receives heat δQ_R from the reservoir and supplies heat δQ_{system} to the system. The system is at an instantaneous uniform temperature of $T(t)$ at that point in time t. Let the work produced by the Carnot engine be δW_{Carnot} and the work performed by the system be δW_{System}. Now denote the *combined system* (the system–with–reversible-Carnot-machine) as C, which is identified by the dotted-line box in Fig. 5.1. The combined work output of the Carnot engine and the system will be expressed as δW_C

$$\delta W_{Carnot} + \delta W_{System} = \delta W_C$$

Applying the energy balance to the combined system C yields

$$\delta W_C = \delta Q_R - dU_C$$

Since the Carnot machine undergoes a cyclic process, there is no change in the internal energy of its working fluid; the internal energy change of the system dU_{System} in the above equation represents a change in the internal energy of the combined system C.

$$\delta W_C = \delta Q_R - dU_{System}$$

Applying either Eq. (46) or Eq. (55) to the Carnot machine

$$\frac{\delta Q_R}{T_R} = \frac{\delta Q_{System}}{T'}$$

where T' represents the temperature of the source of the heat quantity δQ_{system} that the Carnot machine surrenders to the system. Note that for reversible processes, heat exchange necessarily takes place over infinitesimally small ΔT, i.e., $T' = T_{system}$. Substituting this relation into the energy balance equation

$$\delta W_C = T_R \frac{\delta Q_{System}}{T'} - dU_{System}$$

Repeating the infinitesimal steps so that the system is made eventually to undergo a complete cyclic process, while the Carnot machine undergoes through a sequence of corresponding microcycles.[1] We shall refer to this as a "CYCLIC PROCESS of the combined system". Carrying out the cyclic integration of the above equation and noting that the term involving system energy vanishes, we obtain

$$W_C = T_R \oint \frac{\delta Q_{System}}{T'}$$

Note that

$$W_C = \oint \delta Q_R$$

As represented by the direction of the arrows of Fig. 5.1, a positive W_C would indicate a positive work performed by the combined system going through the CYCLIC PROCESS by converting an amount of heat from a single source $\oint \delta Q_R$ completely into work W_C. This would be in violation of the Kelvin–Planck statement of the second law. W_C must, therefore, be negative, i.e.,

$$\oint_{OriginalCYCLE} \frac{\delta Q_{System}}{T'} \leq 0 \tag{59}$$

Conversely, it can be readily shown that when the *reverse* of the original reversible CYCLIC PROCESS is considered, all terms in the energy balance equation retain the same relationship with a consistent sign convention. Thus, the same inequality must be true

$$\oint_{ReverseCYCLE} \frac{\left(\delta Q_{System}\right)_{Reverse}}{\left(T'\right)_{Reverse}} \leq 0$$

[1]The Carnot machine operates as a Carnot heat engine during some microcycles and as a Carnot heat pump during the other microcycles.

Since both CYCLIC PROCESSES are reversible, we have the following two *reversible relations* (i) and (ii) between the two cycles:

$$\left(\delta Q_{system}\right)_{\mathrm{REVERSE}} = -\left(\delta Q_{system}\right)_{\mathrm{ORIGINAL}} \quad \text{(i)}$$

$$(T')_{\mathrm{REVERSE}} = T_{system}(\equiv T) = T' \quad \text{(ii)}$$

It follows, therefore,

$$\oint_{ReverseCYCLE} \frac{\left(\delta Q_{System}\right)_{Reverse}}{T} = -\oint_{OriginalCYCLE} \frac{\delta Q_{System}}{T} \leq 0$$

That is to say, "Kelvin–Planck" also requires

$$\oint_{OriginalCYCLE} \frac{\delta Q_{System}}{T} \geq 0 \quad (60)$$

The only way in which both requirement (59) and requirement (60) can be met is if the equality sign holds, therefore,

$$\oint_{REVERSIBLE} \frac{\delta Q_{System}}{T_{System}} = 0 \quad (58\text{ or }61)$$

This conclusion is known as the *first Clausius theorem.*

5.3 The Entropy, a New State Variable

Consider two end equilibrium states, A and B, of a system. Consider two arbitrarily chosen reversible paths I and II connecting A to B (Fig. 5.2). Let path I be represented by AP_IB and path II by $AP_{II}B$. Now, imagine a cyclic process of $AP_IBP_{II}A$ and consider the cyclic integration Eq. (58 or 61) along this cycle, the result of which is

$$0 = \oint_{AP_IBP_{II}A} \frac{\delta Q}{T} = \int_{AP_IB} \frac{\delta Q}{T} + \int_{BP_{II}A} \frac{\delta Q}{T}$$

Since

$$\int_{BP_{II}A} \frac{\delta Q}{T} = -\int_{AP_{II}B} \frac{\delta Q}{T},$$

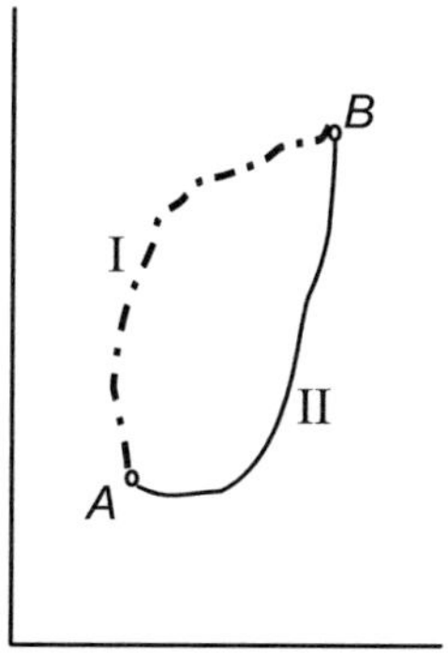

Fig. 5.2 Depiction of a cyclic process: the first leg may be reversible (as considered in this section) or irreversible, but the second leg is reversible as considered here as well as in Sect. 5.4

the first Clausius theorem yields

$$\int_{AP_IB} \frac{\delta Q}{T} = \int_{AP_{II}B} \frac{\delta Q}{T}$$

along any arbitrary reversible path from A to B. That is to say, there exists a *state variable* denoted by S and known as *entropy* which is defined by

$$S_B - S_A = \left(\int_A^B \frac{\delta Q}{T} \right)_{Reversible} \tag{62}$$

Correspondingly,

$$dS = \left(\frac{\delta Q}{T} \right)_{Reversible} \tag{62A}$$

or,

$$(\delta Q)_{reversible} = TdS$$

Like any thermodynamic state variable, the **state-variable *entropy*** can be expressed in the form of functions of state: for instance, $S = S(T, V)$ or $S = S(T, p)$ or $S = S(p, V)$. If the two end states differ from each other only infinitesimally, the change of dS may be expressed as

$$dS = \left(\frac{\partial S}{\partial T}\right)_V dT + \left(\frac{\partial S}{\partial V}\right)_T dV \tag{63A}$$

$$dS = \left(\frac{\partial S}{\partial T}\right)_p dT + \left(\frac{\partial S}{\partial p}\right)_T dp \tag{63B}$$

5.3.1 Gibbs U-V-S Surface

An important result is obtained by the substitution of Eq. (62A) into (25)

$$dU = TdS - pdV \tag{64}$$

At first sight, it might seem that this equation is restricted to *reversible* infinitesimal changes of state only. However, this is not the case: While, individually, Eq. (62A) is restricted to reversible paths and Eq. (25) is restricted to quasi-static paths, Eq. (64), in which δQ and δW both disappear, is strictly a differential relation of state variables independent of the paths between the states. The original path restriction on the reversibility in Eq. (62A), or quasi-staticity of paths (or internal reversibility) in Eq. (25), is irrelevant! As will be discussed in Chap. 9, Eq. (64) implies that U can be regarded as a function of S and V, corresponding to a U-V-S surface, called Gibbs U-V-S surface

$$U = U(S, V) \tag{64A}$$

5.3.2 Entropy Change in Isobaric Processes

A useful result from Eq. (64) is obtained for an isobaric process

$$dH = d(U + pV) = TdS + Vdp = TdS$$

It follows, therefore,

$$C_p = \left(\frac{\partial H}{\partial T}\right)_p = T\left(\frac{\partial S}{\partial T}\right)_p$$

Equation (63B) for isobaric processes assumes the form

$$dS = \frac{C_p(T)}{T} dT$$

That is, for isobaric processes we have the general expression

$$S_B - S_A = \int_A^B \frac{C_p(T)}{T} dT \tag{63C}$$

We shall consider the full implication of this *general equation* in detail in the following chapters, especially in Chap. 9.

5.3.3 The Entropy of Ideal Gases

Here, we consider its application to ideal gases to determine the ideal gases entropy expression in terms of *T*, *p*, and *V*. Rewrite Eq. (64)

$$dS = \frac{dU + pdV}{T} = \frac{dU + d(pV) - Vdp}{T} = \frac{dH - Vdp}{T} \tag{65}$$

Substitution of Eqs. (34) and (35), $p/T = NR/V$, and $V/T = NR/p$ into the equation above leads to the following results, respectively,

$$dS = Nc_V \frac{dT}{T} + NR\frac{dV}{V} \tag{66}$$

$$dS = Nc_p \frac{dT}{T} - NR\frac{dp}{p} \tag{67}$$

Integration of both equations yields

$$S(T, V) = Nc_V lnT + NRlnV + const \tag{68}$$

$$S_B - S_A = Nc_V ln\frac{T_B}{T_A} + NRln\frac{V_B}{V_A} \tag{68A}$$

and

$$S(T, p) = Nc_p lnT - NRlnp + const \tag{69}$$

$$S_B - S_A = Nc_p ln\frac{T_B}{T_A} - NRln\frac{p_B}{p_A} \tag{69A}$$

Making use of the thermal equation of state of an ideal gas, either Eq. (68A) or (69A) yields

$$S_B - S_A = Nc_V ln \frac{p_B}{p_A} + Nc_p ln \frac{V_B}{V_A} \tag{70}$$

5.3.4 The Entropy of Liquids/Solids, An Approximate Formula

As examples of systems of both ideal gases and liquids and solids will be considered, an approximate formula for liquids and solids is given without derivation, which will be given in Chap. 9. The simple formula is

$$S_B - S_A = \int_A^B \frac{C_p(T)}{T} dT \cong C_p ln \frac{T_B}{T_A} \tag{71}$$

It is noted that while (63C) and (71) are of the same expression, (63C) applies to all substances under the isobaric condition whereas (71) applies to liquids and solids under general conditions, i.e., for liquids and solids entropy dependency on pressure is negligible.

5.4 Entropy Change in a System Undergoing an Irreversible Process

Let us use the same setup of Fig. 5.1 and consider now that the system, instead of undergoing a reversible change, undergoes a sequence of *quasi-static* changes along the *same* trajectory of the original reversible thought experiment. (Note that a quasi-static process may or may not be reversible [6], an issue to be studied further in Chap. 6.) The same conclusion of Eq. (59) holds in this case

$$\oint \frac{\delta Q_{System}}{T'} \leq 0 \tag{72}$$

While the quasi-static change of the system can be reversed along the same trajectory in the reversed direction as long as the full interaction means are available for such purpose, the *reversible relations* (i) and (ii) do not hold any longer between the two opposite-direction cycles. Values of heat, δQ_{system}, and T' for each quasi-static step of the system in the reverse direction operation are different from those values in the original operation. There is, thus, no counterpart to (72) [as that to (59) in the form of (60)] in this irreversible case. Without the counterpart, the inequality (72) alone holds, which is known as the *second Clausius theorem* or *Clausius' Inequality.*

It should be especially noted that T' represents the temperature of the source of the heat quantity δQ_{system} that the Carnot engine surrenders to the system, and <u>is not</u>, except in the reversibility limit, <u>equal to the temperature of the system (or any part of the system)</u> along the quasi-static trajectory [5: p. 48, in a Footnote],

$$T' \neq T_{System}$$

For the present quasi-static cycles, (72) cannot be replaced by

$$\oint \delta Q_{System} / T_{System} \leq 0$$

Therefore, a serious ambiguity exists in the application of Inequality since ***T′* in Eq.** (72) **is ill-defined**.

We may now establish the corresponding entropy change for the general thermodynamic process of a system following a similar reasoning in Sect. 5.3: Consider again two end equilibrium states of a system, *A* and *B*, and two different paths I and II, each connecting *A* to *B* (using the same Fig. 5.2)—note, however, here only path II is restricted to be reversible. Let path I be represented by AP_IB (which is irreversible) and path II by $AP_{II}B$ (which is reversible). Now, imagine a cyclic process of $AP_IBP_{II}A$ and consider the cyclic integration along this cycle

$$\oint_{AP_IBP_{II}A} \frac{\delta Q}{T'} = \int_{AP_IB} \frac{\delta Q}{T'} + \int_{BP_{II}A} \frac{\delta Q}{T} \leq 0$$

Since path II is reversible, the following equality holds:

$$\int_{BP_{II}A} \frac{\delta Q}{T} = - \int_{AP_{II}B} \frac{\delta Q}{T}$$

Therefore,

$$\int_{AP_IB} \frac{\delta Q}{T'} - \int_{AP_{II}B} \frac{\delta Q}{T} \leq 0$$

i.e.,

$$\int_{AP_IB} \frac{\delta Q}{T'} \leq \int_{AP_{II}B} \frac{\delta Q}{T}$$

Since the integral along the reversible path $AP_{II}B$ equals the entropy change $S_B - S_A$

$$S_B - S_A \geq \left(\int_A^B \frac{\delta Q}{T'} \right)_{quaisstatic} \tag{73}$$

Correspondingly,

$$dS \geq \frac{\delta Q}{T'} \tag{73A}$$

The ambiguity with regard to T' remains in Eqs. (73) and (73A). A new interpretation of the term which removes the ambiguity will be given in Chap. 6 in terms of modern formalism.

5.5 The Principle of the Increase of Entropy

This ambiguity (the ill-defined T') is avoided when Clausius' inequality is applied to a thermally isolated system, resulting in the PRINCIPLE OF THE INCREASE OF ENTROPY (short as the entropy principle):

For any transformation i → f occurring in a thermally isolated system, the entropy of the final state can never be less than that of the initial state, i.e.,

$$\begin{aligned} &S(f) - S(i) \geq 0 \\ &\text{or,} \\ &(\Delta S)_{iso-sys} \geq 0 \end{aligned} \tag{74}$$

This principle was noted by Clausius himself when he wrote in 1865 "The entropy of the universe strives to attain a maximum value" [4], which became the starting point of Gibbs' formulation of Gibbsian thermodynamics [7]. However, Clausius' own commitment to the idea of entropy was ambivalent, as Uffink reported, "the remarkable fact that in the [his 1876 revised edition of] book…from his collected articles [*The Mechanical Theory of* Heat], every reference to the entropy of the universe and even to the idea that entropy never decreases in irreversible processes in adiabatically isolated systems is deleted!" [8]. The current status of the entropy principle, instead, owns a great deal to Planck, [9] who was credited to be the first to use the phrase of the *principle of the increase of entropy* and propagated the view that the essence of the second law lies in the principle of the increase of entropy, Eq. (74), as the second universal principle of thermodynamics, *the* counterpoint to the principle of conservation of energy

$$U(f) - U(i) = Q - W \tag{22}$$

5.5.1 Examples of the Application of the Entropy Principle

Since a single system and its surroundings together can always be defined as an isolated combined system, the law of the increase of entropy can be applied to the combined system even if the individual systems are not isolated. This principle is a powerful instrument of drawing quite *general* inference from the second law. Note, however, valid and useful inference can only be drawn in cases that such a system, when it is "defined" to be isolated, remains meaningful, not when the system has its fundamental characteristic changed when it is so defined. For example, the planet Earth would lose its meaning as a system if it is so defined!

The fact that all spontaneous transformations in an isolated system proceed in such a direction dictated by (74) can be illustrated by two simple examples. As the first example, we consider a dispersal irreversible process of a composite system, the exchange of heat by heat conduction between two bodies, A_1 and A_2. Let T_1 and T_2 be the temperatures of these two bodies, respectively, and let $T_1 < T_2$. Since heat flows by conduction from the hotter body to the colder body, the hotter body A_2 gives up a quantity of heat δQ which is absorbed by A_1. Thus, the entropy of A_1 changes by an amount of $\delta Q/T_1$, while that of A_2 by an amount of $-\delta Q/T_2$. The entropy change of the isolated composite system is, accordingly,

$$\frac{\delta Q}{T_1} - \frac{\delta Q}{T_2}$$

Since $T_1 < T_2$, this change of entropy for the isolated composite system, $\frac{\delta Q}{T_1} - \frac{\delta Q}{T_2}(>0)$, is positive.

Now, let us consider the continuing heat exchange between the two parts eventually leading to thermal equilibrium between A_1 and A_2, Suppose changes in the temperatures of A_1 and A_2 are related to heat exchange by

$$\delta Q = C_{p1} dT_1$$

$$\delta Q = -C_{p2} dT_2$$

The final temperature at thermal equilibrium, therefore, is

$$T_{final} = \frac{C_{p1}T_1 + C_{p2}T_2}{C_{p1} + C_{p2}} = \frac{T_1 + (C_{p2}/C_{p1})T_2}{1 + C_{p2}/C_{p1}}$$

Correspondingly, the total entropy change of the composite system is (assuming pressure remaining constant and constant $C_p's$) obtained from the application of Eq. (63C) to each part

$$\Delta S = C_{p1} ln \frac{T_{final}}{T_1} + C_{p2} ln \frac{T_{final}}{T_2} = C_{p1} ln \frac{(T_{final})^{1+C_{p2}/C_{p1}}}{T_1\left[(T_2)^{C_{p2}/C_1}\right]}$$

This quantity can be shown to be positive, consistent with the entropy principle.

As a second example, we consider the production of heat in a body by friction, generated by a work input, $-W$. The body is again kept under isobaric condition. This frictional irreversible process also results in an increase of entropy: Input to the body is in the form of workflow and the part of the body that is affected by frictional entropy production transformation of work into heat will experience increase in temperature

$$-W = U_B - U_A = \int_A^B \left(\frac{\partial U}{\partial T}\right)_p dT$$

It will be shown in Chap. 9 that the following property relation applies:

$$\left(\frac{\partial U}{\partial T}\right)_p = C_p - pV\beta \approx C_p,$$

where the approximation applies for the body of substance with a negligible thermal expansion coefficient β. The final temperature is then

$$T_B = T_A + \frac{-W}{C_p}$$

Applying Eq. (63C)

$$\Delta S = \int_A^B \frac{C_p}{T} dT = C_p ln \frac{T_A + \frac{-W}{C_p}}{T_A} (>0)$$

which is positive as long as work is done to the body, i.e., W is negative. A possibility of spontaneous work derived from the body is, of course, impossible in violation of the entropy principle. Note that a workflow is not associated with an entropy flow, this increase of entropy is not compensated by an entropy outflow from (thus, a decrease of entropy in) the system of work source. This increase of entropy in affected part of the body is the total change of entropy for the "composite system of body and work-source."

5.6 The Definition of Heat

The entropy principle holds that all spontaneous transformations in an isolated system proceed in the direction of increasing entropy, such transformations manifest in either the dispersal of gradients or the dissipation of high-grade energy into low-grade energy eventually into heat energy. With this understanding, we can investigate how heat and work, though both are energy in transit and both are equivalent to each other according to the MEH, are differentiated.

Work is commonly defined as

> *Work is energy in transit.*

In association with MEH, there has been a mistaking attempt to treat heat and work, Q and W, in Eq. (22), in a parallel fashion. For instance, Newburgh and Leff wrote "It is no more appropriate to speak of heat in a body than work in a body. Both statements are not sensible" [10].

Objection to the use of heat is also made for a second reason: For instance, a 2004 paper reports: "When it comes to language about heat, there is consensus that physicists' language can be misleading, but little agreement about why it is misleading or how it can be corrected" wrote Brookes et al. [11]. They concluded their study

> Although physicists are quite aware that heat should be thought of as a *process* rather than a *substance*, our coding shows that their language does not reflect this understanding.... Physicists talk about heat predominately as if it were a substance.

Romer [12] made similar point

> *Heat* presents one of our most series linguistic problems. Not only is it a common word in the outside world, but in addition its frequent misuse within physics reinforces ancient and erroneous views of the physical world [the caloric theory]...surely the experiments of Joule and the careful thinking of the thermodynamicians of the second half of the nineteenth century should have put it [i.e., the caloric theory] to rest forever. Yet, we continue to hear vestiges of caloric theory in common talk...

Let us examine the first objection: Both heat exchange and work exchange are energy-in-transition processes. It is true that we never speak of the work in a body. But, work and heat are fundamentally different kinds of processes. Whereas heat exchange is a one-step process (see below), work exchange other than compression work is a two-step process. Only the first step of which is the process of mechanical energy in transition across the system boundary of a receiving body, which naturally results in a gain of mechanical energy in the receiving body. Then, the internal frictional dissipation—dissipatively converting mechanical energy into heat energy—of the second step takes place resulting in a gain of thermal internal energy in the body, which of course is neither "work" nor "mechanical energy" anymore. That is why we do not speak of the work (or mechanical energy, except in the short duration immediately after the exchange across the boundary) in a body.

In contrast, heat exchange is a one-step heat energy in transition, which naturally results in gain of heat energy, i.e., thermal internal energy, in a body. There is no corresponding reason as in the case of the two-step "work → mechanical energy in the body → heat energy in the body," for avoiding the language of heat energy in a body. The same heat energy in transition remains heat energy in the body with its form largely intact. This is because heat energy is the final form of energy for all energy transformation processes. The same term of heat can be used for both heat exchange, Q, and heat energy, a special (thermal) form of internal energy, U.

A simple way to address this difference between work and heat is that work exchange is not associated with entropy flow (see Fig. 4.5), while heat flow is always associated with entropy flow (see Eq. [62A], and *definition of heat below*): explicit considerations of how entropy[2] are involved is the best way to differentiate energy exchange in the form of work from energy exchange in the form of heat.

The second objection, if valid, is a much more serious objection. As it was noted in Table 3.1 summarizing the energy conservation principle, it originated from the conceptual differentiation of caloric into heat exchange and heat energy. By using the same term, heat, when one speaks of heat exchange and heat energy, one is guilty of denying the conceptual difference between the two. But, this is simply not the case.

The conceptual differentiation of caloric, as it was discussed in Chaps. 3 and 4, did not stop at heat energy. Heat energy, or thermal internal energy, is a special form of internal energy. The introduction of *internal energy* by Clausius and *energy* (=internal energy + mechanical energy) by Kelvin (see Chap. 4) are, therefore, additional steps beyond the initial conceptual differentiation phase, #1 in Table 3.1. The complete conceptual differentiation consists of steps #1, #3, and #5 in Sect. 3.3 together. When one uses heat to mean heat energy, there is no risk in treating heat as a substance *unless* one denies the transformation of energy from one form to another. It is safe to note that the triumph of MTH over the caloric theory is safe that *everyone* accepts the existence of energy transformation, even though the triumph is tainted for a different reason as I shall address that issue in Chap. 8.

This subtle point is critical for sorting out the way how physicists and engineers use the term "heat" including specific heats, latent heats, waste heat, etc.

Our serious linguistic dilemma resulted from the inability of defining *heat* on the one hand, and the drastic step of stripping *heat* down to its barebone Q on the other hand. A middle ground can be found between the broadest meaning of *heat* and the narrowest interpretation of heat as Q by defining heat as "a process *and* as a state function corresponding to natural end states of spontaneous changes":

> **Definition of Heat**: Heat, as denoted by Q, is energy and entropy in transit; waste heat or *heat* in a body (i.e., heat used as short for thermal internal energy) is high-entropy, i.e., lowest-grade, form of energy.

This is how we use the term of heat, and there should be no objection to this use: any remaining objection will be addressed in Chap. 8. How we use the term energy will be discussed in Chap. 7 when we consider energy together with the concept of exergy.

[2]Take the example of the system of a cooking vessel, which is brought to a very high temperature in two ways of cooking. Electric resistive coil cooking is an example of heat-exchange process carrying both energy flow and its associated entropy flow; electric induction cooking is an example of (electromagnetic field energy) work-exchange process carrying no entropy flow—with heat generation entropy-production process taking place inside cooking vessel. Entropy gain in the former case results from entropy in-flow across the boundary of the vessel and in the latter case from entropy production within the boundary.

5.7 Statistical Mechanics Formula of Boltzmann

The fact that the entropy of an isolated system can never decrease during any transformation has a clear interpretation from the statistical point of view. Boltzmann calculated the phase volume W of an ideal gas of $\mathcal{N}$ atoms on volume V, for which the energy lies in (E, $E + dE$):

$$W = \int_{\mathcal{R}} d^3x_1 \cdots d^3x_{\mathcal{N}} d^3p_1 \cdots d^3p_{\mathcal{N}} = CV^N E^{\frac{3\mathcal{N}}{2}-1} dE$$

where the region $\mathcal{R}$ of integration is those points for which all coordinates are within a volume V and momenta p_i satisfy

$$E \leq \sum p^2/2m \leq E + dE$$

The constant C is independent of V and E.

Now, the entropy S of an ideal gas of constant molar heat is $S(T, V) = Nc_V lnT + NRlnV + const$. It is evident that $\ln W$, or $k\ln W$, has the same volume and energy dependence as the entropy of the gas. Such a relationship between W and S was established by Boltzmann, who proved

$$S = klnW \tag{75}$$

where k is a constant called the *Boltzmann constant* and is equal to the ratio of the gas constant to Avogadro's number, $R/\mathcal{N}_A$. The number W, i.e., phase volume or the volume of the *phase space*, is the number of dynamical states or microstates that correspond to the given thermodynamic state.

The entropy principle can now be interpreted according to the *equal* a priori *probability postulate*, which states

> For an isolated system with an exactly known energy and exactly known composition, the system can be found with equal probability in any microstate consistent with that knowledge.

Equation (75) states that a larger entropy corresponds to a larger volume of phase space (a larger number of microstates). The postulate means that the system will not spontaneously move into a restricted region of the phase space (i.e., a lower S) by avoiding the rest region of the phase space, i.e., the entropy of an isolated system never decreases. On the other hand, given the availability of larger phase space, the system will always avail itself to every corner of the space, i.e., the tendency of the entropy of a system to increase.

For the first time since Newton, we have a new kind of driving force that exists in the statistical realm of ensemble and probability, instead of the classical realm of masses and physical forces.

5.8 Isentropic Processes and Carnot Cycles

A Carnot gas cycle's four steps cyclic processes $1 \rightarrow 2 \rightarrow 3 \rightarrow 4 \rightarrow 1$ are plotted in the *p*-*V* diagram as shown in Fig. 4.4. The reversible adiabatic steps of $2 \rightarrow 3$ and $4 \rightarrow 1$ correspond to constant entropy steps according to Eq. (62) called isentropic steps or processes. In Fig. 4.4, *p* and *V* variations of the isentropic steps are related according to Eq. (70), or Eq. (41).

As shown in Fig. 5.3, the same Carnot gas cycle is plotted in a *T*-*S* diagram. The four steps of $1 \rightarrow 2 \rightarrow 3 \rightarrow 4 \rightarrow 1$ are shown as a series of horizontal line, downward vertical line, horizontal line, and upward vertical line: the two isothermal expansion/compression steps are shown as horizontals and the two isentropic steps as verticals in the *T*-*S* diagram.

In terms of falling of either the caloric or the entropy flow of the reversible Carnot heat engine, and the corresponding heat flow, their relation becomes straightforward in Fig. 5.3. The entropy flow through the heat engine per cycle is $\Delta S = S_2 - S_1 = S_3 - S_4$: the same entropy flows into the engine each cycle as it flows out from the engine. The corresponding heat flowing into the engine is $T_1(S_2 - S_1) = T_H\Delta S$, and heat flowing out from the engine is $T_4(S_3 - S_4) = T_C\Delta S$, i.e., respectively, the area defined by the $1 \rightarrow 2 \rightarrow C \rightarrow D \rightarrow 1$ and the shaded area in blue defined by the $3 \rightarrow 4$ isothermal, $3 \rightarrow 4 \rightarrow D \rightarrow C \rightarrow 3$.

Work output per cycle is thus the area of the rectangular 1-2-3-4-1, the difference between the large shaded area of $Q_H = T_H\Delta S$ and the blue shaded area of $Q_C = T_C\Delta S$.

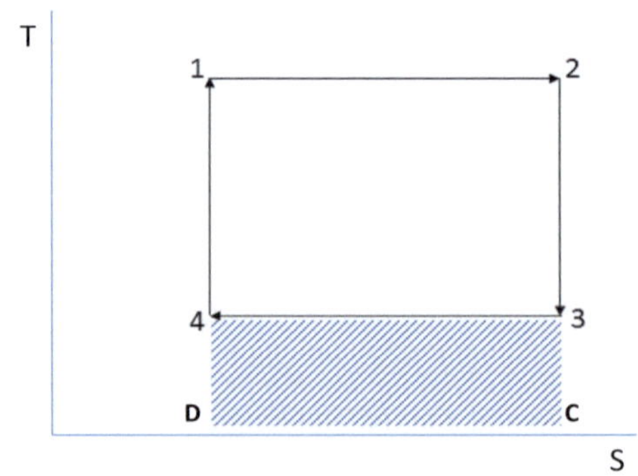

Fig. 5.3 Carnot gas cycle (see Fig. 4.3 for comparison): Step $1 \rightarrow 2$ is isothermal expansion; Step $2 \rightarrow 3$ is isentropic expansion; Step $3 \rightarrow 4$ is isothermal compression; Step $4 \rightarrow 1$ is isentropic compression

An Example of Carnot Vapor Cycle

A gas cycle, such as an internal combustion piston engine cycle, i.e., IC engine, or an internal combustion gas turbine cycle, e.g., a jet engine, can be a closed cycle (the former case of IC engine) or a steady-flow cycle (the latter case of jet engine). Consider the example of a steady-flow Carnot cycle using water as the working fluid. Water changes from saturated liquid to saturated vapor as heat is transferred to it from a source at 250 °C. Heat rejection takes place at a pressure of 20 kPa. Show the cycle on a *T*-*s* diagram relative to the saturation lines, and determine for per kg mass of water and per cycle (a) the thermal efficiency, (b) the amount of heat rejected, and (c) the net work output.

Solution

The cycle is shown in Fig. 5.4 on a *T*-*s* diagram with state 1 to be saturated liquid water at 250 °C, which, according to Table 5.1, corresponds to a saturation pressure of 3976.2 kPa. That is,

State 1: 250 °C, 3976.2 kPa

Saturated liquid enthalpy, $(h_f)_1 = 1085.7\,\text{kJ/kg}$

Saturated liquid entropy, $(s_f)_1 = 2.7933\,\text{kJ/kg}\cdot\text{K}$

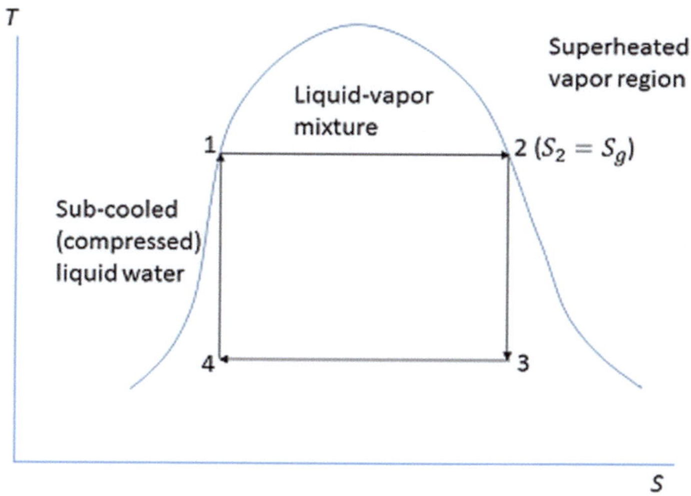

Fig. 5.4 Carnot steam cycle: 1 → 2 is isothermal expansion, 2 → 3 isentropic expansion, 3 → 4 isothermal compression, 4 → 1 isentropic compression

The isothermal process in the mixed liquid–vapor region coincides with the isobaric process. Thus, from Table 5.1, one finds.

Table 5.1 Saturated water–temperature table

Table A-4												
Temp.,	Sat. press,	Specific volume. m^3/kg		Internal energy, kJ/kg			Enthalpy, kJ/kg			Entropy, kJ/kg · K		
		Sat. liquid.	Sat. vapor,	Sat. liquid.	Evap.,	Sat. vapor,	Sat. liquid.	Evap.,	Sat. vapor,	Sat. liquid,	Evap.,	Sat. vapor,
T °C	P_{sat}, kPa.	v_f	v_g	u_f	u_{fg}	u_g	h_f	h_{fg}	h_g	s_f	s_{fg}	s_g
0.01	0.6117	0.001000	206.00	0.000	2374.9	2374.9	0.001	2500.9	2500.9	0.0000	9.1556	9.1556
5	0.8725	0.001000	147.03	21.019	2360.8	2381.8	21.020	2489.1	2510.1	0.0763	8.9487	9.0249
10	1.2281	0.001000	106.32	42.020	2346.6	2388.7	42.022	2477.2	2519.2	0.1511	8.7488	8.8999
15	1.7057	0.001001	77.885	62.980	2332.5	2395.5	62.982	2465.4	2528.3	0.2245	8.5559	8.7803
20	2.3392	0.001002	57.762	83.913	2318.4	2402.3	83.915	2453.5	2537.4	0.2965	8.3696	8.6661
25	3.1698	0.001003	43.340	104.83	2304.3	2409.1	104.83	2441.7	2546.5	0.3672	8.1895	8.5567
30	4.2469	0.001004	32.879	125.73	2290.2	2415.9	125.74	2429.8	2555.6	0.4368	8.0152	8.4520
35	5.6291	0.001006	25.205	146.63	2276.0	2422.7	146.64	2417.9	2564.6	0.5051	7.8466	8.3517
40	7.3851	0.001008	19.515	167.53	2261.9	2429.4	167.53	2406.0	2573.5	0.5724	7.6832	8.2556
45	9.5953	0.001010	15.251	188.43	2247.7	2436.1	188.44	2394.0	2582.4	0.6386	7.5247	8.1633
50	12.352	0.001012	12.026	209.33	2233.4	2442.7	209.34	2382.0	2591.3	0.7038	7.3710	8.0748
55	15.763	0.001015	9.5639	230.24	2219.1	2449.3	230.26	2369.8	2600.1	0.7680	7.2218	7.9898
60	19.947	0.001017	7.6670	251.16	2204.7	2455.9	251.18	2357.7	2608.8	0.8313	7.0769	7.9082
65	25.043	0.001020	6.1935	272.09	2190.3	2462.4	272.12	2345.4	2617.5	0.8937	6.9360	7.8296
70	31.202	0.001023	5.0396	293.04	2175.8	2468.9	293.07	2333.0	2626.1	0.9551	6.7989	7.7540
75	38.597	0.001026	4.1291	313.99	2161.3	2475.3	314.03	2320.6	2634.6	1.0158	6.6655	7.6812
80	47.416	0.001029	3.4053	334.97	2146.6	2481.6	335.02	2308.0	2643.0	1.0756	6.5355	7.6111
85	57.868	0.001032	2.8261	355.96	2131.9	2487.8	356.02	2295.3	2651.4	1.1346	6.4089	7.5435
90	70.183	0.001036	2.3593	376.97	2117.0	2494.0	377.04	2282.5	2659.6	1.1929	6.2853	7.4782
95	84.609	0.001040	1.9808	398.00	2102.0	2500.1	398.09	2269.6	2667.6	1.2504	6.1647	7.4151
100	101.42	0.001043	1.6720	419.06	2087.0	2506.0	419.17	2256.4	2675.6	1.3072	6.0470	7.3542

(continued)

Table 5.1 (continued)

Table A-4

Temp.,	Sat. press.,	Specific volume. m^3/kg		Internal energy, kJ/kg			Enthalpy, kJ/kg			Entropy, kJ/kg · K		
		Sat. liquid.	Sat. vapor,	Sat. liquid.	Evap.,	Sat. vapor,	Sat. liquid.	Evap.,	Sat. vapor,	Sat. liquid,	Evap.,	Sat. vapor,
T °C	P_{sat}, kPa.	v_f	v_g	u_f	u_{fg}	u_g	h_f	h_{fg}	h_g	s_f	s_{fg}	s_g
105	120.90	0.001047	1.4186	440.15	2071.8	2511.9	440.28	2243.1	2683.4	1.3634	5.9319	7.2952
110	143.38	0.001052	1.2094	461.27	2056.4	2517.7	461.42	2229.7	2691.1	1.4188	5.8193	7.2382
115	169.18	0.001056	1.0360	482.42	2040.9	2523.3	482.59	2216.0	2698.6	1.4737	5.7092	7.1829
120	198.67	0.001060	0.89133	503.60	2025.3	2528.9	503.81	2202.1	2706.0	1.5279	5.6013	7.1292
125	232.23	0.001065	0.77012	524.83	2009.5	2534.3	525.07	2188.1	2713.1	1.5816	5.4956	7.0771
130	270.28	0.001070	0.66808	546.10	1993.4	2539.5	546.38	2173.7	2720.1	1.6346	5.3919	7.0265
135	313.22	0.001075	0.58179	567.41	1977.3	2544.7	567.75	2159.1	2726.9	1.6872	5.2901	6.9773
140	361.53	0.001080	0.50850	588.77	1960.9	2549.6	589.16	2144.3	2733.5	1.7392	5.1901	6.9294
145	415.68	0.001085	0.44600	610.19	1944.2	2554.4	610.64	2129.2	2739.8	1.7908	5.0919	6.8827
150	476.16	0.001091	0.39248	631.66	1927.4	2559.1	632.18	2113.8	2745.9	1.8418	4.9953	6.8371
155	543.49	0.001096	0.34648	653.19	1910.3	2563.5	653.79	2098.0	2751.8	1.8924	4.9002	6.7927
160	618.23	0.001102	0.30680	674.79	1893.0	2567.8	675.47	2082.0	2757.5	1.9426	4.8066	6.7492
165	700.93	0.001108	0.27244	696.46	1875.4	2571.9	697.24	2065.6	2762.8	1.9923	4.7143	6.7067
170	792.18	0.001114	0.24260	718.20	1857.5	2575.7	719.08	2048.8	2767.9	2.0417	4.6233	6.6650
175	892.60	0.001121	0.21659	740.02	1839.4	2579.4	741.02	2031.7	2772.7	2.0906	4.5335	6.6242
180	1002.8	0.001127	0.19384	761.92	1820.9	2582.8	763.05	2014.2	2777.2	2.1392	4.4448	6.5841
185	1123.5	0.001134	0.17390	783.91	1802.1	2586.0	785.19	1996.2	2781.4	2.1875	4.3572	6.5447
190	1255.2	0.001141	0.15636	806.00	1783.0	2589.0	807.43	1977.9	2785.3	2.2355	4.2705	6.5059
195	1398.8	0.001149	0.14089	828.18	1763.6	2591.7	829.78	1959.0	2788.8	2.2831	4.1847	6.4678
200	1554.9	0.001157	0.12721	850.46	1743.7	2594.2	852.26	1939.8	2792.0	2.3305	4.0997	6.4302
205	1724.3	0.001164	0.11508	872.86	1723.5	2596.4	874.87	1920.0	2794.8	2.3776	4.0154	6.3930

(continued)

Table 5.1 (continued)

Table A-4

Temp.,	Sat. press,	Specific volume. m^3/kg		Internal energy, kJ/kg			Enthalpy, kJ/kg			Entropy, kJ/kg · K		
		Sat. liquid.	Sat. vapor,	Sat. liquid.	Evap.,	Sat. vapor,	Sat. liquid.	Evap.,	Sat. vapor,	Sat. liquid,	Evap.,	Sat. vapor,
T °C	P_{sat}, kPa.	v_f	v_g	u_f	u_{fg}	u_g	h_f	h_{fg}	h_g	s_f	s_{fg}	s_g
210	1907.7	0.001173	0.10429	895.38	1702.9	2598.3	897.61	1899.7	2797.3	2.4245	3.9318	6.3563
215	2105.9	0.001181	0.094680	918.02	1681.9	2599.9	920.50	1878.8	2799.3	2.4712	3.8489	6.3200
220	2319.6	0.001190	0.086094	940.79	1660.5	2601.3	943.55	1857.4	2801.0	2.5176	3.7664	6.2840
225	2549.7	0.001199	0.078405	963.70	1638.6	2602.3	966.76	1835.4	2802.2	2.5639	3.6844	6.2483
230	2797.1	0.001209	0.071505	986.76	1616.1	2602.9	990.14	1812.8	2802.9	2.6100	3.6028	6.2128
235	3062.6	0.001219	0.065300	1010.0	1593.2	2603.2	1013.7	1789.5	2803.2	2.6560	3.5216	6.1775
240	3347.0	0.001229	0.059707	1033.4	1569.8	2603.1	1037.5	1765.5	2803.0	2.7018	3.4405	6.1424
245	3651.2	0.001240	0.054656	1056.9	1545.7	2602.7	1061.5	1740.8	2802.2	2.7476	3.3596	6.1072
250	3976.2	0.001252	0.050085	1080.7	1521.1	2601.8	1085.7	1715.3	2801.0	2.7933	3.2788	6.0721
255	4322.9	0.001263	0.045941	1104.7	1495.8	2600.5	1110.1	1689.0	2799.1	2.8390	3.1979	6.0369
260	4692.3	0.001276	0.042175	1128.8	1469.9	2598.7	1134.8	1661.8	2796.6	2.8847	3.1169	6.0017
265	5085.3	0.001289	0.038748	1153.3	1443.2	2596.5	1159.8	1633.7	2793.5	2.9304	3.0358	5.9662
270	5503.0	0.001303	0.035622	1177.9	1415.7	2593.7	1185.1	1604.6	2789.7	2.9762	2.9542	5.9305
275	5946.4	0.001317	0.032767	1202.9	1387.4	2590.3	1210.7	1574.5	2785.2	3.0221	2.8723	5.8944
280	6416.6	0.001333	0.030153	1228.2	1358.2	2586.4	1236.7	1543.2	2779.9	3.0681	2.7898	5.8579
285	6914.6	0.001349	0.027756	1253.7	1328.1	2581.8	1263.1	1510.7	2773.7	3.1144	2.7066	5.8210
290	7441.8	0.001366	0.025554	1279.7	1296.9	2576.5	1289.8	1476.9	2766.7	3.1608	2.6225	5.7834
295	7999.0	0.001384	0.023528	1306.0	1264.5	2570.5	1317.1	1441.6	2758.7	3.2076	2.5374	5.7450
300	8587.9	0.001404	0.021659	1332.7	1230.9	2563.6	1344.8	1404.8	2749.6	3.2548	2.4511	5.7059
305	9209.4	0.001425	0.019932	1360.0	1195.9	2555.8	1373.1	1366.3	2739.4	3.3024	2.3633	5.6657
310	9865.0	0.001447	0.018333	1387.7	1159.3	2547.1	1402.0	1325.9	2727.9	3.3506	2.2737	5.6243

(continued)

Table 5.1 (continued)

Table A-4

Temp.,	Sat. press,	Specific volume. m³/kg		Internal energy, kJ/kg			Enthalpy, kJ/kg			Entropy, kJ/kg · K		
		Sat. liquid.	Sat. vapor,	Sat. liquid.	Evap.,	Sat. vapor,	Sat. liquid.	Evap.,	Sat. vapor,	Sat. liquid,	Evap.,	Sat. vapor,
T °C	P_{sat}, kPa.	v_f	v_g	u_f	u_{fg}	u_g	h_f	h_{fg}	h_g	s_f	s_{fg}	s_g
315	10,556	0.001472	0.016849	1416.1	1121.1	2537.2	1431.6	1283.4	2715.0	3.3994	2.1821	5.5816
320	11,284	0.001499	0.015470	1445.1	1080.9	2526.0	1462.0	1238.5	2700.6	3.4491	2.0881	5.5372
325	12,051	0.001528	0.014183	1475.0	1038.5	2513.4	1493.4	1191.0	2684.3	3.4998	1.9911	5.4908
330	12,858	0.001560	0.012979	1505.7	993.5	2499.2	1525.8	1140.3	2666.0	3.5516	1.8906	5.4422
335	13,707	0.001597	0.011848	1537.5	945.5	2483.0	1559.4	1086.0	2645.4	3.6050	1.7857	5.3907
340	14,601	0.001638	0.010783	1570.7	893.8	2464.5	1594.6	1027.4	2622.0	3.6602	1.6756	5.3358
345	15,541	0.001685	0.009772	1605.5	837.7	2443.2	1631.7	963.4	2595.1	3.7179	1.5585	5.2765
350	16,529	0.001741	0.008806	1642.4	775.9	2418.3	1671.2	892.7	2563.9	3.7788	1.4326	5.2114
355	17,570	0.001808	0.007872	1682.2	706.4	2388.6	1714.0	812.9	2526.9	3.8442	1.2942	5.1384
360	18,666	0.001895	0.006950	1726.2	625.7	2351.9	1761.5	720.1	2481.6	3.9165	1.1373	5.0537
365	19,822	0.002015	0.006009	1777.2	526.4	2303.6	1817.2	605.5	2422.7	4.0004	0.9489	4.9493
370	21,044	0.002217	0.004953	1844.5	385.6	2230.1	1891.2	443.1	2334.3	4.1119	0.6890	4.8009
373.95	22,064	0.003106	0.003106	2015.7	0	2015.7	2084.3	0	2084.3	4.4070	0	4.4070

Source Tables A-4 through A-8 are generated using the Engineering Equation Solver (EES) software developed by S. A. Klein and F. L. Alvarado. The routine used In calculations is the highly accurate Steam_JAPWS, which incorporates the 1995 formulation for the Thermodynamic Properties of Ordinary Water Substance for General and Scientific Use, issued by 'International Association for the Properties of Water and Steam (IAPWS). This formulation replaces the 1984 formulation of Haar, Gallagher, and Kell (NBS/NRC Steam Tables. Hemisphere Publishing Co 1984), which is also available in EES as the routine STEAM. The new formulation is based on the correlations of Saul and Wagner (J. Phys. Chem. Ref. Data. 16. 893, 1987) with modifications to adjust to the International Temperature Scale of 1990. The modifications are described by Wagner and Pruss (J. Phys. Chem. Ref. Data. 22, 783. 1993). The properties of ice are based on Hyland and Wexler. "Formulations tor the Thermodynamic Properties of the Saturated Phases of H_2O from 173.15 to 473.15 K," *ASHRAE Trans.*, Part 2A. Paper 2793, 1983`

State 2: 250 °C, 3976.2 kPa

$$\text{Saturated vapor enthalpy } (h_g)_2 = 2801.0\,\text{kJ/kg}$$
$$\text{Saturated vapor entropy } (s_g)_2 = 6.0721\,\text{kJ/kg}\cdot\text{K}$$

The isentropic condition of 2 → 3 determines $s_3 = s_2 = 6.0721\,\text{kJ/kg}\cdot\text{K}$.
Note from Table 5.2: $s_g(20\,\text{kPa}) = s_f + s_{fg} = 0.8320 + 7.0752 = 7.9073$
Therefore,

$$s_3 = s_f + x_3 s_{fg}, \text{i.e.}, 6.0721 = 0.8320 + x_3 \times 7.0752\,\text{yields},$$

Quality at state 3 is 0.74063.
It follows

State 3: 60.06 °C, 20 kPa (see Table 5.2)

$$x_3 = 0.74063$$

$$h_3 = h_f + x_3 \times h_{fg} = 251.42 + 0.74063 \times 2357.5 = 1997.45\,\text{kJ/kg}$$

(see Table A-5 for the above enthalpy values)

Use the entropy value at state 1 and the same procedure of determining state 3, i.e.,

$$2.7933 = 0.8320 + x_4 \times 7.0752$$

to find the properties at state 4.

State 4: 60.06 °C, 20 kPa (see Table 5.2)

$$x_4 = 0.27721$$

$$h_4 = h_f + x_4 \times h_{fg} = 251.42 + 0.27721 \times 2608.9 = 904.937\,\text{kJ/kg}$$

Answers to (b), (c), and (a): Heat and work are related to enthalpy changes for steady-flow processes:

(b) Heat rejected = $h_3 - h_4 = 1092.5\,\text{kJ/kg}$ each cycle
(c) Net work output = $(h_2 - h_3) - (h_1 - h_4) = 622.787\,\text{kJ/kg}$
(a) $\eta_{th} = \frac{NetWorkOutput}{h_2 - h_1} = \frac{622.787}{1715.3} = 0.363$

Which of course checks with the Kelvin formula

$$\eta_{th} = 1 - \frac{T_C}{T_H} = 1 - \frac{60.06 + 273.15}{250 + 273.15} = 0.363$$

Table 5.2 Saturated water–pressure table

Press.,	Sat. temp.,	Specific volume, m^3/kg		Internal energy, kJ/kg			Enthalpy, kJ/kg			Entropy, kJ/kg · K		
		Sat. liquid,	Sat. vapour,	Sat. liquid,	Evap.,	Sat. vapour,	Sat. liquid,	Evap.,	Sat. vapour,	Sat. liquid,	Evap.,	Sat. vapour,
P kPa	T_{sat} °C	v_f	v_g	u_f	u_{fg}	u_g	h_f	h_{fg}	h_g	s_f	s_{fg}	s_g
1.0	6.97	0.001000	129.19	29.302	2355.2	2384.5	29.303	2484.4	2513.7	0.1059	8.8690	8.9749
1.5	13.02	0.001001	87.964	54.686	2338.1	2392.8	54.688	2470.1	2524.7	0.1956	8.6314	8.8270
2.0	17.50	0.001001	66.990	73.431	2325.5	2398.9	73.433	2459.5	2532.9	0.2606	8.4621	8.7227
2.5	21.08	0.001002	54.242	88.422	2315.4	2403.3	88.424	2451.0	2539.4	0.3118	8.3302	8.6421
3.0	24.08	0.001003	45.654	100.98	2306.9	2407.9	100.98	2443.9	2544.8	0.3543	8.2222	8.5765
4.0	28.96	0.001004	34.791	121.39	2293.1	2414.5	121.39	2432.3	2553.7	0.4224	8.0510	8.4734
5.0	32.87	0.001005	28.185	137.75	2282.1	2419.8	137.75	2423.0	2560.7	0.4762	7.9176	8.3938
7.5	40.29	0.001008	19.233	168.74	2261.1	2429.8	168.75	2405.3	2574.0	0.5763	7.6738	8.2501
10	45.81	0.001010	14.670	191.79	2245.4	2437.2	191.81	2392.1	2583.9	0.6492	7.4996	8.1488
15	53.97	0.001014	10.020	225.93	2222.1	2448.0	225.94	2372.3	2598.3	0.7549	7.2522	8.0071
20	60.06	0.001017	7.6481	251.40	2204.6	2456.0	251.42	2357.5	2608.9	0.8320	7.0752	7.9073
25	64.96	0.001020	6.2034	271.93	2190.4	2462.4	271.96	2345.5	2617.5	0.8932	6.9370	7.8302
30	69.09	0.001022	5.2287	289.24	2178.5	2467.7	289.27	2335.3	2624.6	0.9441	6.8234	7.7675
40	75.86	0.001026	3.9933	317.58	2158.8	2476.3	317.62	2318.4	2636.1	1.0261	6.6430	7.6691
50	81.32	0.001030	3.2403	340.49	2142.7	2483.2	340.54	2304.7	2645.2	1.0912	6.5019	7.5931
75	91.76	0.001037	2.2172	384.36	2111.8	2496.1	384.44	2278.0	2662.4	1.2132	6.2426	7.4558
100	99.61	0.001043	1.6941	417.40	2088.2	2505.6	417.51	2257.5	2675.0	1.3028	6.0562	7.3589
101.325	99.97	0.001043	1.6734	418.95	2087.0	2506.0	419.06	2256.5	2675.6	1.3069	6.0476	7.3545
125	105.97	0.001048	1.3750	444.23	2068.8	2513.0	444.36	2240.6	2684.9	1.3741	5.9100	7.2841
150	111.35	0.001053	1.1594	466.97	2052.3	2519.2	467.13	2226.0	2693.1	1.4337	5.7894	7.2231
175	116.04	0.001057	1.0037	486.82	2037.7	2524.5	487.01	2213.1	2700.2	1.4850	5.6865	7.1716
200	120.21	0.001061	0.88578	504.50	2024.6	2529.1	504.71	2201.6	2706.3	1.5302	5.5968	7.1270

(continued)

Table 5.2 (continued)

Press.,	Sat. temp.,	Specific volume, m^3/kg		Internal energy, kJ/kg			Enthalpy, kJ/kg			Entropy, kJ/kg · K		
		Sat. liquid,	Sat. vapour,	Sat. liquid,	Evap.,	Sat. vapour,	Sat. liquid,	Evap.,	Sat. vapour,	Sat. liquid,	Evap.,	Sat. vapour,
P kPa	T_{sat} °C	v_f	v_g	u_f	u_{fg}	u_g	h_f	h_{fg}	h_g	s_f	s_{fg}	s_g
225	123.97	0.001064	0.79329	520.47	2012.7	2533.2	520.71	2191.0	2711.7	1.5706	5.5171	7.0877
250	127.41	0.001067	0.71873	535.08	2001.8	2536.8	535.35	2181.2	2716.5	1.6072	5.4453	7.0525
275	130.58	0.001070	0.65732	548.57	1991.6	2540.1	548.86	2172.0	2720.9	1.6408	5.3800	7.0207
300	133.52	0.001073	0.60582	561.11	1982.1	2543.2	561.43	2163.5	2724.9	1.6717	5.3200	6.9917
325	136.27	0.001076	0.56199	572.84	1973.1	2545.9	573.19	2155.4	2728.6	1.7005	5.2645	6.9650
350	138.86	0.001079	0.52422	583.89	1964.6	2548.5	584.26	2147.7	2732.0	1.7274	5.2128	6.9402
375	141.30	0.001081	0.49133	594.32	1956.6	2550.9	594.73	2140.4	2735.1	1.7526	5.1645	6.9171
400	143.61	0.001084	0.46242	604.22	1948.9	2553.1	604.66	2133.4	2738.1	1.7765	5.1191	6.8955
450	147.90	0.001088	0.41392	622.65	1934.5	2557.1	623.14	2120.3	2743.4	1.8205	5.0356	6.8561
500	151.83	0.001093	0.37483	639.54	1921.2	2560.7	640.09	2108.0	2748.1	1.8604	4.9603	6.8207
550	155.46	0.001097	0.34261	655.16	1908.8	2563.9	655.77	2096.6	2752.4	1.8970	4.8916	6.7886
600	158.83	0.001101	0.31560	669.72	1897.1	2566.8	670.38	2085.8	2756.2	1.9308	4.8285	6.7593
650	161.98	0.001104	0.29260	683.37	1886.1	2569.4	684.08	2075.5	2759.6	1.9623	4.7699	6.7322
700	164.95	0.001108	0.27278	696.23	1875.6	2571.8	697.00	2065.8	2762.8	1.9918	4.7153	6.7071
750	167.75	0.001111	0.25552	708.40	1865.6	2574.0	709.24	2056.4	2765.7	2.0195	4.6642	6.6837
800	170.41	0.001115	0.24035	719.97	1856.1	2576.0	720.87	2047.5	2768.3	2.0457	4.6160	6.6616
850	172.94	0.001118	0.22690	731.00	1846.9	2577.9	731.95	2038.8	2770.8	2.0705	4.5705	6.6409
900	175.35	0.001121	0.21489	741.55	1838.1	2579.6	742.56	2030.5	2773.0	2.0941	4.5273	6.6213
950	177.66	0.001124	0.20411	751.67	1829.6	2581.3	752.74	2022.4	2775.2	2.1166	4.4862	6.6027
1000	179.88	0.001127	0.19436	761.39	1821.4	2582.8	762.51	2014.6	2777.1	2.1381	4.4470	6.5850
1100	184.06	0.001133	0.17745	779.78	1805.7	2585.5	781.03	1999.6	2780.7	2.1785	4.3735	6.5520
1200	187.96	0.001138	0.16326	796.96	1790.9	2587.8	798.33	1985.4	2783.8	2.2159	4.3058	6.5217

(continued)

Table 5.2 (continued)

Press.,	Sat. temp.,	Specific volume, m³/kg		Internal energy, kJ/kg			Enthalpy, kJ/kg			Entropy, kJ/kg · K		
		Sat. liquid,	Sat. vapour,	Sat. liquid,	Evap.,	Sat. vapour,	Sat. liquid,	Evap.,	Sat. vapour,	Sat. liquid,	Evap.,	Sat. vapour,
P kPa	T_{sat} °C	v_f	v_g	u_f	u_{fg}	u_g	h_f	h_{fg}	h_g	s_f	s_{fg}	s_g
1300	191.60	0.001144	0.15119	813.10	1776.8	2589.9	814.59	1971.9	2786.5	2.2508	4.2428	6.4936
1400	195.04	0.001149	0.14078	828.35	1763.4	2591.8	829.96	1958.9	2788.9	2.2835	4.1840	6.4675
1500	198.29	0.001154	0.13171	842.82	1750.6	2593.4	844.55	1946.4	2791.0	2.3143	4.1287	6.4430
1750	205.72	0.001166	0.11344	876.12	1720.6	2596.7	878.16	1917.1	2795.2	2.3844	4.0033	6.3877
2000	212.38	0.001177	0.099587	906.12	1693.0	2599.1	908.47	1889.8	2798.3	2.4467	3.8923	6.3390
2250	218.41	0.001187	0.088717	933.54	1667.3	2600.9	936.21	1864.3	2800.5	2.5029	3.7926	6.2954
2500	223.95	0.001197	0.079952	958.87	1643.2	2602.1	961.87	1840.1	2801.9	2.5542	3.7016	6.2558
3000	233.85	0.001217	0.066667	1004.6	1598.5	2603.2	1008.3	1794.9	2803.2	2.6454	3.5402	6.1856
3500	242.56	0.001235	0.057061	1045.4	1557.6	2603.0	1049.7	1753.0	2802.7	2.7253	3.3991	6.1244
4000	250.35	0.001252	0.049779	1082.4	1519.3	2601.7	1087.4	1713.5	2800.8	2.7966	3.2731	6.0696
5000	263.94	0.001286	0.039448	1148.1	1448.9	2597.0	1154.5	1639.7	2794.2	2.9207	3.0530	5.9737
6000	275.59	0.001319	0.032449	1205.8	1384.1	2589.9	1213.8	1570.9	2784.6	3.0275	2.8627	5.8902
7000	285.83	0.001352	0.027378	1258.0	1323.0	2581.0	1267.5	1505.2	2772.6	3.1220	2.6927	5.8148
8000	295.01	0.001384	0.023525	1306.0	1264.5	2570.5	1317.1	1441.6	2758.7	3.2077	2.5373	5.7450
9000	303.35	0.001418	0.020489	1350.9	1207.6	2558.5	1363.7	1379.3	2742.0	3.2866	2.3925	5.6791
10,000	311.00	0.001452	0.018028	1393.3	1151.8	2545.2	1407.8	1317.6	2725.5	3.3603	2.2556	5.6159
11,000	318.08	0.001488	0.015988	1433.9	1096.6	2530.4	1450.2	1256.1	2706.3	3.4299	2.1245	5.5544
12,000	324.68	0.001526	0.014264	1473.0	1041.3	2514.3	1491.3	1194.1	2685.4	3.4964	1.9975	5.4939
13,000	330.85	0.001566	0.012781	1511.0	985.5	2496.6	1531.4	1131.3	2662.7	3.5606	1.8730	5.4336
14,000	336.67	0.001610	0.011487	1548.4	928.7	2477.1	1571.0	1067.0	2637.9	3.6232	1.7497	5.3728
15,000	342.16	0.001657	0.010341	1585.5	870.3	2455.7	1610.3	1000.5	2610.8	3.6848	1.6261	5.3108
16,000	347.36	0.001710	0.009312	1622.6	809.4	2432.0	1649.9	931.1	2581.0	3.7461	1.5005	5.2466

(continued)

Table 5.2 (continued)

Press.,	Sat. temp.,	Specific volume, m³/kg		Internal energy, kJ/kg			Enthalpy, kJ/kg			Entropy, kJ/kg · K		
		Sat. liquid,	Sat. vapour,	Sat. liquid,	Evap.,	Sat. vapour,	Sat. liquid,	Evap.,	Sat. vapour,	Sat. liquid,	Evap.,	Sat. vapour,
P kPa	T_{sat} °C	v_f	v_g	u_f	u_{fg}	u_g	h_f	h_{fg}	h_g	s_f	s_{fg}	s_g
17,000	352.29	0.001770	0.008374	1660.2	745.1	2405.4	1690.3	857.4	2547.7	3.8082	1.3709	5.1791
18,000	356.99	0.001840	0.007504	1699.1	675.9	2375.0	1732.2	777.8	2510.0	3.8720	1.2343	5.1064
19,000	361.47	0.001926	0.006677	1740.3	598.9	2339.2	1776.8	689.2	2466.0	3.9396	1.0860	5.0256
20,000	365.75	0.002038	0.005862	1785.8	509.0	2294.8	1826.6	585.5	2412.1	4.0146	0.9164	4.9310
21,000	369.83	0.002207	0.004994	1841.6	391.9	2233.5	1888.0	450.4	2338.4	4.1071	0.7005	4.8076
22,000	373.71	0.002703	0.003644	1951.7	140.8	2092.4	2011.1	161.5	2172.6	4.2942	0.2496	4.5439
22,064	373.95	0.003106	0.003106	2015.7	0	2015.7	2084.3	0	2084.3	4.4070	0	4.4070

This example of Carnot cycle demonstrates how the introduction of entropy by the second law facilitates idealized power-cycle analyses. Various degrees of removal of idealization to representing real cycles can be considered. In those considerations, the departure of real power-cycle steps from idealized isentropic power-cycle steps is associated with entropy increase. Thus, universal entropy increase as predicted by the entropy law is usually interpreted to be the central meaning of the second law as *the law of dissipation*. But, there is a more important meaning of the second law as *the law of driving force*—as discussion in the two sections below suggests.

5.9 Mixtures of Ideal Gases and Their Properties

The specific internal energy and the enthalpy of ideal gases are functions of temperature alone. The total values of internal energy and enthalpy for a mixture can be expressed as the sum of the products of the mole fraction and the specific internal energy or enthalpy of each component in the mixture

$$U(T) = \sum_{i=1}^{n} N_i u_i = N \sum_{i=1}^{n} x_i u_i(T) \tag{76}$$

$$H(T) = \sum_{i=1}^{n} N_i h_i = N \sum_{i=1}^{n} x_i h_i(T) \tag{77}$$

The entropy of each ideal gas depends on temperature and volume or temperature and pressure. The determination of the total value for a mixture requires the application of Gibbs' theorem.

Gibbs' theorem is demonstrated by a thought experiment: Consider an isolated composite system made of one stationary cylinder- chamber of volume V and its extension chamber of equal volume, and one sliding cylinder (on the right in the figure on next page) of equal volume (see Figs. 5.5 and 5.6). The stationary cylinder chamber with its extension chamber is made of three walls: two end walls and a middle partition, which separates gas A in the stationary cylinder chamber from the extension chamber. Initially, the sliding cylinder is occupied with gas B at the same temperature of gas A. Of the three fixed walls, the middle partition is a semi-permeable partition permeable to gas B but impermeable to gas A, while both end walls are impermeable. Of the two walls of the sliding cylinder, the left wall is a semi-permeable membrane permeable to A, but impermeable to B, while the right wall is an impermeable wall.

Initially, the sliding cylinder and the extension chamber are coincident in space with gas B inside the cylinder (as shown in the top view of Fig. 5.5). Reversible mixing can take place by sliding the sliding cylinder into the stationary cylinder. Imagine an intermediate position of the sliding cylinder in the middle of the process as shown in Fig. 5.6. And imagine that there are four spaces (designated as α, β, γ,

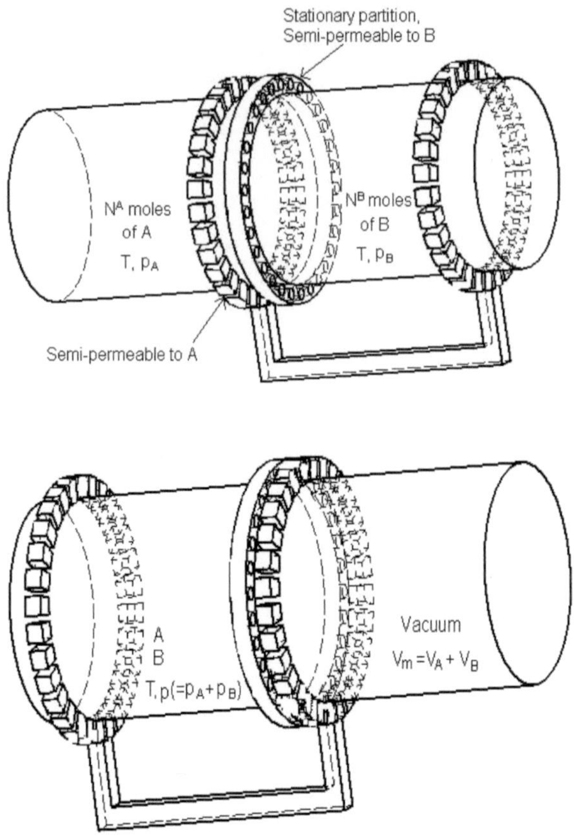

Fig. 5.5 Reversible mixing of two ideal gases with a sliding cylinder shown in its initial position at top drawing, and its final position at the lower drawing

δ): α-space is the space to the left of moving wall; it is a space occupied by gas A because the moving wall is impermeable to B. The β-space, as shown in Fig. 5.6, is between the moving wall (permeable to A) and the stationary partition (permeable to B) and it is occupied by the mixture of A and B (A from diffusion across the moving wall on the left and B from diffusion across the stationary partition). The γ-space is between the stationary partition and the right moving wall with gas B in the space, and the δ-space is vacuum space, since the right moving wall is impermeable to gases (assuming the sliding surfaces are gas tight that is impervious to any leakage). At the end of the quasi-static reversible process, the sliding cylinder coincides with the space of the stationary cylinder, i.e., the β-space occupies a

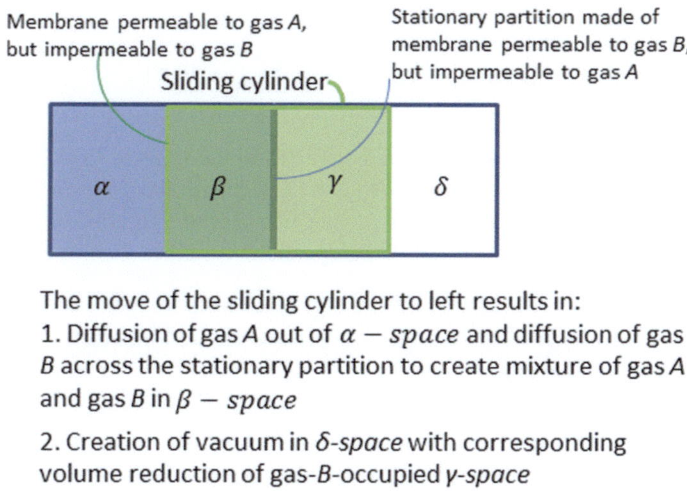

Fig. 5.6 Reversible mixing of two ideal gases, gas A and gas B, with a sliding cylinder in an intermediate position

volume V which is filled with the mixture of A and B. The process involves zero Q. It also requires no W: The forces against the motion on the left moving wall (according to Dalton's law) are

$$[p_A - (p_A + p_B)] \cdot (cylinder\_cross \cdot section\_area)$$

and the force on the right moving wall is

$$p_B \cdot (cylinder\_cross \cdot section\_area)$$

Which are balanced resulting in zero net force and consequently zero work. Whence $\Delta U = 0$ and T remains constant.

That is, the temperature of the mixture in the final β-space is the same T as that of gas A in the initial α-space and gas B in the initial γ-space. So also are the volumes of the three spaces, the final β-space occupied by mixture, initial α-space occupied by gas A, and initial γ-space occupied by gas B, equal to one another. In addition to these equalities, the reversible adiabatic mixing involves no entropy change, i.e., the final entropy of the mixture at T and V equals the sum of the initial entropy of gas A and the initial entropy of gas B under the same T and V. The inference of the thought experiment is, therefore,

Gibbs' theorem: The entropy of a mixture of ideal gases is the sum of the entropies that each gas would have if it alone were to occupy the volume V at temperature T.

That is,

$$S(T,V)=\sum_{i=1}^{n}N_i s_i=N\sum_{i=1}^{n}x_i s_i(T,V) \tag{78}$$

Despite the similarity in the form of Eq. (78) and the form of Eqs. (76) and (77), they are quite different because Eqs. (76) and (77) remain the same for either *T-V* or *T-p* as the set of independent variables, while Eq. (78) will assume a different form if it is expressed in terms of the set of independent variables *T-p*, as shown below.

5.9.1 Entropy and Specific Gibbs Function of Mixture in Terms of T-p

According to Dalton's law, when a gas of specie i alone occupies the volume V at temperature T, the pressure is equal to the partial pressure p_i of the specie in the mixture. Therefore,

$$S=N\sum_{i=1}^{n}x_i s_i(T,V)=N\sum_{i=1}^{n}x_i s_i(T,p_i) \tag{79}$$

Note that this is different from

$$S=N\sum_{i=1}^{n}x_i s_i(T,p)$$

where p is the pressure of the mixture.

Recall from Eq. (67)

$$dS=Nc_p\frac{dT}{T}-NR\frac{dp}{p}$$

For ideal gases, therefore,

$$s(T,p)-s_0=\int_{T_0}^{T}\frac{c_p(T')}{T'}dT'-R\ln\frac{p}{p_0}$$

The molar entropy of an ideal gas in terms of *T-p* is

$$\begin{aligned}s_i(T,p_i)&=s_{i0}+\int_{T_0}^{T}\frac{c_{pi}(T)}{T}dT-R\ln\frac{p_i}{p_0}=s_{i0}+\int_{T_0}^{T}\frac{c_{pi}(T)}{T}dT-R\ln\frac{p_i}{p}\frac{p}{p_0}\\&=s_{i0}+\int_{T_0}^{T}\frac{c_{pi}(T')}{T'}dT'-R\ln\frac{p}{p_0}-R\ln\frac{p_i}{p}\end{aligned}$$

Substitution of the molar entropy expression into (79) yields

$$S(T,p) = N\sum_{i=1}^{n} x_i s_i(T,p_i)$$
$$= \sum_i N_i s_{i0} + \sum_i N_i \int_{T_0}^{T} \frac{c_{pi}(T')}{T'} dT' - NR\ln\frac{p}{p_0} + NR\sum_i x_i\left(-\ln\frac{p_i}{p}\right) \tag{80}$$

That is, from Eqs. (8) and (80) becomes,

$$S(T,p) = N\sum_{i=1}^{n} x_i s_i(T,p) + NR\sum_i x_i(-\ln x_i) \tag{80A}$$

When a gas A and a gas B, both at the same pressure p, are mixed to form a mixture at p, the entropy increase, as a result of irreversible mixing, equals $NR(-x_A lnx_A - x_B lnx_B)$, as shown in the last term of Eq. (80A).

An Example of the Mixing of Ideal Gases

Consider an oxygen gas which is kept in a container of 0.6818 m^3 at 300 K and 220 kPa, and in a separate container of 0.8182 m^3 is kept nitrogen gas at 300 K and 220 kPa. Determine the mass and kmol of the oxygen and nitrogen gases. Now open the connection between the two containers so that the gases are mixed under the constant pressure of 220 kPa. Determine the entropy gain as a result of irreversible mixing. (Note the final volume of the mixture is 0.6818 + 0.8182 = 1.5 m^3.)

a. $N_{O_2} = pV_{O_2}/RT = 220 \cdot 0.6818/8.31447 \cdot 300 = 0.060136\,\text{kmol}$

$N_{N_2} = pV_{N_2}/RT = 220 \cdot 0.8182/8.31447 \cdot 300 = 0.072163\,\text{kmol}$

Correspondingly,

$m_{O_2} = 0.060136\,\text{kmol} \times 31.999\text{kg/kmol} = 1.9243\,\text{kg}$

$m_{N_2} = 0.072163\,\text{kmol} \times 28.013\,\text{kg/kmol} = 2.0215\,\text{kg}$

b. $S_{mixture}[300\,\text{K}, 220\,\text{kPa}] - (N_{O_2}s_{O_2}[300\,\text{K}, 220\,\text{kPa}] + N_{N_2}s_{N_2}[300\,\text{K}, 220\,\text{kPa}])$

$= NR\left(-\frac{N_{O_2}}{\sum N_i}\log_e\frac{p_{O_2}}{220} - \frac{N_{N_2}}{\sum N_i}\log_e\frac{p_{N_2}}{220}\right)$

$= NR(-x_{O_2}\log_e x_{O_2} - x_{N_2}\log_e x_{N_2}) = 0.7579\,\text{kJ/K}$

5.10 The Examples of Reversibly Controlled "Free Expansion" and Reversible Mixing of Ideal Gases: Why Kelvin's Second *General Conclusion* Is Not True?

We now consider the same oxygen and nitrogen systems at the same 300 K temperature but kept in different containers ($V_{O_2} = 0.5\,\mathrm{m}^3$ and $V_{N_2} = 1\,\mathrm{m}^3$). Correspondingly, the pressures are 300 kPa and 180 kPa, respectively. These initial conditions are chosen so that when the two gases are mixed to become a mixture in a container of 1.5 m^3, the final state of the mixture is the same final state in the above example: 300 K, 220 kPa (as it can be readily confirmed by the application of Dalton's law).

The entropy of the mixture is the same as in the example above. The entropy of the initial oxygen and nitrogen systems is different, thus, the entropy gain of this mixing process will be different, which can be readily calculated by using the Gibbs theorem and Eq. (68A) or (69A)

$$\begin{aligned} S_{mixture} - [S_{O_2} + S_{N_2}]_{initial} &= \left[S_{O_2}\left(300\,\mathrm{K}, 1.5\,\mathrm{m}^3\right) - S_{O_2}\left(300\,\mathrm{K}, 0.5\,\mathrm{m}^3\right)\right] \\ &+ \left[S_{N_2}\left(300\,\mathrm{K}, 1.5\,\mathrm{m}^3\right) - S_{N_2}\left(300\,\mathrm{K}, 1\,\mathrm{m}^3\right)\right] \end{aligned}$$

Recall Eq. (68A)

$$S_B - S_A = Nc_V ln\frac{T_B}{T_A} + NRln\frac{V_B}{V_A}$$

Thus, as the mole numbers of the two gases are determined in the aforementioned example, and there is no change in temperature

$$\begin{aligned} S_{mixture} - [S_{O_2} + S_{N_2}]_{initial} &= \left[S_{O_2}\left(300\,\mathrm{K}, 1.5\,\mathrm{m}^3\right) - S_{O_2}\left(300\,\mathrm{K}, 0.5\,\mathrm{m}^3\right)\right] \\ &+ \left[S_{N_2}\left(300\,\mathrm{K}, 1.5\,\mathrm{m}^3\right) - S_{N_2}\left(300\,\mathrm{K}, 1\,\mathrm{m}^3\right)\right] \\ &= 0.0601 \cdot 8.31447 \log_e 3 + 0.722 \cdot 8.31447 \log_e 1.5 \\ &= 0.7926\,\mathrm{kJ/K} \end{aligned}$$

We now consider the thought experiment of a reversible change of the oxygen and nitrogen systems from their initial states to the final mixture state as defined in Fig. 5.7. The reversible change takes place as follows: Assuming availability of a vacuum of total 1.5 m^3, one part of which 1 m^3 is to be used for the first step of the oxygen system and the second part of which 0.5 m^3 is to be used for the first step of the nitrogen system.

5.10.1 Controlled Expansion of the Oxygen System/Vacuum System

As shown in Fig. 5.8, the oxygen system and the vacuum system are a composite system with a piston separating the two and are submerged in a heat reservoir at 300 K. Let the piston undergo a slow expansion of the oxygen gas until the piston reaches the opposite end of the vacuum. We have then the following result:

$$\Delta S_{O_2} = N_{O_2} R \log_e 1.5/0.5 = 0.0601 \cdot 8.31447 \log_e 3 = 0.5493\,\mathrm{kJ/K}$$

The expansion extracts from the heat reservoir heat in the amount,
$T_0 \cdot \Delta S_{O_2} = 300[K] \cdot \Delta S_{O_2} = 164.79\,\mathrm{kJ}$, which equals the work of isothermal expansion

$$W_{reversible-expansion} = \int p dV = N_{O_2} RT \int \frac{dV}{V} = 0.0601 \cdot 8.31447 \cdot 300 \cdot \log_e 1.5/0.5$$
$$= 164.79\,\mathrm{kJ}$$

5.10.2 Controlled Expansion of the Nitrogen System/Vacuum System

Similarly, as shown in Fig. 5.9, combine the nitrogen system and the vacuum system into a composite system with a piston separating the two, and submerge the composite system in a heat reservoir of 300 K. The piston undergoes a slow expansion of the nitrogen gas until it reaches the opposite end of the vacuum. We have the following results:

$$\Delta S_{N_2} = N_{N_2} R \log_e 1.5\,/1.51\ - 1 = 0.0722 \cdot 8.31447 \log_e 1.5 = 0.2433\,\mathrm{kJ\,/K}$$

$$300[K] \cdot \Delta S_{N_2} = 72.9837\,\mathrm{kJ}$$

$$W_{reversible-expansion} = \int p dV = N_{N_2} RT \int \frac{dV}{V} = 0.0722 \cdot 8.31447 \cdot 300 \cdot \log_e 1.5$$
$$= 72.98\,\mathrm{kJ}$$

We are now in the position of bringing the systems to the final state of the mixture, as shown in Fig. 5.7.

5.10.3 Reversible Mixing of the $1.5\,\mathrm{m}^3$ Oxygen and the $1.5\,\mathrm{m}^3$ Nitrogen Systems

Reversible mixing of the $1.5\,\mathrm{m}^3$ oxygen and the $1.5\,\mathrm{m}^3$ nitrogen systems results in the creation of a vacuum system of $1.5\,\mathrm{m}^3$ as shown in Fig. 5.10. This is exactly the same process shown in Fig. 5.6, where the created vacuum space is the final δ-space.

5.10.4 In Sum

The change of the above processes associated with Figs. 5.8, 5.9, and 5.10 collectively corresponds to change from the initial oxygen and the initial nitrogen systems to their mixed final state as shown in Fig. 5.7. This change can be the result of either irreversible mixing as usually implied by Fig. 5.7 or, the result of reversible steps as shown in Figs. 5.8—5.10. The difference between the two events is that, in the latter case, heat totaling (164.79 + 72.98 =) 237.78 kJ is absorbed from the two heat baths in Figs. 5.8 and 5.9, and that heat is converted into mechanical work; whereas, in the former case, the mixing is usually taking place adiabatically without heat absorption or dissipation. Note that there is no change in the vacuum space: the required vacuum initially used in expansion processes is restored during the reversible mixing. Nor is there any change in energy, or energy degradation, of the oxygen gas and the nitrogen gas. Significantly, the only change occurs in entropy of the gases.

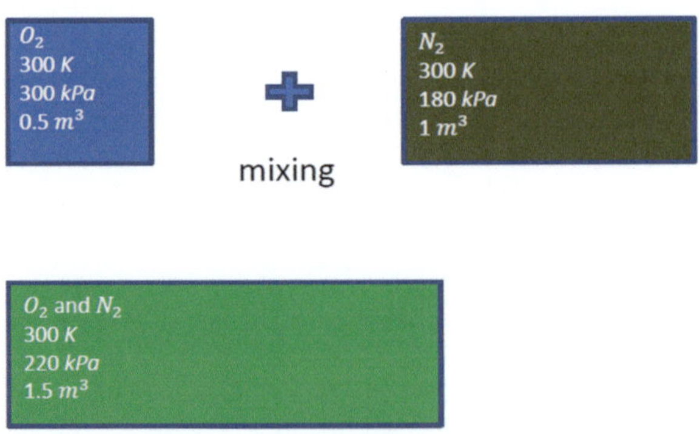

Fig. 5.7 The initial state and the final state of a mixing process: as it is discussed here the mixing process can be an irreversible change or a reversible change of states

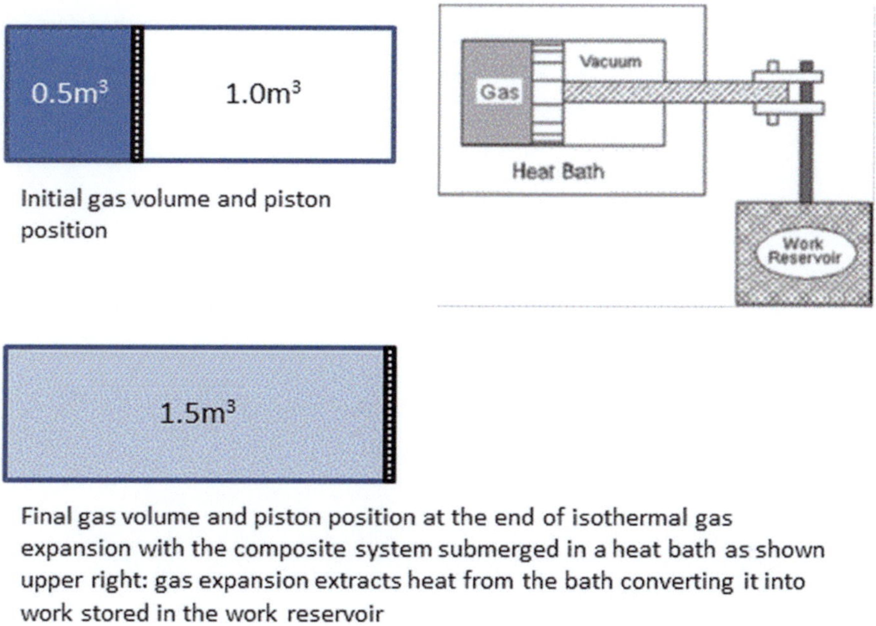

Fig. 5.8 Depiction of a reversibly controlled expansion of a gas–vacuum composition system from the initial state (left top) to final state (left bottom). Upper right depicts a schematic arrangement of the expansion process being driven by an infinitesimally small net force

5.10.5 Kelvin's Energy Principle

What about the self-evident truth of the universal dissipation of mechanical energy. There is no doubt that mechanical energy dissipates spontaneously. The key claim of universal dissipation is Kelvin's second *general conclusion*

> Any *restoration* of mechanical energy [of a given amount], without [the expenditure of] more than an equivalent [amount] of [mechanical energy resulting in its] dissipation, is impossible in inanimate material processes…

That is, restoration or creation of mechanical energy (or, energy of certain "grade") of a given amount requires the *expenditure* of mechanical energy (or, the energy of the same grade or higher grade) of a greater amount. Since energy is conserved and no energy is ever lost or spent, the expenditure of a given amount of energy can only be understood in terms of occurrence in the form-change of the given amount of energy.

The above example of reversible mixing involves no change in forms of oxygen and nitrogen, both of which remain at the same temperature; there is, thus, no energy expended in the creation (or, restoration) of mechanical energy. Kelvin's second general conclusion is false: mechanical energy dissipates spontaneously, not universally.

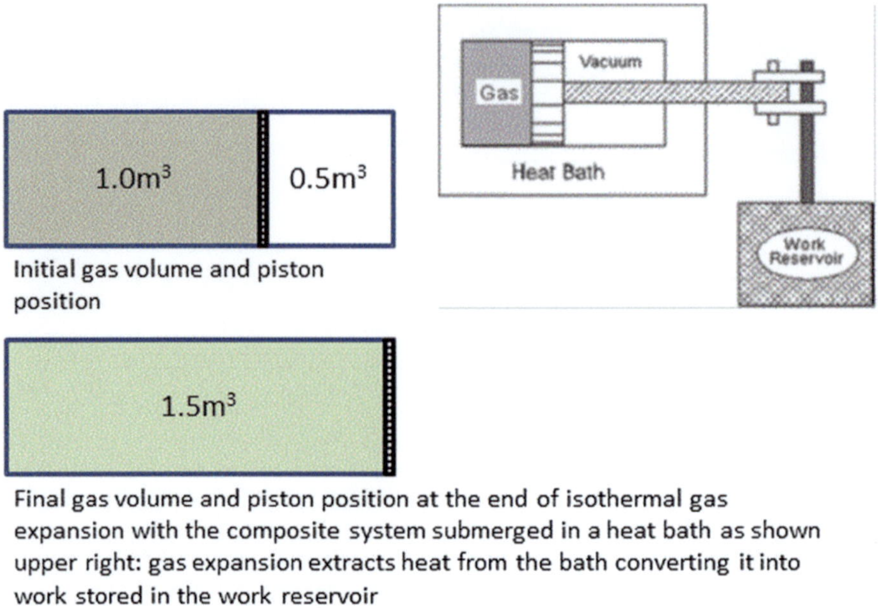

Fig. 5.9 Depiction of a reversibly controlled expansion of a gas (nitrogen)–vacuum composition system from the initial state (left top) to the final state (left bottom). Upper right depicts a schematic arrangement of the expansion process being driven by an infinitesimally small net force

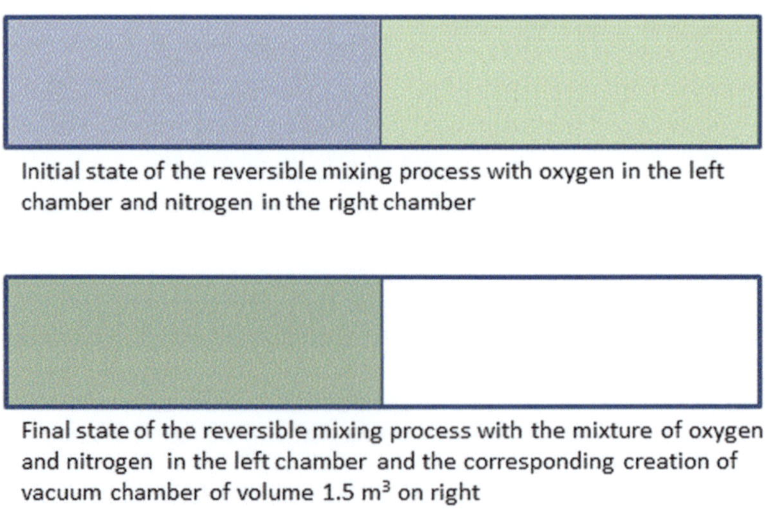

Fig. 5.10 Initial state of separate oxygen and nitrogen of equal volumes and final state of oxygen–nitrogen mixture

The production of mechanical energy, in that case, results from the growth of entropy, which is the driving force of that process. Such condition was pointed out by Planck some time ago:

> The real meaning of the second law has frequently been looked for in a "dissipation of energy." This view, proceeding, as it does, from the irreversible phenomena of conduction and radiation of heat, presents only one side of the question. There are irreversible processes in which the final and initial states show exactly the same form of energy, e.g. the diffusion of two perfect gases or further dilution of a dilute solution. Such processes are accompanied by no perceptible transference of heat, nor by external work, nor by any noticeable transformation of energy. They occur only for the reason that they lead to an appreciable increase of the entropy. [1: 103–104]

Unfortunately, in the general thermodynamics literature, this insightful point has been overlooked and we have the bizarre situation among students of thermodynamics that while the entropy law is accepted as the official second law, the law is almost universally understood in terms of Kelvin's *general conclusions* of universal dissipation: "entropy and the dissipation of energy are as inseparable as Siamese twins in the thought of every student of thermodynamics" [13]. The demonstration in the above provides details to Planck's observation so that it cannot be overlooked anymore—as well as necessity, as it'll be carried out, in a critical assessment of the energy principle, the entropy principle and the real meaning of the second law.

It suffices to note here that the critical assessment, as it will be fully developed in Chaps. 7 and 8, will conclude that energy is not the sole driver of change, but instead, energy is the proxy of entropy growth (more precisely, entropy growth potential to be introduced in Chap. 8), which is the *universal* driving force of all processes in nature. That lesson, rather than entropy as a merely dissipative mechanism, is the true lesson we should take from the second law.

5.11 Concluding Remarks: Applications to Special States of Thermodynamic Equilibrium

Figure 5.11 suggests that the formulation of the entropy principle is a central part of the evolution of thermodynamic thought. First of all, it notes that Kelvin and Clausius formulated two distinctive principles of unidirectionality. The two, though closely connected, are by no means identical (see Sect. 5.10). There is no doubt, especially in view of the statistical mechanical interpretation being directly and successively linked to the entropy principle (while its linkage to the dissipation proposition is not as clear), that the entropy principle is *the* counterpoint to the first law. Its privileged position accords it, like the first law, to be considered as one of the universal principles of nature.

The position of the dissipation proposition (the energy principle), though already repudiated in Sect. 5.10, remains to be further examined and, certainly, its importance in thermodynamic thought warrants a more detailed assessment of it in Chap. 8.

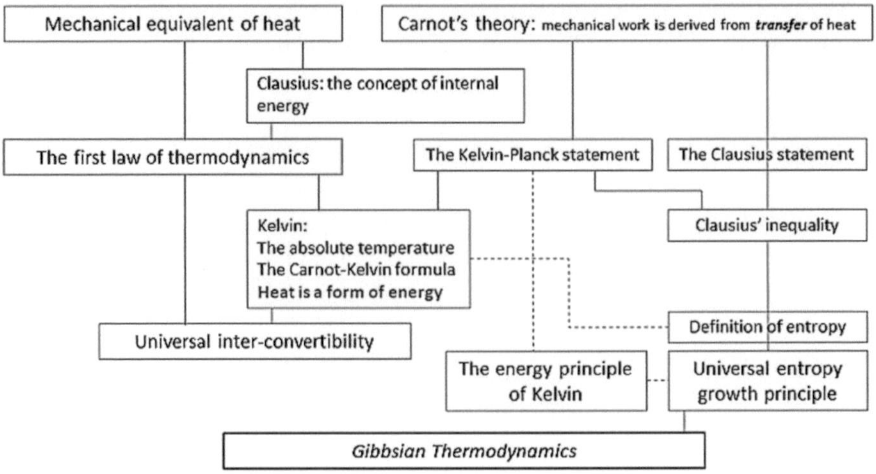

Fig. 5.11 Evolution of thermodynamic thought 2_The entropy principle and equilibrium thermodynamics: The figure summarizes Clausius' contributions to the MTH: the introduction of internal energy and the formulation of the 1st law, the Clausius statement and Clausius' inequality, the introduction of entropy, and his formulation of the second law in terms of universal growth of entropy

The establishment of the entropy principle signaled the beginning of a new era—in which thermodynamics, which had been originated as a branch of engineering knowledge, separated into two distinctive streams.

The engineering stream, one that originated from the source of heat and work transformations, continued the course of its development. The central position of universal interconvertibility in the MTH remained intact and engineering practice was still based on the Carnot–Kelvin formula as a manifestation of *heat's apparent utility*. The contribution of the establishment of the entropy principle to the engineering stream was limited to the availability of *properties of working fluids* for application in engineering systems analysis, while the long-term *conceptual* contribution derived from the entropy principle would not materialize fully until the development of the theory of exergy (see Chap. 7).

A milestone in this chapter is the formulation of a definition of heat in Sect. 5.6 that reflects how physicists and engineers actually use the term heat. This is a partial step toward our understanding of heat and energy, a definition of which will be given only in Chap. 7. As well as the fact that it remains, in Chap. 8, to remove the impossible burden universal interconvertibility placed on heat per se as a driving force.

The real short-term, significant contribution the entropy principle brought about was the creation of a new stream, the scientific stream. The establishment of the entropy principle made it possible to develop a complete thermodynamic formalism of thermodynamic equilibrium states, which is called Gibbsian Thermodynamics (briefly summarized in Chap. 9). This development corresponded to a shift from the British engineering tradition (Joule and Thomson) [3] to the German science tradition (Clausius, Boltzmann, and Gibbs). Unlike the messy, incoherent engineering thermodynamics, Gibbsian Thermodynamics, i.e., *equilibrium thermodynamics*, is a coherent system and, as a branch of theoretical physics, is, furthermore, amenable to reformulation into elegant systems with postulational formulations of the entropy growth principle (of various versions) instead of the empirical treatment as shown in Fig. 5.11 [7, 15–18].

In the meantime, *atomism* (and *mechanism* and *reductionism*) came to dominate science at the end of the nineteenth century with the presupposition that, in the naturalistic study of a system, laws should be centered on a natural system in itself without having to involve a man-made machine. Such formulations do not address engineering applications by leaving out operational details, and as a result, the definitions of work and heat, which are "central to the formulation of the first law of thermodynamics," remain unsettled and contentious topics, as Gislason and Craig noted (to their surprise and disappointment) in 2005 [19]. Chapter 6 presents an assessment of this naturalistic project of classical formalism. Which is then followed with an introduction of the modern formalism. The modern formalism with its *second entropy principle* provides a formalism that can incorporate operational consideration of the system in interaction with its surroundings and opens the door for restoring once again, in Chaps. 7 and 8, thermodynamics as the synthesis of Carnot's theory and MEH.

A note on "restoring…the synthesis": the MTH resulted from the Kelvin–Clausius "synthesis" of Carnot's theory and the MEH. But, a case can be made that the impression of the Kelvin–Clausius synthesis' success is formed from its success in producing a coherent system of equilibrium thermodynamics, not in resulting in a coherent system of engineering stream of thermodynamics, the failure of which is reflected in the fact that engineering thermodynamics cannot even formulate meaningful definitions of heat and energy as noted by Gislason and Craig. Chaps. 7 and 8 aim for restoring engineering thermodynamics as the true synthesis of Carnot's theory and the MEH by according equal status to the first law and the second law in its complete form.

[3]The engineering connection is particularly strong in North British: "North British group of scientists and engineers, including James Joule, James Clerk Maxwell, William and James Thomson, Fleeming Jenkin, and P. G. Tait, developed energy physics to solve practical problems encountered by Scottish shipbuilders and marine engineers" noted Smith [14].

Problems

5.1 The heat capacity of a body is given by

$$C_p = a + bT$$

where a = 20.35 JK^{-1} and b = 0.20 JK^{-2}. Calculate the change in entropy when the temperature of this body is raised from 298.15 K to 304.0 K under constant pressure.

$$1.5654\,\mathrm{JK}^{-1}$$

5.2 What is the entropy of 1.0 L of gaseous N_2 at T = 350 K and p = 2.0 bar given that s_0 = 191.61 JK^{-1} $gmol^{-1}$ at T_0 = 298.15 K and p_0 = 1 bar?

$$0.0687\,\mathrm{gmol};\,13.094\,\mathrm{J/K}$$

5.3 The heat of fusion of ice at 1 atm and 0 °C is l_{fu} = 6013.5 kJ/kmol and the heat of vaporization of water at 1 atm and 100 °C is $l_{va} = h_{fg}$ = 40683.6 kJ/kmol. Assuming an average molar heat capacity of 75.56 kJ/°C-kmol at 1 atm between 0 and 100 °C for water, calculate the difference of the entropy of one k mol of ice at 1 atm and 0 °C, and the entropy of one kmol of steam at 1 atm and 100 °C.

$$154.61\,\mathrm{kJ/kmol} - \mathrm{K}(22.02 + 23.57 + 109.03)$$

5.4 Two blocks A and B are initially at 100 and 500 °C, respectively. They are brought together and isolated from the surroundings. They are allowed to reach a final state of internal thermal equilibrium. Determine the final equilibrium temperature of the blocks and the entropy change of each block and of the whole isolated system. Block A is aluminum $[c_p = 0.900\,\mathrm{kJ/kg \cdot K}]$ with m_A = 0.5 kg and block B is copper $[c_p = 0.386\,\mathrm{kJ/kg \cdot K}]$ with m_B = 1.0 kg.

$$557.839\,\mathrm{K};\,0.0549\,\mathrm{kJ/K}$$

5.5 An ideal gas of 0.1 kmol at the initial state of 298.15 K and 303.9 kPa occupies one chamber of an insulated composite system. The other chamber (of a double volume of the first) contains a vacuum. Determine the volume of the first chamber V_1. After the removal of the partition between the two chambers, the ideal gas undergoes an adiabatic free expansion (see Problem 4.2) from its initial volume V_1 to its final volume 3 V_1. Explain why the gas remains at the same temperature at its final state (hint: note the values of heat exchange Q and work W in this case; give a **precise** reason for the W value you assume). And determine the entropy change of the free expansion.

$$0.816\,\mathrm{m}^3, 0.91\,\mathrm{kJ/K}$$

References

1. Planck M (1969) Treatise on Thermodynamics, 3rd edition. Dover, New York
2. Jaynes ET (1984) The evolution of Carnot's principle. EMBO Workshop on Maximum-Entropy Methods (Orsay, France, April 24–28, 1984. Reprinted in Ercksen & Smith [1988], ***1***:267-282)
3. Smith C, Wise MN (1989) Energy and Empire: A Biographical Study of Lord Kelvin. Cambridge Univ Press
4. Clausius R (1865;1867) On Different Forms of the Fundamental Equations of the Mechanical Theory of Heat and Their Convenience for Application. Presented to *Zuricher naturforschende Gesellschaft* on April 24, 1865. Published in *Abhandlungen uber die mechanische Warmetheories* **2**:1–44 (1867)
5. Fermi E (1956) Thermodynamics. Dover (p. 48)
6. Wang LS (2007) The nature of spontaneity-driven processes. International J of Ecodynamics 2: 231–244
7. Gibbs JW (1961) The Scientific Papers of J. W. Gibbs, Vol. 1: Thermodynamics. Dover (p. 354)
8. Uffink J (2001) Bluff your way in the Second Law of Thermodynamics. Stud Hist Phil Sci Part B: 2001 Studies in Hist and Philo of Modern Physics 32(No.2):305–395
9. Planck M (1887) Uber das Princip der Vermehrung der Entropie, in 3 parts. Ann. D. Phys. 30:562–582; 31:189-203; 32:462-503
10. Newburgh R, Leff H S (2011) The Mayer-Joule Principle: The foundation of the first law of thermodynamics. The Physics Teacher 49(November):484–487
11. Brookes D, Horton G, Van Heuvelen A, Etkina E (2005) Concerning scientific discourse about heat. AIP Conference Proceedings 790:149–152.
12. Romer RH (2001) Heat is not a noun. Am J Phys 69(2):107–109
13. Daub EE (1970) Entropy and dissipation. Historical Studies in the Physical Sciences 2:321–354
14. Smith C (1991) Lord Kelvin: Scientist of energy. Supercond Sci Technol 4:502–506
15. Callen HB (1st edition, 1960; 2nd edition, 1985) Thermodynamics and an Introduction to Thermostatistics. Wiley, New York
16. Born M (1949) Natural Philosophy of Cause and Chance. Oxford Univ Press (p. 38)
17. Carathéodory C (1909) Untersuchungen uber die Grundlagen der Thermodynamik. Math. Ann. 67: 355–386. A translated text "Investigation into the foundations of thermodynamics" is collected in J. Kestin, *The Second Law of Thermodynamics* (Dowden, Hutchinson and Ross, 1976). Pages referred to this reference *in-text* are that of the translated text.
18. Landsberg PT (1961) Thermodynamics. Interscience, New York (p. 94)
19. Gislason EA, Craig NC (2005) Cementing the foundations of thermodynamics: comparison of system-based and surroundings-based definitions of work and heat. J. Chem. Thermodynamics 37:954–966

Reversible Processes Versus Quasi-static Processes, and the Condition of Internal Reversibility

6

Abstract

A presupposition in the naturalistic sciences is that laws should be centered on a natural object in itself without having to involve a man-made machine. That is the legacy of Newtonian science of mechanical objects as mass bodies. A strong movement in equilibrium thermodynamics initiated by Caratheodory has been based on this presupposition. This disquisition rejects naturalistic presupposition and the Caratheodory formalism by pointing out the unique nature of thermodynamic objects as systems with essential design characteristics. This chapter applies Prigogine's modern formalism for the effective treatment of system–surroundings interactions for a better understanding of thermodynamic processes.

Keywords

Caratheodory formulation of the second law · Quasi-static processes · Quasi-static heat and work · Local thermodynamic equilibrium · Brussels school formalism · Quasi-staticity versus internal reversibility versus reversibility · Nonreversible processes · Reversible-like processes · Joule free expansion

L.-S. Wang, *A Treatise of Heat and Energy*, Mechanical Engineering Series,
https://doi.org/10.1007/978-3-030-05746-6_6

6.1 The Project of Classical Formalism

The treatment in Chaps. 4 and 5 will be referred to as the Carnot and CKP (Clausius, Kelvin, and Planck [1])[1] tradition. Born [2] spoke for all physicists who subscribe to scientism[2] and reductionism when he wrote that the use of "the conception of idealized thermal machines" by Carnot and the CKP tradition for the formulation of the second law of thermodynamics and the establishment of the concepts of entropy and absolute temperature "deviated too much from the ordinary methods of physics"—the methods of mechanistic physics which came down from Galileo and Newton. (The prevailing view is that the doctrine of *mechanism* survived the challenge from the science of heat unscathed and, in fact, rose triumphantly with the rise of atomism but in the twentieth century faced and remains facing the new, stronger challenge from the quantum revolution.) At Born's suggestion, Carathéodory [3] developed an alternative formulation of the second law, the classical formalism, which made no use of *reversible* "idealized thermal machines" in the Carnot and CKP tradition. That is what Carathéodory, Born, and generations of physicists since thought: the project of classical formalism succeeded in removing reversible machines from the formulation of thermodynamic theory by transforming the science of heat into a typical theory of mechanistic physics, which is based on *causal laws*, e.g., the typical laws of physics such as Maxwell's equations, the heat transfer equation with Fourier's law of heat conduction, the Navier–Stokes equations, etc.

There are two problems with this view. First, the first law and the second law of thermodynamics are *not* causal laws [4]. That is, they are not associated with *governing equations* such as the Maxwell equations and Navier–Stokes equations, and we shall study the implications of this point in Chaps. 8 and 10. The second problem is one that is argued here: the formulation of Carathéodory's principle (the project of classical formalism) did not succeed in removing reversibility. Reversibility and the idea of machine or design, even if they do not appear explicitly, are tacitly included in the classical formalism (see Sect. 6.3). Without them, we are not certain why we can apply the doctrine of latent and sensible heats and why we can use mathematical expressions such as the quasi-static work by Eq. (81), the validity of which is taken, mistakenly, for granted in the classical formalism.

Let us first grant the rationale of the classical formalism that it is possible to study a system in and of itself without having to refer to a reversible machine or/and a reservoir. The formalism did give rise to Gibbsian thermodynamics. That was indeed an extraordinarily successful first step. The conventional view is that

[1]Uffink famously in Ref. [1] made this characterization of CKP, "the unargued statements of Kelvin, the bold claims of Clausius and the strained attempts of Planck." While CKP may not be perfect in their logic or elegance (there is an opinion that P's logic is better than that of C and K), their scientific judgment is superior in my mind to that of Caratheodory and Uffink. Certainly, their scientific legacy supports this opinion.

[2]I use the term here somewhat like reductionism to mean the position that any concept that is not in the governing equations of physics such as purpose, action, or the operation of thermal machines has no scientific meaning.

Gibbsian thermodynamics, by limiting consideration to quasi-equilibrium processes as "idealized reversible processes", can serve as the starting point of the theoretical development toward understanding irreversibility. This limitation does succeed in achieving a high degree of elegance in Gibbsian thermodynamics as well as making it highly useful in thermo-physics and physical chemistry (Chap. 9). However, it will be shown in this chapter that quasi-equilibrium processes, or quasi-static processes, cannot represent idealized reversible processes, and therefore Gibbsian thermodynamics cannot be the platform toward understanding irreversibility.

Constatin Carathéodory (1873–1950)

6.2 Quasi-static Processes and the Classical (Caratheodory) Formalism

Carathéodory [3] set out to develop an alternative formulation to the definition of entropy, (62A), in Chap. 5, without using the *reversible* thermal machines. He began by noting the application of the first law, $dU = \delta Q - \delta W$, to an adiabatic composite system (which consists of subsystems, e.g., two subsystems "1" and "2") separated diathermanously undergoing quasi-static adiabatic processes

$$\delta Q = dU + \delta W = dU_1 + dU_2 + \delta W$$

Because of being diathermanously related, the two subsystems are at the same temperature.

By “quasi-static” he meant, “A process that occurs so slowly that the difference between the work performed externally and the preceding limit, *pdV*, is smaller than the uncertainty of our measurements” [3:238]. A quasi-static process is an infinitely slow process.

On the same page, however, he added, “a quasistatic adiabatic process can be regarded as a *series of* [infinitely dense succession of] *equilibrium states*” [3:238]. Callen [5:96] also defined quasi-static processes to be “in terms of a dense succession of equilibrium states.” I make the interpretation here that both Carathéodory and Callen equated infinitely dense processes with infinitely slow processes as synonymous terms. In the following, a restated version is used for discussion in this chapter as the definition of QUASI-STATIC PROCESS (QUASI-EQUILIBRIUM PROCESS) set:

> The set consists of a subset-series of infinitely dense succession of equilibrium states *and* the corresponding subsets of all spontaneous transient states between each pairs of equilibrium states in the series.

The inclusion in the above definition of “transient states” will become clear, the existence of which in fact was pointed out by Callen in his penetrating but inconsistent treatment (see below).

Furthermore, Carathéodory added “quasistatic adiabatic processes of a simple system are ‘reversible’” [3:239]. I shall come back to both issues of whether infinitely dense is the same as infinitely slow (in Sect. 6.3) and whether quasi-static processes are reversible (in Sect. 6.5). For the purpose of summarizing the rest of Carathéodory’s reasoning, let us accept for now the crucial assertion of Carathéodory, Born, and in fact almost all physicists with one notable exception, [6] that *quasi-static work* is

$$(\delta W)_{Quasistatic} = pdV \tag{81}$$

With Eq. (81), the above first law equation for the composite system becomes

$$\delta Q = \left(\frac{\partial U_1}{\partial V_1} + p_1\right)dV_1 + \left(\frac{\partial U_2}{\partial V_2} + p_2\right)dV_2 + \left(\frac{\partial U_1}{\partial \vartheta} + \frac{\partial U_2}{\partial \vartheta}\right)d\vartheta \tag{82}$$

where ϑ is the *empirical temperature* [2:36] of the system. In mathematics, Eq. (82) is known as a Pfaffian system

$$\delta Q = Xdx + Ydy + Zdz \tag{82A}$$

of three variables: $x = V_1$, $y = V_2$, $z = \vartheta$.

Pfaffians have the following mathematical property: If a Pfaffian in an *xyz*-space has an integrating denominator λ

$$\frac{Xdx + Ydy + Zdz}{\lambda} = d\phi,$$

then, the *xyz*-space is made of nonintersecting constant ϕ surfaces. Conversely, if in the neighborhood of a point $P_0(x, y, z)$ in the *xyz*-space there are points P which cannot be connected to P_0 by a line satisfying $\delta Q = 0$, then, there exists λ. Carathéodory made a connection of this mathematical property with the physical concept, which was adopted as the second law Axiom of Carathéodory

> In every arbitrary close neighborhood of a given initial state there exist states that cannot be approached arbitrarily closely by adiabatic processes.

The Axiom entails the existence of the integrating denominator of the Pfaffian and, therefore, the existence of ϕ, i.e., the entropy

$$(\delta Q)_{Quasistatic} = \lambda d\phi = TdS \tag{83}$$

Equation (83) is shown as Eq. (5.28) in [2:42].

The concept of quasi-static processes is thus thought to be the central concept in the classical formalism of thermodynamics: "The concept of *reversible processes*, which plays an essential role in many expositions of thermodynamics [in the CKP tradition], is not required in the present [Carathéodory's] approach" noted Landsberg [7]. Callen [5] presented a well-received postulatory formulation, which is closely related to Carathéodory's and the Gibbsian approach, and noted

> The postulatory formulation of thermodynamics features states, rather than processes, as fundamental constructs. Statements about Carnot cycles and about the impossibility of perpetual motion of various kinds do not appear in the postulates, but state functions, energy, and entropy become the fundamental concepts. An enormous simplification in the mathematics is obtained, for processes [i.e., quasistatic processes] then enter simply as differentials of the state functions. [5(1960):viii].

Callen's words express the universal view of physicists who subscribe to the scientism that the book of nature is written in the language of mathematics; it is little wonder that the acceptance of quasi-static work, Eq. (81), is complete (see below for the somewhat different acceptance of quasi-static heat [8]).

So much so that Callen devoted an interesting discussion on the contradiction of the "continuous free expansion" process with these comments

> Whether this atypical 'continuous free expansion' process should be considered as quasi-static is a delicate point. On the positive side is the observation that the terminal states of the infinitesimal expansions can be spaced as closely as one wishes along the locus. On the negative side is the realization that the system necessarily passes through

> nonrepresentable nonequilibrium states during each expansion; the irreversibility of the microexpansions is essential and irreducible. The fact that $dS > 0$ whereas $dQ = 0$ is inconsistent with presumptive applicability of the relation $dQ = TdS$ to all quasi-static processes. We *define* (by somewhat circular logic!) the continuous free expansion process as being "essentially irreversible" and *non-quasi-static* [5:99].

That micro-free-expansion is irreversible is a fact; that it can be idealized by quasi-static process is a matter of definition (by Callen's own definition). We have a syllogism of a major premise (the applicability of Eq. (81)) and a subsidiary premise (the definition) linked to a conclusion, and the conclusion is not supported by the fact. Either the major premise or the subsidiary premise must be rejected. Callen, in rejecting the subsidiary premise definition "micro-free-expansion = quasi-static", did not, in fact, commit logical error but made the wrong choice. He reached the wrong conclusion by not questioning the validity of Eqs. (81) and (83) and, instead, made an unusual move by abandoning the definition.

The universal premise *quasi-static* $\equiv$ *reversible* in the context of Eq. (83) has been questioned by a number of physicists (the latest example is in a paper by Samiullah, who wrote, "although not all quasi-static processes are reversible, the converse statement that all reversible processes are quasi-static is true" [8:24]).

In contrast, the equally mistaken presumptive applicability of Eq. (81) for processes of infinitely dense equilibrium states has never been critiqued! It must be noted that since "if Eq. (81), then Eq. (83)," the rejection of Eq. (83) must be followed by the rejection of Eq. (81) according to *modus tollens*. For engineering problems of heat and work, we need a more precise condition for the applicability of Eqs. (81) and (83) as it'll be given in Sect. 6.5.

The classical formalism and Gibbsian thermodynamics represented the mechanistic capture of the mechanical theory of heat, which views thermodynamics as the naturalistic science of energy rather than the engineering science of heat and energy. Though the move had been initiated by Kelvin, Kelvin himself retained viewing energy's role to be its engineering capacity for doing work grounded in the Carnot–Kelvin tradition of heat, power, and reversible processes. That tradition underwent transformation post-Kelvin to become the Gibbs, Carathéodory, and Callen problem of energy, entropy, and quasi-static processes of today. In that modern view, thermodynamic events are captured completely by "differentials of the state functions" [5(1960):viii]. This *mechanistic* interpretation of the mechanical theory of heat was probably the inevitable result of the growing popularity of atomism in the late 19th century. The classical formalism finds its great success in application to physical and chemical problems as it'll be treated in Chap. 9.

6.3 Infinitely Dense State Function Does Not Always Equal to Infinitely Slow Process

In the application to engineering problems of heat and work, however, the fundamental mistake of the classical formalism is the presumption that a series of *infinitely dense* equilibrium states (see Fig. 6.1 and Problem 6.2) is synonymous with an *infinitely slow* process (see Fig. 6.2 and Problem 6.3). Making a process to be a series of infinitely dense equilibrium states does not ensure that the process also proceeds infinitely slow. A set of very closely spaced constraints for the piston placement of a gas-filled piston cylinder will ensure the system equilibrium states (resulting from relaxation of constraints [see Ref. [5] for definition] as shown in

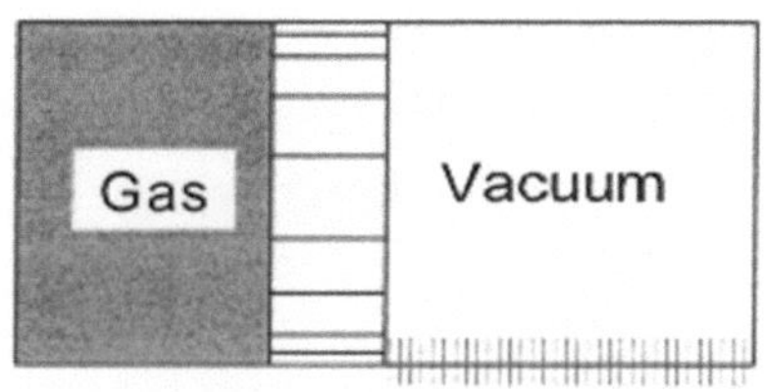

Fig. 6.1 Quasi-static process resulting from relaxation of constraints, which is a set of infinitely dense equilibrium states (this figure represents the same process in Fig. 1.6)

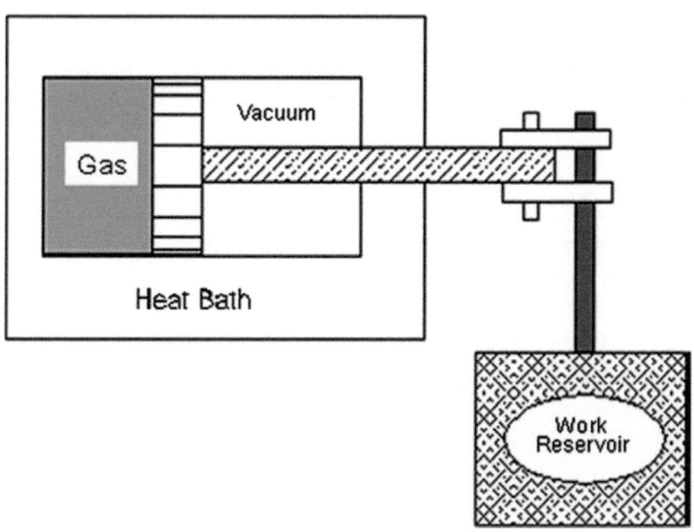

Fig. 6.2 Reversible process resulting from the infinitesimally small net force, which is a set of equilibrium states moving infinitely slow from one equilibrium state to the next equilibrium state (this figure represents the same process in Fig. 1.5)

Fig. 6.1) to be closely positioned in the thermodynamic space. But, unless an additional condition that the *net force* acting on the piston is infinitesimal also prevails through a **design arrangement of balancing all forces** (Fig. 6.2 and Problem 6.3), the changes at the relaxation of each constraint will be sudden, not slow. *This force-balance design arrangement is present either explicitly or tacitly when one applies* Eq. (81). In absence of such arrangement, during each transition between neighboring equilibrium states "the system necessarily passes through non-equilibrium states" [5:99] as noted Callen, no matter how dense the set of the equilibrium states is; for such cases Eq. (81) is not applicable.

The idea of *design* (force-balance design) is always present, tacitly or explicitly, in the classical formalism of thermodynamics of Gibbs, Carathéodory, and Callen. What is absent is *useful* work (definition in Sect. 6.5), not the idea of design or machines.

Internal reversibility (see Sects. 6.4 and 6.5), not quasi-static processes, should be the central concept in classical formalism as we shall see in the following.

Max Planck

6.4 Local Thermodynamic Equilibrium and the Modern (Brussels School) Formalism

The synthesis of Carnot–Kelvin–Clausius was presented by Planck in an authoritative exposition [9]. Planck puts the second law, the concept of entropy, and irreversibility in nonequilibrium systems at the very center of thermodynamics. As a working knowledge, however, thermodynamics was restricted to handle problems of equilibrium systems. How to incorporate the handling of irreversibility satisfactorily into thermodynamic theory remained unresolved in both the CKP approach and the classical formalism approach. The step taken by "modern thermodynamics"—the classical formalism of thermodynamics of Gibbs [10], Carathéodory [3], and Callen [5]—actually represented a step backward by removing the importance of universal irreversibility's role in thermodynamic theory [1] and its corresponding implication in engineering applications.

Ilya Prigogine (1917–2003)

6.4.1 The Entropy Principle of the Modern Formalism

Significantly, another line of research advance took place in the twentieth century: De Donder, Onsager, and Prigogine developed the *modern formalism of thermodynamics*, which is a theory of irreversible processes. Its central message is that the universe is fundamentally irreversible [11, 12]. The modern formalism approach introduces the assumption of *local* thermodynamic equilibrium (LTE, see definition in Sect. 1.3) [11, 13, 14], which offers a new "mind's eye" opening door for resolving some of the ambiguous questions in the CKP tradition and the classical formalism approach. I shall first here apply the new mind's eye to remove the ambiguity in Clausius' inequality (Sect. 5.4), and the ambiguity in the definition of entropy: is it defined in terms of Eq. (83) in this chapter or the original Eq. (62A) under the more strict condition of reversibility?

With LTE, the entropy principle is not restricted to systems being in equilibrium as a whole. When a whole system cannot be specified by single-valued *thermostatic* variables as long as LTE prevails at every point within the system, local thermodynamic variables can be defined within every *elemental volume* of the system. All thermodynamic relations remain valid for relating local variables, both intensive variables and molar-based or mass-based specific extensive variables. An extensive variable of the system is specified by the integrals of local, e.g., molar, extensive variables. Consider the molar entropy,

$$s = s(T[x, y, z, t], p[x, y, z, t]) \text{ or } s = s(T[x, y, z, t], v[x, y, z, t])$$

The system entropy is

$$S_{System}(t) = \int_{V_{System}} \rho_N(x, y, z, t) s(x, y, z, t) dV(x, y, z, t),$$

where ρ_N (x, y, z, t) is the molar density (*mole number per unit volume*).

LTE assumption is a necessary condition for the *continuum hypothesis*, e.g., as it is applied to mass and energy balances. We apply in the following the continuum hypothesis to the consideration of entropy balance at every local point

$$\rho_N \frac{Ds}{Dt} = -div J_S + \sigma \tag{84}$$

where the local molar entropy change is balanced with the convergence of the entropy flux, J_S, and the local entropy production, σ. LTE makes it possible to relate J_Q, the heat flux, to J_S, the entropy flux [15:345]

$$\frac{J_Q(x, y, z, t)}{T(x, y, z, t)} = J_S(x, y, z, t) \tag{85}$$

(In the case of an incoming radiative heat flux, J_Q, LTE requires that the temperature of J_Q must be infinitesimally close to T of the system.) With the substitution of Eq. (85), the volume integration of Eq. (84) in which the divergence theorem of Gauss is applied to the first term on the RHS yields the rate of *system* entropy *change*,

$$\frac{dS(t)}{dt} = \int_V \left(-div \frac{J_Q}{T}\right) dV + \int_V \sigma dV = \int_A \frac{-J_Q}{T} dA + \int_V \sigma dV \tag{84A}$$

Introduce the following notations of system entropy change due to entropy exchange, $\delta_E S$, and system entropy change due to growth (or production), $\delta_G S$:

$$\delta_E S = \left[\int_A \frac{-J_Q(x,y,z,t)}{T(x,y,z,t)} dA \right] dt \tag{86}$$

$$\delta_G S = \left[\int_V \sigma dV \right] dt \tag{87}$$

Equation (84A) becomes [16; 17:410; 15:88],[3]

$$dS = \delta_E S + \delta_G S \tag{84B}$$

Or in the cases of finite changes $1 \to 2$

$$\Delta S = S_2 - S_1 = \Delta_E S + \Delta_G S \tag{84C}$$

where $\Delta_E S = \int_{t_1}^{t_2} \delta_E S$ is the system (net) entropy exchange, and $\Delta_G S = \int_{t_1}^{t_2} \delta_G S$ the system entropy growth (production).

The system entropy change, dS or ΔS, is the result of two parts: entropy exchange associated with the heat flow across the system boundary and entropy growth or generation due to irreversible processes of nonuniform temperature and other affinity gradients in the *interior* of the system.

The first entropy principle of the second law (the principle of the increase of entropy), Eq. (74) in Chap. 5, can now be restated in its most general and elegant form, the second entropy principle of the second law [17:409]: Entropy *growth (generation) is always nonnegative or entropy cannot be destroyed*, i.e.,

$$\sigma \geq 0 \tag{88A}$$

$$\delta_G S \geq 0 \tag{88B}$$

Equation (88B) holds for all systems at all times regardless of whether the system is open or closed and regardless of whether the system is natural or artificial [15:90]. Equation (88A) holds in the macroscopic universe at all times and at all places.

---

[3] In the article, Coveney [17] wrote

> From an epistemological viewpoint, the contributions of Prigogine's Brussels School are unquestionably of signal importance. The myth of a completely timeless, deterministic Universe is henceforth replaced by a world in which static affairs are enlarged to embrace the probabilistic kinetics of process; in which reversibility and irreversibility are accorded equal objectivity; and in which the notions of 'being' and 'becoming' are unified within a single conceptual framework. (p. 414)

6.4.2 Temperature *of Clausius' Inequality*

Substituting inequality Eq. (88B) into (84B) yields

$$dS \geq \delta_E S = \left[\int_A \frac{-J_Q}{T} dA \right] dt \tag{89A}$$

If one defines an effective mean temperature T_{mean} in accordance with

$$T_{mean} \equiv \frac{\int_A J_Q dA}{\int_A \frac{J_Q}{T} dA} \tag{90}$$

inequality Eq. (89A) becomes (note $\left[\int_A -J_Q dA \right] dt = \delta Q$)

$$dS \geq \frac{\left[\int_A -J_Q dA \right] dt}{T_{mean}} = \frac{\delta Q}{T_{mean}} \tag{89B}$$

$$S(B) - S(A) \geq \left(\int_A^B \frac{\delta Q}{T_{mean}} \right)_{Quasistatic} \tag{89C}$$

Inequality Eq. (89C) in terms of an effective mean temperature defined by Eq. (90) supersedes the ill-defined Eq. (73).

6.4.3 Internal Reversibility as the Condition for Defining Entropy

In the case of vanishing entropy production, $\delta_G S = 0$ in (84B), the system entropy change dS is balanced solely with the system entropy exchange, $\delta_E S$. That is, the condition for equality

$$dS = \left[\int_A \frac{-J_Q}{T} dA \right] dt \tag{91}$$

depends on the *interior condition* of the system alone—*independent* of any other specifics concerning how the system interacts with its surroundings. We call this interior condition of vanishing entropy production within the system by the name *internal reversibility* (see Sect. 6.5 for a precise definition).

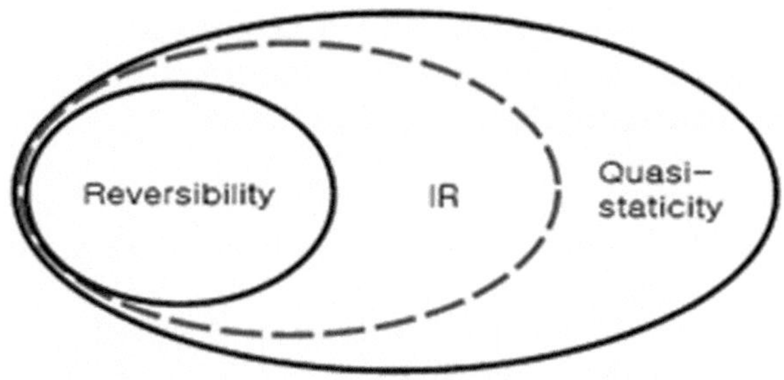

Fig. 6.3 Venn diagram of the conditions of reversibility, internal reversibility (IR), and quasi-staticity

When the condition of internal reversibility prevails within the system, the temperature gradient within the system must become infinitesimal and $T(x, y, z, t)$ becomes the single-valued *thermostatic* temperature of the system, i.e.,

$$\left[\int_A \frac{-J_Q}{T} dA\right] dt = \frac{\left[\int_A -J_Q dA\right] dt}{T} = \frac{\delta Q}{T}$$

Equation (91) takes the form

$$dS = \left(\frac{\delta Q}{T}\right)_{InternalReversible} = \frac{dU + pdV}{T} \tag{92}$$

I shall note in Sect. 6.5 that both reversibility and internal reversibility (IR) are subsets of the set of quasi-staticity (see also Fig. 6.3). Correspondingly, the applicability of (92) [18] is based on.

a. reversibility as the sufficient condition,
b. internal reversibility as the *necessary and sufficient condition*, and,
c. quasi-staticity (see Sect. 6.5 for definition) as the necessary condition.

Equation (92) is the precise definition of the entropy, superseding the classical formalism definition, Eq. (83) in this chapter and the Clausius definition, Eq. (62A) in Chap. 5, as it will be summarized in Sect. 6.5.

6.5 Useful Work and Action, Which Are What Distinguishes Reversible-Like Processes from Spontaneous Natural Processes

The condition for the applicability of Eq. (81) is also internal reversibility. When a system changes infinitely slowly as a result of an infinitesimal net force, the change satisfies internal reversibility condition.

However, internal reversibility is not reversibility. Reversibility describes nature not as it is but as the idealized version it could be. Reversibility could be realized only if an ideal machine is used to bring about *reversible* work (corresponding to a perfectly controlled system change), which is stored in a *work reservoir* so that this exact amount of reversible work will be able to return the system and its surroundings to their original "entropy states" [8]. In reversible-like processes (see below), the external force that balances the internal system force must be used to produce USEFUL WORK (which can be defined as work *that can be gainfully used for a purpose or stored in work reservoir for later use*), which can partially return the system to its original state. Work, as machines, is necessarily an anthropogenic or organismic concept. Work is a defining characteristic of organismic existence: what characterizes life systems are "work cycles in a web of propagating organization of processes" [19] argued Kauffman.

Having nothing to do with work reservoirs, a quasi-static process in the strict sense can never be a reversible process. Quasi-static processes and reversible processes represent two fundamentally different classes of spontaneity-driven processes [18]: spontaneous natural processes and reversible-like processes (see below). A quasi-static process is an idealization of a spontaneous natural process. Recall its definition in Sect. 6.2, "*the set consists of a subset-series of infinitely dense succession of equilibrium states and the corresponding subsets of all the transient states between each pairs of equilibrium states in the series*." Consider such an idealized representation of a system undergoing a natural change. The system, in dividing its change into small steps and representing the system at the end of each step in terms of its equilibrium states, passes through a series of equilibrium states (the first quasi-equilibrium subset), *and in addition*, between each pair of equilibrium states of the first subset, passes through transient states, which may be "non-representable, non-equilibrium states" [5:97 and 99] (the second transient subsets).

We may thus define QUASI-STATICITY as the condition that a process consists of *the set of a series of infinitely dense succession of equilibrium states*—i.e., the erstwhile first subset. Note the difference of this definition from the earlier definition of quasi-static process repeated above.

Like quasi-staticity, INTERNAL REVERSIBILITY (IR) is not a process operation, but a condition that a process satisfies: defined as the condition that *transient states in every second-subset of transient states become "at all times infinitesimally near"*

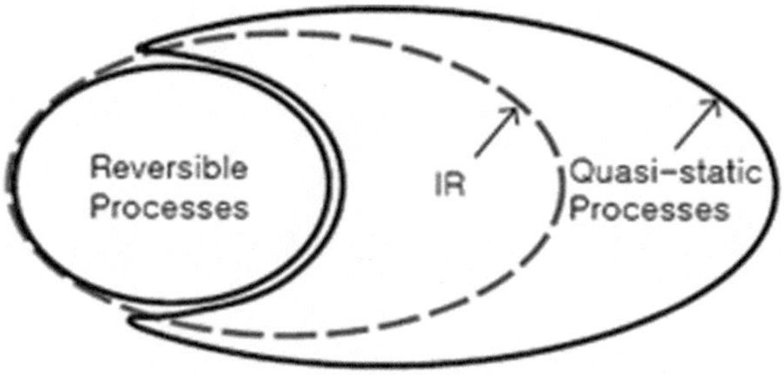

Fig. 6.4 Venn diagram of reversible processes and quasi-static processes, with the IR condition superimposed

[20:43] *their corresponding pairs of equilibrium states of the first subset*. A set of infinitesimally close constraints only ensures that the first set is infinitely dense, not this condition of transient states being infinitesimally near equilibrium. The *merging* of the second sets into the first set, i.e., the whole process in totality being infinitely slow, requires that the process be brought about by infinitesimal affinity, not merely its constraints being infinitesimally close.

The second sets of a quasi-static process sometimes, e.g., thermal processes, merge into the first set, but in other instances, e.g., mechanical *free expansion* processes [5:99] and chemical processes, they do not merge into the first set and the transient states remain non-representable and nonequilibrium.

It is useful to review the three conditions of (a) reversibility, (b) IR, and (c) quasi-staticity. Reversibility is synonymous with reversible processes, which are ensured by machines (and work reservoir) that keep all affinity (driving forces) differences balanced and all affinity gradients infinitesimally small. Reversibility means having this condition both within the system (internally reversible) and outside the system (externally reversible).

The condition of reversibility, thus, is a subset of IR (which is defined above)—see Fig. 6.3. Quasi-staticity is the condition that a system passes through a series of equilibrium states, i.e., the first quasi-equilibrium set of a quasi-static process. IR is a subset of quasi-staticity (see Fig. 6.3). Quasi-staticity (in Fig. 6.3) is not synonymous with a quasi-static process (in Fig. 6.4). Quasi-static processes, as it was noted, can never be a reversible process; we are reminded in Fig. 6.4 that reversible processes and quasi-static processes are fundamentally different classes of processes. The condition of IR is superimposed on the two distinctive processes in Fig. 6.4 to show that while a reversible process always meets the IR condition, some quasi-static processes do [(81) and (83) are applicable] and others do not [(81) and (83) are invalid].

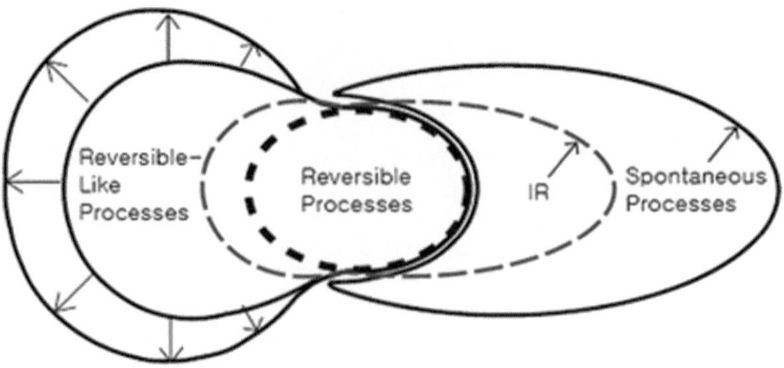

Fig. 6.5 Venn diagram of reversible-like processes and non-reversible (spontaneous) processes. Reversible-like processes are shown to be an expanding domain, signifying that new reversible-like processes are continuously created and constructed. IR is shown to overlap a part of reversible-like processes to represent idealized practical reversible-like processes, such as an ideal Rankine cycle which is discussed in engineering thermodynamics books

6.5.1 Nonreversible Processes and Reversible-like Processes

It is also useful to introduce the concepts of *nonreversible processes* and *reversible-like processes* [18]. All real processes are irreversible. It is, therefore, pointless to characterize a real process irreversible. Spontaneous natural processes ("nature as it is": a room being warmed up in the summer) are irreversible and nonreversible, i.e., irreversible processes that are not "managed". In contrast, reversible-like processes ("nature as it can be": a room is *made* colder in the summer) are irreversible but not nonreversible, i.e., irreversible processes that are "managed" or are made to happen. The distinction between nonreversible processes and reversible-like processes serves a useful purpose, e.g., in Chap. 10 an important distinction is made between heat transfer processes/heat production processes, which are nonreversible spontaneous natural processes, and heat extraction processes, which are reversible-like.

Again, spontaneous natural processes and reversible-like processes are the two fundamental classes of spontaneity-driven macroscopic processes (Fig. 6.5). Quasi-static processes are the idealization of spontaneous natural processes (as shown by comparing Figs. 6.4 and 6.5), whereas reversible processes are the perfection of reversible-like processes (Fig. 6.5).

For all processes, IR is the necessary and sufficient condition for the applicability of

$$(\delta W)_{IR} = pdV \tag{93}$$

$$(\delta Q)_{IR} = TdS \tag{92}$$

6.6 Internal Reversibility and the $C_p - C_V$ Question in Sect. 2.3

In the caloric theory, the doctrine of latent and sensible heats of Sect. 2.3 often gives the impression that heat, Q, is treated as a function, e.g., $Q(V, T)$

$$\delta Q = C_T^{(V)}(V,T)dV + C_V^{(T)}(V,T)dT \tag{12}$$

or, as a function of p and T,

$$\delta Q = C_T^{(p)}(p,T)dp + C_p^{(T)}(p,T)dT \tag{13}$$

That impression is incorrect. The meaning of δQ in Eqs. (12) and (13), in fact, should be interpreted strictly as TdS in accordance with Eq. (92) and correspondingly, Eqs. (12) and (13) are valid only if they are applied to processes that meet the internal reversibility condition. That is, Eqs. (12) and (12) are understood exclusively as

$$TdS = T\left(\frac{\partial S}{\partial V}\right)_T dV + T\left(\frac{\partial S}{\partial T}\right)_V dT = T\left(\frac{\partial S}{\partial V}\right)_T dV + C_V dT \tag{94}$$

$$TdS = T\left(\frac{\partial S}{\partial p}\right)_T dp + T\left(\frac{\partial S}{\partial T}\right)_p dT = T\left(\frac{\partial S}{\partial p}\right)_T dp + C_p dT \tag{95}$$

The latent heat with respect to volume and the latent heat with respect to pressure are correspondingly

$$C_T^{(V)}(V,T) = T\left(\frac{\partial S}{\partial V}\right)_T \tag{94A}$$

$$C_T^{(p)}(p,T) = T\left(\frac{\partial S}{\partial p}\right)_T \tag{95A}$$

Similarly, Eq. (15) used by Poisson in his handling of adiabatic heating in Sect. 2.4

$$\delta Q = \left(\frac{\delta Q}{\partial p}\right)_V dp + \left(\frac{\delta Q}{\partial V}\right)_p dV$$

is subject to the same interpretation. That is, it is understood exclusively as

$$TdS = T\left(\frac{\partial S}{\partial p}\right)_V dp + T\left(\frac{\partial S}{\partial V}\right)_p dV = T\left(\frac{\partial S}{\partial T}\right)_V \left(\frac{\partial T}{\partial p}\right)_V dp + T\left(\frac{\partial S}{\partial T}\right)_p \left(\frac{\partial T}{\partial V}\right)_p dV$$

or

$$TdS = C_V\left(\frac{\partial T}{\partial p}\right)_V dp + C_p\left(\frac{\partial T}{\partial V}\right)_p dV$$

The introduction of the concept of internal reversibility condition resolves one mystery why the caloric theory of heat as used by Laplace and Poisson was so successful. It may be suited here to repeat the comment made in Sect. 2.4,

> It is noted that the δQ Eqs. (12), (13), and (15), are valid under the condition that the material media are internally reversible—a notion that [is discussed in this chapter]. This does not infer that $Q(T, V)$ itself is a state function: while the condition that $Q(T, V)$ is a state function infers that $\left(\frac{\delta Q}{\partial T}\right)_V$ and $\left(\frac{\delta Q}{\partial T}\right)_p$ are state functions, the opposite inference—that $\left(\frac{\delta Q}{\partial T}\right)_V$ and $\left(\frac{\delta Q}{\partial T}\right)_p$ are state functions infers that $Q(T, V)$ is a state function—is not true. Still, we have the intriguing possibility that the fact that $\left(\frac{\delta Q}{\partial T}\right)_V$ and $\left(\frac{\delta Q}{\partial T}\right)_p$[are widely and successfully treated for producing valid results] might have led caloricists to their erroneous belief that the heat content, Q, of a substance is a state function too…That mistaking inference might in turn strengthen their belief in the ontological status of caloric as matter-like.

6.7 Conclusion: Nature as It Is and It Can Become

A reversible machine remains the best or natural approach to start the consideration of the concept of entropy, Eq. (62A),

$$(\delta Q)_{Reversible} = TdS$$

Once the introduction is made, classical formalism is correct in pointing out that reversibility is a too restrictive condition for defining entropy. Classical formalism is mistaken, however, in replacing reversibility with quasi-staticity. The modern formalism shows that quasi-staticity in the classical formalism, Eq. (83),

$$(\delta Q)_{quasistatic} = TdS,$$

is *in fact* internal reversibility, which is the necessary and sufficient condition for the definition of entropy, Eq. (92),

$$(\delta Q)_{IR} = TdS$$

Why was it necessary for Carnot to make use of idealized machines? One reason is that Nature does not readily disclose how she is and even less in what she can become.

> Nature loves to hide.—Heraclitus

Scientific method was developed for decoding nature. It combined the Platonic–Cartesian–Galilean tradition of mathematical laws and the Aristotelian tradition of the rules of reasoning based on the universal premises [19:232]. This pre-Carnot tradition emphasized the power of the mathematical language for decoding a *TRI dynamical* *nature* as it is.

When Carnot noted that the production of power is always accompanied by a circumstance of "the re-establishing of equilibrium in the caloric," he pointed out the centrality of irreversibility in the universe—the concept of irreversibility was important to Kelvin and Clausius, and stressed by Planck in his exposition, but has been neglected by "modern physicists" who are still steeped in the pre-Carnot mathematical language tradition.

The irreversible universe of thermodynamic systems is also an interconnected universe. *No thermodynamic system is an island.* Reversibility or idealized machines is the new dialectic (irreversibility–reversibility) language to "decode" the irreversible, interconnected universe, in which it is no longer possible to study a thermodynamic system in itself, without referring to a reversible machine and its work reservoir, and a surrounding reservoir (see Fig. 6.6; [21]; Chap. 8). This new language, which prescribes/constructs nature as it could and can become (one example is the ever-growing operating principles of reversible-like processes (Fig. 6.5) through the *action* or *design* of engineers), is the majestic, paradigm-breaking contribution of Carnot.

Surprisingly, the contribution of Carnot is almost universally misunderstood and the not-yet-discovered gem of Carnot's theory will be unfolded in Chap. 8.

Problems

6.1 An ideal gas initially at T_1 expands adiabatically and reversibly from V_1 to V_2 ($= 2\ V_1$). Determine the final temperature in terms of the initial temperature, and the final pressure in terms of the initial pressure by first using the ideal gas equation of state, $p_2 = NRT_2/V_2$.

6.2 *Joule free expansion* (an example of a process of infinitely dense equilibrium states): Consider a composite system consisting of two compartments of equal volume (the two compartments are separated by a piston with perfect seal—see Fig. 6.1, note though that the figure incorrectly shows compartments of different volumes). The first compartment is filled with an ideal gas at T_1 and the second is evacuated at vacuum. If the ideal gas is permitted to (to push against the piston and) expand into the evacuated compartment region, thereby

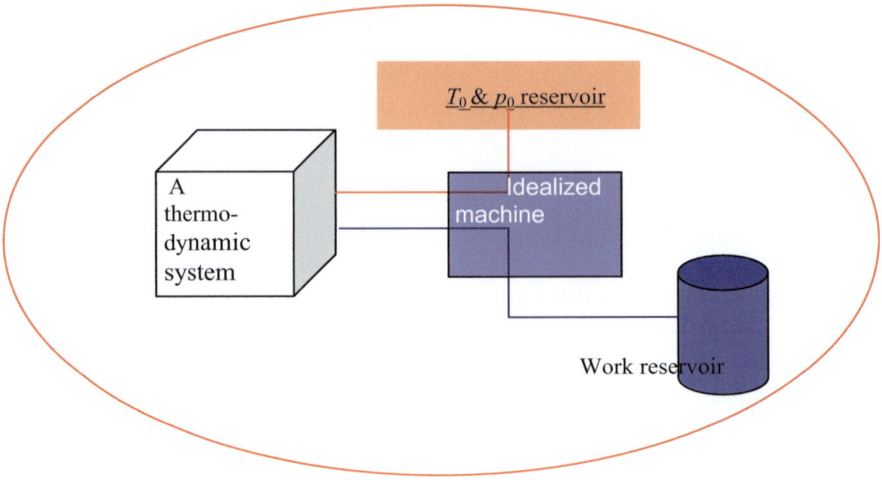

Fig. 6.6 Standard schematic of thermodynamic investigation (Fig. 1.7b is reproduced here with the addition of an idealized machine. *Note* Professor Bent [21] made the case for the operational definitions of heat and work and he wroteIn thermodynamics one usually needs to keep track of three things: a system, its thermal surroundings, and its mechanical surroundings…The [proposed operational] notation…offers one the option of purging from one's thermodynamic vocabulary the often troublesome terms "heat" and "work," by substituting, respectively, these operationally more expressive, if grammatically less succinct, phrases: "energy lost by the thermal surroundings"; "energy gained by the mechanical surroundings" [i.e., work reservoir]

increasing its volume from V_1 to $2V_1$, and if the walls of the composite system are rigid and adiabatic,

(2a) What is the ratio of the initial and final pressures? What is the ratio of the initial and final temperatures? What is the difference of the initial and final entropies?

(2b) Note that "quasi-static" work and heat expressions

$$\delta W = P\mathrm{d}V$$
$$\delta Q = T\mathrm{d}S,$$

are not applicable in this case of quasi-static expansion. Give the precise reason why they are not applicable.

(2c) What is the entropy change of the system and surroundings?

6.3 An example of infinitely slow process: (see Fig. 6.2: note, though, that both it and Fig. 6.1 incorrectly show compartments of different volumes) Consider the composite system of Problem 6.2 again: the piston is now connected to a mechanism that balances the force exerted by the gas on the piston and is equipped with work storage capacity—and the whole composite system is submerged in a heat reservoir/bath at T_1. Repeat the consideration of gaseous expansion to double the initial volume. The near balance of net force on the piston results in slow expansion of gas along the same quasi-static path of Problem 6.2. (Note that the process in the present case is reversible due to the balanced net force and the slowness of expansion and heat extraction from the bath.)

(3a) What is the ratio of the initial and final pressures? What is the ratio of the initial and final temperatures? What is the difference between the initial and final system entropies?

(3b) What is the useful work stored away in the work storage?

(3c) What is the heat extracted from the bath by the expanding gas? Is the heat completely transformed into work?

(3d) What is the total entropy change of the composite system and the bath?

(3e) This process has exactly the same path as that of Problem 6.2. Explain why the two processes are fundamentally different while they share the same path.

References

1. Uffink J (2001) Bluff your way in the Second Law of Thermodynamics. Stud Hist Phil Sci Part B: 2001 Studies in Hist and Philo of Modern Physics 32(No. 2): 305–395.
2. Born M (1949) Natural Philosophy of Cause and Chance. Oxford Univ Press (p. 38).
3. Carathéodory C (1909) Untersuchungen uber die Grundlagen der Thermodynamik. Math. Ann. 67: 355–386. A translated text "Investigation into the foundations of thermodynamics" is collected in J. Kestin, The Second Law of Thermodynamics (Dowden, Hutchinson and Ross, 1976). Pages referred to this reference *in-text* are that of the translated text.
4. Wang LS (2006) The auxiliary components of thermodynamic theory and their non-empirical, algorithmic nature. Physics Essays 19(2):174–199.
5. Callen HB (1st edition, 1960; 2nd edition, 1985) Thermodynamics and an Introduction to Thermostatistics. Wiley, New York.
6. Plank M (1932) Theory of Heat, Volume V of Introduction to Theoretical Physics. Macmillan, London. In which on p. 51, Planck wrote, "the content of the second law is not exhausted if—as was done occasionally by Clausius and later with renewed emphasis by Ostwald—every process in nature is resolved into a series of energy transformations [i.e., a series of steps in a quasi-static process]."
7. Landsberg PT (1961) Thermodynamics. Interscience, New York (p. 94).
8. Samiullah M (2007) What is a reversible process? Am J Physics 75: 609.
9. Planck M (1969) Treatise on Thermodynamics, 3rd edition. Dover, New York.
10. Gibbs JW (1961) The Scientific Papers: Thermodynamics. Vol. 1. Dover
11. Prigogine I (1949) Le domaine de validité de la thermodynamique des phénomènes irréversibles. Physica 15: 272.
12. Planck M (1887) Uber das Princip der Vermehrung der Entropie, in 3 parts. Ann. D. Phys. 30:562–582; 31:189–203; 32:462–503.
13. De Groot SR, Mazur P (1969) Non-Equilibrium Thermodynamics. North Holland, Amsterdam.
14. Butler HW (2007) Tracing the second law. Mechanical Engineering, ASME, July 2007. (pp. 38–40).

15. Kondepudi D, Prigogine I (1998) Modern Thermodynamics: From Heat Engines to Dissipative Structures. Wiley, New York.
16. Prigogine I (1947) Etude Thermodynamique des Phénomènes Irréversibles. Dunod, Paris.
17. Coveney PV (1988) The second law of thermodynamics: entropy, irreversibility and dynamics. Nature 333(No. 2): 409–415.
18. Wang LS (2007) The nature of spontaneity-driven processes. International J of Ecodynamics 2: 231–244.
19. Kauffman SA (2008) Reinventing the Sacred. Perseus Books (pp. 88–91).
20. Zemansky MK (1943) Heat and Thermodynamics, 2nd edition. McGraw-Hill, New York.
21. Bent HA (1972) A note on the notation and terminology of thermodynamics. J of Chem Education 49(No. 1):44–46.

Free Energy, Exergy, and Energy: The Exergetic Content of Energy

7

Abstract

Kelvin had introduced the general idea of available energy. The advent of the entropy principle led to, in quick succession, the formulation of equilibrium thermodynamics by Gibbs—which was the prized fruit of the Kelvin–Clausius synthesis. One of its important results was the concept of free energies, which provided the general idea of available energies with specific examples—especially the Gibbs free energy of electrochemical energy→electrical work. Further development of available energy has evolved into the theory of exergy. Significantly, both the first law and the theory of exergy can be formulated for an engineering device's control volume. In that formulation, the exergy treatment is reducible to the Carnot–Kelvin formula, a key tool in mechanical engineering, as well as to the Gibbs free energy, a key tool in chemical engineering.

Keywords

Thermodynamic potentials · Helmholtz free energy · Gibbs free energy · Exergy · Material exergy · Energy equation for open systems · Exergy equation for open systems · Definition of energy · Spontaneous energy conversion · Mechanical theory of heat (MTH)

7.1 Thermodynamic Potentials and Free Energies

This chapter introduces the concepts of free energy and exergy. Beginning with Chap. 4, we have been centrally concerned with the phenomena of the natural tendency of the material world and how this tendency is intimately related to the production of work. Both of the above concepts help to deepen our understanding of this issue. It turns out that the only way to deliberate about energy meaningfully is through thinking in terms of entropy and exergy. In this section, we begin with free energies.

L.-S. Wang, *A Treatise of Heat and Energy*, Mechanical Engineering Series,
https://doi.org/10.1007/978-3-030-05746-6_7

Equation (64) in Chap. 5 is reproduced here

$$dU = TdS - pdV \tag{64}$$

Equation (64) suggests that the set of *U*-*S*-*V* is a unique one. Their functional relationship may be written as

$$f(U, S, V) = 0 \tag{96}$$

Equation (64) is known under the name of *fundamental differential* or *principal exact differential.* Correspondingly, Eq. (96) may be referred to as a *fundamental thermodynamic function of state* in the sense that its differentiation results in Eq. (64).

The fundamental thermodynamic function and the fundamental differential, which "capture the *complete* thermodynamic information about a system" (see Chap. 9), serve as the foundation of equilibrium thermodynamics, which will be treated in Chap. 9. But, it does not capture all aspects of the first and the second laws, which are the foundation of engineering thermodynamics that studies the interaction of a system and its surroundings, as it will be evident in the following treatment of free energy and exergy. Even so, the concept of free energies begins with the concept of thermodynamic potentials, which (the potential expressions themselves) can be conveniently introduced in terms of equilibrium thermodynamics whereas the interpretation of thermodynamic potential as free energies must, of course, involve how the system interacting with its surroundings.

Beginning with the energy representation of the fundamental function, Eq. (96),

$$U = U(S, V) \tag{96A}$$

we compute the first differential

$$dU = \left(\frac{\partial U}{\partial S}\right)_V dS + \left(\frac{\partial U}{\partial V}\right)_S dV \tag{97}$$

Comparing Eq. (97) with Eq. (64)

$$T \equiv \left(\frac{\partial U}{\partial S}\right)_V = T(S, V) \tag{98}$$

$$-p \equiv \left(\frac{\partial U}{\partial V}\right)_S = -p(S, V) \tag{99}$$

Equations (96A), (98), and (99) are called *canonical form*, and *S* and *V* the *generalized coordinates* of the canonical form.

It is useful to choose a different set of independent variables as generalized coordinates, and the question arises whether the resulting *U*-function of the new set of independent variables, e.g., *S*-*p* or *T*-*V* instead of the original *S*-*V*, retains the

Table 7.1 Thermodynamic potentials

Independent variables, X and Y	Dependent thermodynamic functional variable ψ derived from Legendre transformation	Name of the alternative dependent variable ψ	Fundamental function, $\psi = \psi(X, Y)$
S and V	U		U-S-V
S and p	$U + pV$	Enthalpy H	H-S-p
T and V	$U - TS$	Helmholtz function A_H	A_H -T-V
T and p	$H - TS = A_H + pV = U + pV - TS$	Gibbs function G	G-T-p

same usefulness as a fundamental function. In these cases, for the purpose of preserving the all-inclusive information, the dependent variable U can be transformed by Legendre transformation [1:137–149] into a new variable, i.e., an alternative fundamental functional variable, ψ.

Consider the case of the transformation of U-S-V into ψ-S-p, in which ψ is the new dependent variable

$$\psi \equiv U - \left(\frac{\partial U}{\partial V}\right)_{S,N_i} V = U - (-p)V = U + pV \tag{100}$$

which has been identified as the enthalpy, H (Sect. 3.6).

Consider the case of the transformation of U-S-V into ψ-T-V

$$\psi \equiv U - \left(\frac{\partial U}{\partial S}\right)_{V,N_i} S = U - TS \tag{100}$$

which will be called the Helmholtz function, A_H.

Consider next the case of the transformation of A_H-T-V into ψ-T-p, one gets

$$\psi(=G) = A_H - \left(\frac{\partial A_H}{\partial V}\right)_{T,N_i} \cdot V = A_H + pV = U - TS + pV \tag{101}$$

which will be called the Gibbs function, G.

The complete set of new dependent variables is summarized in Table 7.1.

7.1.1 The Extremum Principle for Thermodynamic Equilibriums of Composite Systems

We now consider composite systems approaching internal equilibrium with the possibility of being subject to the constraint of reservoir that the systems interact with.

For a thermally isolated composite system subject to constant system volume, $V = V^{(1)} + V^{(2)} = constant$, its system internal energy U is also constant in accordance with the first law. Consider the entropy representation of the fundamental function

$$S = S(U, V) = S^{(1)}\left(U^{(1)}, V^{(1)}\right) + S^{(2)}\left(U^{(2)}, V^{(2)}\right) \tag{96B}$$

Application of the entropy principle to Eq. (96B) leads to the extremum principle of maximum system entropy at internal thermodynamic equilibrium

$$S = S^{(1)}\left(U^{(1)}, V^{(1)}\right) + S^{(2)}\left(U^{(2)}, V^{(2)}\right) \rightarrow S_{max}$$

That is, the equilibrium values of $U^{(1)}_{equi}, U^{(2)}_{equi}, V^{(1)}_{equi}, V^{(2)}_{equi}$ are determined by

$$dS = 0, \text{ at } U^{(1)} = U^{(1)}_{equi} \; and \; V^{(1)} = V^{(1)}_{equi} \tag{102A}$$

under the constraints

$$U^{(1)} + U^{(2)} = U^{(1)}_{initial} + U^{(2)}_{initial}, a \; constant$$

$$V^{(1)} + V^{(2)} = V^{(1)}_{initial} + V^{(2)}_{initial}, a \; constant$$

Consider a thermally isolated composite system in interaction with a constant-pressure reservoir, i.e., $p^{(1)} = p^{(2)} = p^r$. The first law becomes

$$U_1 - U_2 = Q - \int_1^2 pdV = 0 - p^r(V_2 - V_1)$$

i.e.,

$$(U + p^r V)_2 - (U + p^r V)_1 = H_2 - H_1 = 0$$

The entropy principle, Eq. (74), applies to this thermally isolated system, and we have

$$S = S^{(1)}\left(H^{(1)}, p^r\right) + S^{(2)}\left(H^{(2)}, p^r\right) \rightarrow S_{max} \tag{102B}$$

The equilibrium values of $H^{(1)}_{equi}$ and $H^{(2)}_{equi}$ are

$$\frac{dS}{dH^{(1)}} = \frac{dS^{(1)}}{dH^{(1)}} + \frac{dS^{(2)}}{dH^{(1)}} = \frac{dS^{(1)}}{dH^{(1)}} + \frac{dS^{(2)}}{dH^{(2)}} \cdot \frac{d\left(\left[H_1^{(1)} + H_1^{(2)}\right] - H^{(1)}\right)}{dH^{(1)}}$$
$$= \frac{dS^{(1)}}{dH^{(1)}} - \frac{dS^{(2)}}{dH^{(2)}} = 0, \text{at } H^{(1)} = H_{equi}^{(1)} \tag{102B}$$

under the constraint.

$$H^{(1)} + H^{(2)} = constant\left(= H_1^{(1)} + H_1^{(2)}\right)$$

We now consider an isothermal composite system in interaction with an isothermal heat reservoir at T^r. Denote U, S, A_H, T to be the internal energy, entropy, Helmholtz function, and temperature of the composite system, and U^r, S^r, T^r be the internal energy, entropy, and temperature of the isothermal heat reservoir (Note again, $T = T^r$, and $V^{(1)} + V^{(2)}$ is maintained constant). Consider the totality of the composite system and the heat reservoir noting that the totality is by definition an isolated system of constant total internal energy, i.e.,

$$d(U + U^r) = 0 \tag{103}$$

The entropy principle, in this case, applies to the totality of the system and the reservoir, and therefore,

$$d(S + S^r) = 0, \text{ when the composite system at equilibrium with the reservoir} \tag{102C}$$

Note that a thermal reservoir is defined as a reversible heat source (or sink) that is so large that any heat transfer of interest does not alter the temperature of the reservoir. For such a system, its thermal interaction is reduced to

$$dU^r = \delta Q_{res} - 0 = T^r dS^r$$

It follows, therefore, (103) may be rewritten as

$$d(U + U^r) = dU + T^r dS^r$$

Furthermore, by Eq. (102C), it becomes

$$dU + T^r dS^r = dU - T^r dS = 0, \text{ at equilibrium}$$

That is, the equilibrium criterion of an isothermal composite system in interaction with an isothermal heat reservoir is that the system Helmholtz function assumes minimum value

$$dA_H = d(U - TS) = 0, \text{ at equilibrium} \tag{104}$$

For a chemical composite system in thermal/mechanical interaction with an isothermal heat reservoir (again, the reservoir is so large that any mechanical interaction of interest does not alter the pressure of the reservoir), denote U, S, G, T, p, V to be the internal energy, entropy, Gibbs function, and temperature, pressure, and volume of the composite system, and U^r, S^r, T^r, p^r be the internal energy, entropy, and temperature and pressure of the isothermal heat reservoir.

Consider the totality of the composite system and the heat reservoir, noting again that the totality is by definition an isolated system. The entropy application leads to

$$d(S + S^r) = 0, \tag{102D}$$

when the composite system at equilibrium with the reservoir.

The application of the first law to the system yields

$$dU = \delta Q - pdV = -\delta Q_{res} - p^r dV$$

Again, by $\delta Q_{res} = T^r dS^r$ and with the substitution of Eq. (102D), the system first law balance becomes

$$dU = -\delta Q_{res} - p^r dV = T^r dS - p^r dV$$

That is,

$$d(U - T^r dS + p^r dV) = dG = 0, \text{at equilibrium} \tag{105}$$

The equilibrium criterion of a chemical composite system in thermal/mechanical interaction with a constant temperature and pressure reservoir is the system Gibbs function assumes minimum value.

Equation (105) may be alternatively written as follows. Introduce the thermodynamic potential ψ of a composite system, $d\psi(\xi) = dG(T, p, N_1[\xi], N_2[\xi], \ldots)$, (see Chap. 9) and correspondingly the degree of reaction, ξ. The process of the chemical system can be determined along the quasi-static path [1] of the system approaching the internal equilibrium of the composite system. Equation (105) becomes

$$\left(\frac{\partial \psi}{\partial \xi}\right)_{equili} = 0 \tag{105A}$$

or

$$\left[A\left(\equiv -\frac{\partial \psi}{\partial \xi}\right)\right]_{equili} = 0 \tag{105B}$$

Table 7.2 Equilibrium criteria

Fundamental functional relation	Thermodynamic potential, ψ	Constraint	Equilibrium criterion
U-S-V	$-T_0\ S$	$\sum U_i = const.$ $\sum V_i = const.$	*S* maximization
H-S-p	$-T_0\ S$	$\sum H_i = const.$ $P = const.$	*S* maximization
A_H *-T-V*	A_H	$T = const.$ $\sum V_i = const.$	A_H minimization
G-T-p	*G*	$T\ = const.$ and $p = const.$	*G* minimization

where *A* is *affinity*, which, in the case of chemical examples, is related to chemical potential [2: Chap. 4].

In sum, Eqs. (102A), (102B), (104), and (105) may be summarized as the following expressions and Table 7.2.

$$S \rightarrow S_{max}(U, V) \tag{102A}$$

$$S \rightarrow S_{max}(H, p) \tag{102B}$$

$$A_H \rightarrow (A_H)_{min}(T, V) \tag{104}$$

$$G \rightarrow G_{min}(T, p) \tag{105}$$

7.1.2 Helmholtz Free Energy and Gibbs Free Energy

There is no difficulty in understanding the importance of the inference of the entropy principle on thermodynamic equilibrium and the corresponding formulation of equilibrium thermodynamics (see Chap. 9). The difficulty in understanding the second law lies in its foundation on the one hand and, on the other hand, its engineering application.[1] This essay/disquisition addresses neither the issue of the

[1]If further evidence is needed for the need of "better understanding", I quote two speakers in the Panel Discussion on "Teaching the Second Law" at the 2007 MIT meeting, *Meeting the Entropy Challenge, AIP Conference Proceedings* ***1033*** [12]:

Professor Robert Silbey,

"I have taught thermodynamics in the Chemistry Department here at MIT for about 40 years. And I have to say that whenever we get to the second law I am always very nervous. And so I'm anxious to hear from the panelists their views on this. And I think, it's a little bit like the old story of Toscanini when he had to conduct Beethoven's Ninth Symphony, he always said to the orchestra 'courage, have courage.' And I think when you do the second law it's the same thing."

Professor Yunus Cengel (The following reflection of his showed that there was, in the Panel Discussion little agreement on what the second law is.),

"First, I would like to thank the panel members for sharing their experiences and ideas on

statistical mechanical foundation of the second law nor the axiomatic foundations of the law, but it does consider the engineering application of the second law to be its central goal.

With that in mind, it is important to warn against a dangerous view in engineering that entropy and the second law should be avoided in favor of an energy-centric treatment of engineering thermodynamics [3]. This would have been a grave mistake, the consequence of which would be condemning practitioners of engineering thermodynamics to be the second-class scientists/engineers instead of the true descendants of Carnot.

The purpose of this section is to show that, through the introduction of Helmholtz free energy and Gibbs free energy, the direct application of the entropy principle leads to the idea of useful energy and exergy, the idea that gave meaning to energy as we understand it. The idea, without which modern engineering practice cannot exist.

An isothermal composite system of constant system volume in interaction with an isothermal heat reservoir at T^r, initially at $V_1^{(1)}$ and $V_1^{(2)}$, corresponding to A_{H_1}, will approach equilibrium when,

$$V^{(1)} \rightarrow V_{equili}^{(1)} \text{ and } V^{(2)} \rightarrow \left[V_1^{(1)} + V_1^{(2)}\right] - V_{equili}^{(1)}$$

corresponding to the minimization of the system Helmholtz function, $A_H \rightarrow A_{H_{min}}$. The difference in system Helmholtz function, $A_{H_1} - A_{H_{min}}$, is subject to the following interpretation: Note first the relation

$$A_{H_1} - A_{H_{min}} = (U - T^r S)_1 - (U - T^r S)_{equili} = \left(U_1 - U_{equili}\right) - T^r\left(S_1 - S_{equili}\right)$$

Consider the process to be reversible and we have $Q_{1\rightarrow equili} = T^r\left(S_{equili} - S_1\right)$. Correspondingly,

$$A_{H_1} - A_{H_{min}} = \left(U_1 - U_{equili}\right) + Q_{1\rightarrow equili}$$

By the first law, $\Delta U = Q_{1\rightarrow equili} - W_{1\rightarrow equili}$, it follows

$$A_{H_1} - A_{H_{min}} = \left(U_1 - U_{equili}\right) + Q_{1\rightarrow equili} = \left(W_{1\rightarrow equili}\right)_{rev}$$

That is,

$$Helmholtz\,free\ energy = A_{H_1} - A_{H_{min}} = Reversible\ Useful\ Work \tag{106}$$

teaching the second law. One thing I noticed is that there is little in common in their presentations, which means we have a long way from finding a unified approach in teaching the second law. I was particularly pleased to hear that MIT is team teaching a combined energy-philosophy course, which, I think, is very interesting. I wish Professor Trout the best of luck in this undertaking…"

If the process is not reversible, Eq. (73) or Eq. (89) yields

$$Q_{1\to equili} \le T^r\left(S_{equili} - S_1\right)$$

We find

$$\begin{aligned} W_{1\to equili} &= \left(U_1 - U_{equili}\right) + Q_{1\to equili} \le \left(U_1 - U_{equili}\right) + T^r\left(S_{equili} - S_1\right) \\ &= A_{H_1} - A_{H_{min}} \end{aligned}$$

That is,

$$\textit{Useful Work}_{Constant-T\&V-system} \le \textit{Helmholtz free energy} \tag{107}$$

Closely related to the Helmholtz free energy is the Gibbs free energy. Consider a chemical composite system at thermal/mechanical interaction with a constant pressure (p^r), isothermal (T^r) heat reservoir. A chemical system will approach equilibrium corresponding to the minimization of the Gibbs function, $G_1 \to G_{min}$, which may be written as

$$G_1 - G_{min} = \left(U_1 - U_{equi}\right) + p^r\left(V_1 - V_{equi}\right) - T^r\left(S_1 - S_{equi}\right) \tag{105A}$$

For this chemical system undergoing internally reversible process under constant temperature, T^r, and constant pressure, p^r, the first law equation may be written as

$$U_{equi} - U_1 = Q_{1\to equili} - W_{1\to equili} = Q_{1\to equili} - \left[p^r\left(V_{equi} - V_1\right) + W_{chemical-work}\right]$$

For the case of reversible operation, the equation becomes

$$U_{equi} - U_1 = T^r\left(S_{equili} - S_1\right) - \left[p^r\left(V_{equi} - V_1\right) + \left(W_{chemical-work}\right)_{rev}\right]$$

The last term in the above version of the first law, $\left(W_{chemical-work}\right)_{rev}$, can be identified as chemical work by comparing the above equation to the following equation, (108), as shown in the following paragraph.

Consider Eq. (154B) in Chap. 9

$$dU = TdS - pdV + \sum_i \mu_i dN_i$$

Which may be written under constant temperature, T^r, and constant pressure, p^r, as,

$$U_{equi} - U_1 = T^r\left(S_{equili} - S_1\right) - p^r\left(V_{equi} - V_1\right) + \int \sum_i \mu_i dN_i \tag{108}$$

Comparison of Eq. (108) with the reversible operation version of the first law equation yields

$$(W_{chemical-work})_{rev} = -\int \sum_i \mu_i dN_i$$

By the reversible operation version of the first law equation

$$(W_{chemical-work})_{rev}\left(= -\int \sum_i \mu_i dN_i\right) = (U_1 - U_{equi}) + T^r(S_{equili} - S_1) - p^r(V_{equi} - V_1)$$

which, according to Eq. (105A), becomes

$$(W_{chemical-work})_{rev} = G_1 - G_{min}(\equiv -\Delta G) \tag{109}$$

If the process is not reversible but internally reversible, similarly, in this case, we have by the first law

$$W_{chemical-work} = -\Delta U + Q_{1\to equili} - p^r(V_{equi} - V_1)$$

Note, in this case,

$$Q_{1\to equili} \le T^r(S_{equili} - S_1)$$

Substituting which into the first law yields

$$W_{chemical-work} \le -\Delta U + T^r(S_{equili} - S_1) - p^r(V_{equi} - V_1) = G_1 - G_{min}(\equiv -\Delta G) \tag{110}$$

7.1.3 Example: Thermodynamics of a Battery

The battery in your car produces chemical work via chemical reaction

$$Pb + PbO_2 + 4H^+ + 2SO_4^{2-} \rightarrow 2PbSO_4 + 2H_2O$$

Note that this work corresponds to electrical work. By using the concept of Gibbs free energy, we compute the maximum electrical work that is produced by modeling the battery operation as an isothermal and isobaric electrochemical process.

We can compute the change in the Gibbs function by taking the difference of the formation values: the difference of the Gibbs function at the initial state of the battery subtracted by the final state, the discharged state of the battery. For one mole of lead, we have the following value of $G_{fully-charged} - G_{equi}$ $(-\Delta G)$,

$$-\Delta G = 0 + (-217.3) + 0 + 2(-744.5) - 2(-813.0) - 2(237.1) = 393.9 \text{ kJ}$$

A perfectly reversible battery will produce 394 kJ electrical work per mole of *Pb* element. A real battery will produce a lower value in electrical work

When you charge the battery, you run this electrochemical reaction in reverse. To put the battery back into the fully charged initial state, a minimum value of 393 kJ (per mole of *Pb*) electrical work is required ideally. The charging of a real battery will require a higher value.

Like the Kelvin–Carnot formula, which accounts for the motive power of heat under ideal theoretical operation in an open system setting such as steam engines, the Helmholtz free energy and the Gibbs free energy account for power under ideal operating conditions derived in a closed system setting, such as batteries.

7.2 Engineering Inference of the Entropy-Energy Principles

There are two versions of the second law: one version is discussed in Chap. 4 (as the energy principle, see Sect. 4.7 and Fig. 4.7) and the second version in Chap. 5 (as the entropy principle, see Sects. 5.4–5.5 and Fig. 5.11). The two versions center their formulations, respectively, on the notion of the dissipation of energy and the notion of the growth of entropy. In a comment, cited by Daub, [4], Maxwell had this to say "The doctrine of the dissipation of energy is closely connected with that of the growth of entropy, but is by no means identical with it" [5:192]. This assessment was, of course, repeated by Planck, as was cited in Sect. 5.10 and in a recent paper [6]. To amplify what Maxwell and Planck wrote on this matter, [6] and Sect. 5.10 explain the relation of the two versions in the following sense: the latter is a universal principle while the former is not; in other words, the growth of entropy is not *exhausted* by the dissipation of energy. If the growth of entropy were exhausted by the dissipation of energy, every case of the growth of entropy would have corresponded to an example of the dissipation of energy; discussion in Sect. 5.10 demonstrates that this is not so.

The implication of the Helmholtz free energy and the Gibbs free energy on engineering thermodynamics is this: the original understanding of heat's apparent utility derived from steam engines and Carnot' theory was transformed into energy utility or availability of energy, as Kelvin first articulated. The development was significant because it transformed Kelvin's general idea (his "general conclusions") into concrete terms by the direct application of the entropy principle to the first law.

Without this and the earlier advent of the energy principle, we only understood the apparent utility of heat in terms of Carnot–Kelvin formula. The role of any energy system in the production of mechanical work would have to go through the phase of heat released from the energy system, i.e., only indirect energy conversion would be physically possible. This is clearly not true.

In the literature, the available energy (free energy) concept was, for obvious reason, attributed to Gibbs [7]. But, the well-deserved attribution to Gibbs was often taken as Gibbs as the originator of the concept, instead of one who gave a general idea its concrete form. That perception is contradicted by the analysis in Chap. 4 of this essay/disquisition, as well as other writers who held a real vantage point on this topic: The source of Gibbs' insight is traced by Daub, as he noted in a historical study on entropy and dissipation, to Kelvin, "Although Gibbs never once mentioned Thomson in his work, he was indebted, I believe, to Thomson's concept of dissipation of energy via the good offices of Maxwell and his *Theory of Heat*. Maxwell, in turn, was indebted to Gibbs in the revision of his treatment of available and unavailable energy in his *Theory of Heat*, thereby uniting the two traditions of entropy and dissipation" [4:351]. Daub's reading of the history of thermodynamics is fully consistent with the historical study of Kelvin's biographers [8]. Finally, the clearest attribution to Kelvin is, in a review of Tait's *Thermodynamics*, [9] the statement by Maxwell, "Thomson, the last but not the least of the three great founders [Clausius, Rankine, and Thomson], does not even consecrate a symbol to denote the entropy, but he was the first to clearly define the intrinsic energy of a body, and to him alone are due the ideas and definitions of the available energy and the dissipation of energy."

These conceptual developments contributed to how engineers understand the "heat→mechanical energy" transformation. In this part of thermodynamics history, in the 1940s and 50s, Joseph Keenan assumed the leadership role in applying second law analysis to engineering applications [10]. This eventually became known as the *theory of exergy*, the term was introduced by Rant in his 1956 article, [11].

The theory of exergy was an important phase in the long-term *conceptual* contribution to be derived from the entropy principle to engineering thermodynamics. In a significant way, the theory of exergy represents an outcome resulting from the inference of the combined entropy and energy principles. Scientists, i.e., physicists and chemists, etc., have no use of the energy principle, needing only the conservation of energy and entropy principles. Engineers, however, in formulating the theory of exergy by applying the entropy principle, have very much in their mind that the bedrock of their enterprise is the energy principle, while the entropy principle as incorporated in exergy analysis is only a tool to make their undertaking more effective.

7.2.1 Why Exergy?

My analysis on the development of the idea of available energy may be summarized: Kelvin started the general idea of available energy, Gibbs provided the general idea its concrete form, and, in the following I shall show, the formulation of the theory of exergy transformed/united streams of ideas into one general framework completing a full circle. We shall conclude that engineers find the utility of energy in its exergetic content, i.e., the consideration of exergetic content of energy is the only way to think in terms of energy intelligently. Another way to appreciate usefulness in exergetic thinking is that, via the theory of exergy, a quantitative linkage can be found (see below) between the Carnot–Kelvin formula and Gibbs free energy, giving unity to the typical core knowledge of mechanical engineering and typically what chemists and chemical engineers know. Furthermore, we shall find exergy analysis is particularly effective in application to component processes (the unit operations level) instead of the whole-systems level, and therefore it is a powerful tool in identifying the problem. Even though it, in itself offers no solution, but identification of where a better solution is needed is paramount.

7.2.2 Energy Equation for Open Systems

So far, our discourse has been focused on closed systems since the first law and the second law are introduced naturally based on the consideration of closed systems. This focus will continue in Chaps. 8 and 9. But, beginning with this chapter and in Chap. 10, engineering application, i.e., engineering thermodynamics, will be the main goal of our discourse. In the realm of engineering application, consideration must be made to open systems, as many devices are open systems.

Beginning with the first law, Eq. (23), which is rewritten as

$$\frac{dE}{dt} = \dot{Q} - \dot{W} \tag{23B}$$

the first law equation is shown in Chap. 10 to be developed into an equation applied to a *control volume*, *cv*, defining a device. Key assumptions and steps there (see Chap. 10 for details) are reproduced here.

The system rate equation is related to expressions of the rate change in the corresponding control volume, known as Reynolds' transport theorem

$$\left(\frac{dE}{dt}\right)_{system} = \frac{\partial}{\partial t}\int_{cv} e\rho dV + \int_{cs} e\rho\vec{V}\cdot d\vec{A}$$

The application of Reynolds' transport theorem to the LHS of the equation leads to

$$\frac{\partial}{\partial t}\int_{cv} e\rho d\mathcal{V} + \int_{cs} e\rho\vec{V}\cdot\hat{n}\, dA = \dot{Q} - \dot{W}$$

The work term on the RHS is developed into

$$\dot{W} = \dot{W}_{shaft} + \dot{W}_{surface} = \dot{W}_{shaft} + \int_{cs} p\vec{V} \cdot d\vec{A}$$

The two terms, $\int_{cs} e\rho\vec{V} \cdot \hat{n}\, dA$ and $\int_{cs} p\vec{V} \cdot d\vec{A}$, may be combined becoming

$$\int_{cs} p\vec{V} \cdot d\vec{A} + \int_{cs} e\rho\vec{V} \cdot d\vec{A} = \int_{cs} \rho(pv + e)\vec{V} d\vec{A}$$

Noting that $e = u + \frac{V^2}{2} + gz$, substitution of the $\dot{W}$ expression into the first law equation for cv making use of the combined term leads to

$$\dot{Q}_{cv} - \dot{W}_{shaft} = \frac{\partial}{\partial t}\int_{cv} e\rho d\mathcal{V} + \int_{cs} \rho\left(h + \frac{V^2}{2} + gz\right)\vec{V} \cdot d\vec{A}$$

This is the reproduced version of Eq. (199) in Chap. 10, which is rewritten as

$$\begin{aligned} \frac{\partial E_{cv}}{\partial t} &= \frac{\partial}{\partial t}\int_{cv} e\rho d\mathcal{V} = \dot{Q}_{cv} - \dot{W}_{shaft} - \int_{cs} \rho\left(h + \frac{V^2}{2} + gz\right)\vec{V} \cdot d\vec{A} \\ &= \dot{Q}_{cv} - \dot{W}_{shaft} + \sum_i \dot{m}_i\left(h_i + \frac{V_i^2}{2} + gz_i\right) - \sum_e \dot{m}_e\left(h_e + \frac{V_e^2}{2} + gz_e\right) \end{aligned} \tag{111/199}$$

7.3 A Brief Review of the Concept of Exergy

As Ghoniem noted in the 2007 AIP-MIT Conference, "we don't have an energy challenge, we have an entropy challenge" [12:5]. This entropy challenge is best handled with the introduction of exergy in the analysis of energy, i.e., exergy analysis. In introducing exergy analysis, Bejan et al. in 1996 explained the necessity of doing so as follows.

> Exergy analysis also provides insights that elude a purely first law approach. Thus, from an energy perspective, the expansion of a gas (or liquid) across a valve without heat transfer (throttling process) occurs without loss. That such an expansion is a site of thermodynamic inefficiency is well known, however, and this can be readily quantified by exergy analysis. From an energy perspective, energy transfers to the environment appear to be the only possible sources of power plant inefficiency. On the basis of first law reasoning alone, for example, the condenser of a power plant may be mistakenly identified as the component primarily responsible for the plant's seemingly low overall efficiency. An exergy analysis correctly reveals not only that the steam generator is the principal site of thermodynamic inefficiency owing to irreversibilities within it, but also the condenser [loss] is relatively unimportant. [13]

The development of research on exergy analysis after Keenan led to the modern definition of exergy.

> The exergy of a thermodynamic system S in a certain state S_A is the maximum theoretical useful work obtained if S is brought into thermodynamic equilibrium with the environment by means of ideal processes in which the system interacts only with this environment. [14]

Another influential definition was formulated by Rant [15] and Baehr [16]:

> Exergy is the portion of energy that is entirely convertible into all other forms of energy; the remainder is anergy.

That is,

$$energy = exergy + anergy \tag{112}$$

We shall refer to them as the first exergy definition and the second exergy definition, respectively.

While one defines exergy in terms of "the portion of energy...," and considers the application of exergy analysis, as well as speaks about exergy components, exergy balance, exergy transfer, etc. ([13], Chap. 3) as one does about energy analysis, energy components, energy balance, and energy transfer, one fundamental difference of exergy from a property such as energy, which is defined for a system, is that exergy, in general, can only be defined for a system *and* the environment (heat reservoir) with which the system interacts. This point is explicitly made in the first exergy definition. In the second exergy definition, the point is not explicit, but implicit because the determination of the portion of energy in a thermal energy system that is entirely convertible requires the specification of a heat reservoir (see below, but, of course, that determination for pure exergy energies does not have this requirement).

7.3.1 Exergy Components

The total exergy of a system, E, can be divided into three components: material exergy E^{MTL}, kinetic exergy E^{KN}, and potential exergy E^{PT}

$$Ex = Ex^{\mathrm{MTL}} + Ex^{\mathrm{KN}} + Ex^{\mathrm{PT}} \tag{113}$$

One could choose to call the material exergy the *internal* exergy, similarly as the division between internal energy and kinetic/potential energies. In the case of that usage of the internal energy it is understood that the same term represents thermal internal energy as well as other forms of internal energies including chemical internal energy. We choose here instead not to use *internal* for the important point that exergies other than the kinetic exergy are not defined in terms of system (material system) alone but in reference to the exteriorly defined environment.

There is a second reason for not to use the word *internal*. In the otherwise authoritative 1996 book by Bejan et al. [13], the treatment introduced an additional division of $E^{\mathrm{MTL}} = Ex^{\mathrm{PH}} + Ex^{\mathrm{CH}}$; thus, the total exergy of a system is, accordingly,

$$Ex = Ex^{\mathrm{PH}} + Ex^{\mathrm{KN}} + Ex^{\mathrm{PT}} + Ex^{\mathrm{CH}} \tag{114}$$

We shall comment in Sect. 7.5 on the difference between the two classifications, Eqs. (113) and (114), represent. In that discussion, irreversible processes toward internal equilibrium is a key. Since the term *internal* used in internal exergy would be too closely linked to the idea of internal energy preventing one from consideration of whether the system is in internal chemical equilibrium or not, the different term of material exergy that, though connoting similarly the idea of other than kinetic and potential exergies, is a better choice.

The kinetic and potential energies are in principle fully convertible to work as the system is brought to rest or to its reference level, respectively. Accordingly, for a system of mass m,

$$Ex^{\mathrm{KN}} = KE = \frac{1}{2}mV^2 \tag{115}$$

$$Ex^{\mathrm{PT}} = PE = mgz \tag{116}$$

where v and z denote velocity and elevation relative to the reference level.

7.3.2 Material Exergy

We first consider a material system approaching thermal and mechanical equilibrium with a standard environment. In this case, the material exergy, which is known as physical exergy, of a closed system at a specified state is given by

$$Ex^{\mathrm{MTL}} = (U - U_0) + p_0(V - V_0) - T_0(S - S_0) \tag{117}$$

where U, V, and S denote, respectively, the internal energy, volume, and entropy of the system at the specified state, and U_0, V_0, and S_0 are the values of the same properties when the system is at the *restricted dead state*. The *dead state* of a system is the state of the system when it reaches the conditions of mechanical, thermal, and chemical equilibrium with its environment at T_0, p_0, and a given equilibrium chemical composition, while *restricted dead state* refers to the state of the system reaching a restricted form of equilibrium with its environment, i.e., mechanical and thermal equilibrium only, with the system at the final dead state at T_0 and p_0.

Equation (117) can be derived as follows. Consider a system that interacts with its environment, its surroundings at T_0, p_0. The system undergoes a work producing process defined by its two end states A (initial state) and B (final state, the restricted dead state). The entirety of the work done by the system is not always in a useful form. For example, when a gas in a piston cylinder expands, part of the work done by the gas W is used to push the atmospheric air at p_0 out of the way of the piston motion. This work,

$$W_{Surr} = p_0(V_B - V_A),$$

is not to be recovered for any useful purpose. The difference between the actual gas expansion work W and the work against the surroundings W_{Surr} is called the useful work defined by

$$W_{Useful} = W - W_{Surr} = W - p_0(V_B - V_A)$$

where W according to the first law equation is equal to change in system internal energy and heat exchange, $Q_{A \to B}$,

$$U_B - U_A = Q_{AB} - W_{AB}$$

or

$$\left(W_{Useful}\right)_{AB} = W_{AB} - W_{Surr} = (U_A - U_B) + Q_{AB} - p_0(V_B - V_A)$$

Note that with system heat exchange $Q_{A \to B}$ (Q_{AB}), the corresponding heat exchange of the surroundings is $-Q_{A \to B}$ (assuming that the system and the surroundings form a thermally isolated system), and correspondingly, the entropy change of the surroundings is $-Q_{AB}/T_0$. Therefore, the total entropy change of the universe is

$$(\Delta S)_{AB} = (S_B - S_A) + \frac{-Q_{AB}}{T_0}$$

The entropy principle holds

$$(\Delta S)_{AB} \geq 0$$

It follows

$$Q_{AB} \leq T_0(S_B - S_A)$$

The application of the first law and the entropy principle leads to

$$\left(W_{Useful}\right)_{AB} = (U_A - U_B) + Q_{AB} - p_0(V_B - V_A)$$

$$\leq [U - T_0 S + p_0 V]_A - [U - T_0 S + p_0 V]_B$$

And, the reversible useful work (maximum useful work) for process $A\rightarrow$ *restricted dead state* is

$$[U - T_0S + p_0V]_A - [U - T_0S + p_0V]_B$$

Or,

$$\left(Ex^{MTL}\right)_A = [U - T_0S + p_0V]_A - [U - T_0S + p_0V]_0 \tag{117}$$

7.3.3 Discussion

Clearly, Eq. (117) is based on the first exergy definition and the physical exergy of the system is defined in terms of the system *and* its environment with which the system interacts. With the first exergy definition, the value of physical exergy is not subjected explicitly to the notion that it is a portion of energy as the second exergy definition declares. In the latter case, as Eq. (96) implies *exergy* $\leq$ *energy* (in other words, all three quantities [*energy*, *exergy*, and *anergy*] are positive definite and a negative *anergy* will be senseless). Yet, there are systems and their environments for which *anergys* are found to be negative and, for these instances, the second exergy definition is problematic. However, the problematic second exergy definition is necessary for the concepts of kinetic exergy and potential energy as shown in Eqs. (115) and (116); it captures importantly that kinetic and potential exergies, as examples of *pure exergy energies*, can be completely converted into other forms of pure exergy energy, such as electrical energy. Nonetheless, while both definitions are necessary, there is a conflict between the two definitions in that the first exergy definition allows the possibility of *exergy* $\geq$ *energy* whereas the second exergy definition implies that *exergy*, as a portion of *energy*, is always smaller than *energy*.

Pure exergy energies and physical exergy are core parts of the theory of exergy, as treated by Bejan et al. [13]. Substitution of Eqs. (115), (116), and (117) into (113) yields the exergy change between two states, state 1 and state 2, of a closed system

$$\begin{aligned} Ex_2 - Ex_1 &= (U_2 - U_1) + p_0(V_2 - V_1) - T_0(S_2 - S_1) + (KE_2 - KE_1) + (PE_2 - PE_1) \\ &= (E_2 - E_1) + p_0(V_2 - V_1) - T_0(S_2 - S_1) \end{aligned} \tag{118}$$

7.4 Thermodynamic Processes and Exergy Balance

Consider the application of energy balance and entropy balance to a closed system, respectively,

$$E_2 - E_1 = \int_1^2 \delta Q - \int_1^2 \delta W = \int_1^2 \delta Q - W \tag{22A}$$

$$dS \geq \delta_E S = \left[\int_A \frac{-J_Q}{T} dA \right] dt \tag{89A}$$

which may be written if a single value T is defined for the system as

$$S_2 - S_1 \geq \int_1^2 \frac{\delta Q}{T}$$

One may introduce *entropy generation* or *entropy growth*, S_G, as $S_2 - S_1 = \int_1^2 \frac{\delta Q}{T} + S_G$. The Clausius inequality becomes

$$S_G \geq 0$$

and the system entropy balance becomes

$$S_2 - S_1 = \int_1^2 \left(\frac{\delta Q}{T} \right)_b + S_G \tag{89D}$$

The exergy balance for a closed system is developed by substituting (118) into (22A)

$$(Ex_2 - Ex_1) - p_0(V_2 - V_1) + T_0(S_2 - S_1) = \int_1^2 \delta Q - W$$

With the replacement of $S_2 - S_1$ by using Eq. (89D) in the above equation

$$Ex_2 - Ex_1 = \left\{ \int_1^2 \left(1 - \frac{T_0}{T} \right) \delta Q - [W - p_0(V_2 - V_1)] \right\} - T_0 S_G \tag{119}$$

The term on the left side of Eq. (119) is the exergy change of the closed system. The terms on the right side represent the breakdown of the system exergy change dependent on processes: The first group represents two kinds of exergy transfers. The exergy transfer associated with the transfer of heat, E_q, is

$$Ex_q = \int_1^2 \left(1 - \frac{T_0}{T}\right)\delta Q$$

The exergy transfer associated with the transfer of work, E_w, is,

$$Ex_w = W - p_0(V_2 - V_1)$$

A positive work output of E_w is associated with a decrease in system exergy. The last term on the right side accounts for the destruction of exergy due to irreversibilities within the system as related to the entropy generation or entropy growth

$$Ex_D = T_0 S_G$$

In the literature, the expression of Ex_D is also known as the Gouy–Stodola theorem.

Equation (119) is the exergy balance for a closed system, which serves as the fundamental equation for exergy analysis of closed systems similar to that of Eq. (22) as the fundamental equation for energy analysis of closed systems. Like Eq. (22), which is the starting point of derivation for Eq. (111/119) in Sect. 7.2, Eq. (119) is the starting point of deriving governing equations for control volume balance analysis of open systems.

7.4.1 Control Volume Exergy Balance

Rewrite Eq. (119) as system exergy rate equation

$$\frac{dEx}{dt} = \sum_j \left(1 - \frac{T_0}{T_j}\right)\dot{Q}_j - \left(\dot{W} - p_0\frac{dV}{dt}\right) - \dot{E}x_D \tag{119A}$$

in which Ex is given as (118A)

$$Ex = [U - T_0 S + p_0 V]_A - [U - T_0 S + p_0 V]_0 + \frac{1}{2}mv^2 + mgz \tag{118A}$$

Apply the Reynolds transport theorem to the LHS to express system change in Ex in terms of control volume change in Ex and exergy flow across the boundary, cs, of the control volume,

$$\frac{\partial}{\partial t}\int_{cv} ex \cdot \rho d\mathcal{V} + \int_{cs} ex \cdot \rho\vec{V}\cdot\hat{n}\,dA = \sum_j \left(1 - \frac{T_0}{T_j}\right)\dot{Q}_j - \left(\dot{W} - p_0\frac{dV}{dt}\right) - \dot{E}x_D$$

The work term, $\dot{W}$, is treated, in the context of exergy analysis, as in Sect. 7.2

$$W = W_{shaft} + W_{normal}$$

where W_{normal} is the system mass expansion work. In the context of exergy analysis, we consider the useful work portion of this expansion work by dealing with the combination of W_{normal} and $-p_0\Delta V$. That is, the useful work exerted by the normal surface of a system mass is

$$W_{normal} - p_0\Delta V$$

Now, consider its rate. Between t and $t+\Delta t$, the change in volume of the system mass is

$$\Delta V = m_e v_e$$

and exergy transfer accompanying work is

$$W_{normal} - p_0 m_e v_e$$

Rate of exergy transfer accompanying work is

$$\lim_{\Delta t \to 0} \frac{W_{normal}}{\Delta t} - \lim_{\Delta t \to 0} \frac{m_e}{\Delta t} p_0 v_e$$

Noting that $W_{normal} = m_e p_e v_e$, as $\Delta t \to 0$, we can write "rate of exergy transfer accompanying normal work" as

$$\dot{m}_e(p_e v_e - p_0 v_e)$$

The mass specific exergy equals

$$ex = (u - u_0) + p_0(v - v_0) - T_0(s - s_0) + \frac{1}{2}v^2 + gz$$

Correspondingly,

$$\int_{cs} \rho(ex)\vec{V}d\vec{A} = -\sum_i \dot{m}_i ex_i + \sum_e \dot{m}_e ex_e$$

Note that the combination of $(pv - p_0 v)$ and $[(u - u_0) + p_0(v - v_0)]$ becomes

$$u + pv - (u + pv)_0 = h - h_0$$

The combination of surface force work term and specific exergy term is, therefore, the following known as *flow specific* exergy (*fe*),

$$fe = (h - h_0) - T_0(s - s_0) + \frac{1}{2}v^2 + gz \tag{120}$$

The exergy balance over a control volume is found to be

$$\frac{\partial Ex_{cv}}{\partial t} = \sum_j \left(1 - \frac{T_0}{T_j}\right)\dot{Q}_j - \dot{W}_{shaft} + \sum_i \dot{m}_i fe_i - \sum_e \dot{m}_e fe_e - \dot{E}x_D \tag{121}$$

7.5 Chemical Exergy and Exergy of Heat and Cold

The restriction in the consideration of our material system approaching restricted dead state, instead of dead state, was nowhere explicit in the above derivation. Therefore, both Eqs. (120) and (121) apply to thermal/mechanical processes as well as thermal/mechanical/chemical processes. The situation is exactly similar to the applicability of Eq. (111), which is applicable to all processes. Only in the use of Eqs. (120) and (121), it is necessary to specify what kind of equilibrium states is considered, the *restricted dead state* or *dead state*.

Rather, it is necessary to consider whether the material of flow exergy is already in internal chemical equilibrium or not. If it is, a material input is necessarily away from thermomechanical equilibrium with the surroundings to be useful because otherwise, the material system is in chemical equilibrium and thermomechanical equilibrium, i.e., the dead state.

If it is not, there are two possibilities: the possibility of chemical reaction toward internal chemical equilibrium and the possibility of a mixing process.

In the former case, what can happen is, first, the material system, which can often appropriately be treated as in thermomechanical equilibrium with the environment, will first approach internal chemical equilibrium resulting in thermomechanical disequilibrium. Then, the material in thermomechanical disequilibrium will approach the dead state in the long run. The problem *cannot* be treated as a problem of approaching a restricted dead state! The above equation derivation of approaching a restricted dead state, in fact, is a valid general derivation of approaching dead state.

There is one example of internal chemical disequilibrium that approaches thermomechanical equilibrium with the environment. That is, systems in chemical metastable equilibrium, e.g., a battery that discharges slowly resulting in small thermal disequilibrium, which would approach thermal equilibrium before the continuing chemical process building up thermal disequilibrium. That is, a slow chemical process-driven process toward a restricted dead state. This, though, is not an important application of Eq. (121).

The only approximation of approaching restricted dead state applies to the latter case of the mixing of nonreactive material components at thermomechanical disequilibrium. In this situation, it may be a useful approximation to consider the process of approaching a restricted dead state in the device defining control volume with individual components exiting from the device where, once outside the control volume, mixing of individual components can take place.

The required treatment-distinction between a nonchemical system and a chemical system is not by classifying physical exergy and chemical exergy according to Eq. (114), but interpreting $(h - h_0) - T_0(s - s_0)$, accordingly, dependent on whether the system is in internal chemical equilibrium or not. We reject Eq. (114), for such classification of exergy is philosophically problematic and pedagogically futile.

7.5.1 Energy and Exergy Equations for a Control Volume

We have then the general energy equation and the general exergy equation:

$$\frac{\partial E_{cv}}{\partial t} = \dot{Q}_{cv} - \dot{W}_{shaft} + \sum_i \dot{m}_i\left(h_i + \frac{V_i^2}{2} + gz_i\right) - \sum_e \dot{m}_e\left(h_e + \frac{V_e^2}{2} + gz_e\right) \tag{111}$$

$$\frac{\partial Ex_{cv}}{\partial t} = \sum_j \left(1 - \frac{T_0}{T_j}\right)\dot{Q}_j - \dot{W}_{shaft} + \sum_i \dot{m}_i fe_i - \sum_e \dot{m}_e fe_e - \dot{Ex}_D \tag{121}$$

where

$$fe = (h - h_0) - T_0(s - s_0) + \frac{1}{2}v^2 + gz$$

Equations (111) and (121) are the fundamental tools for engineering thermodynamic analysis. Equation (111), of course, is a universally used tool of mechanical engineers representing the first law of thermodynamics. Equation (121) is the corresponding version representing the entropy principle.

To the extent that "the goal of this essay-disquisition is to accord equal status to the first law and the second law," the fact that Eqs. (111) and (121) are similar in form suggests that one pedagogical way of achieving the goal is developing the practice of using Eq. (121) whenever Eq. (111) is used.

7.5.2 Relation of Eqs. (118A) and (121) to the Gibbs Free Energy

We shall first demonstrate the central role of Eqs. (111) and (121) in thermodynamic analysis by the latter equation providing unifying linkage between chemical engineering thermodynamics and mechanical engineering thermodynamics. Consider, first, the system rate version of Eq. (119), i.e., Eq. (119A),

$$\frac{dEx}{dt} = \sum_j \left(1 - \frac{T_0}{T_j}\right)\dot{Q}_j - \left(\dot{W} - p_0 \frac{dV}{dt}\right) - \dot{E}x_D$$

where

$$Ex = [U - T_0S + p_0V]_A - [U - T_0S + p_0V]_0 + \frac{1}{2}mv^2 + mgz$$

Specifically, we have a chemical closed system at $T_A = T_0$ and $p_A = p_0$, and the system is subject to the additional assumptions:

$$\dot{Q}_j = 0 \,(\text{thermally isolated})$$

$$dV/dt = 0 \,(\text{constant volume})$$

$$\frac{1}{2}mv^2 = 0$$

$$mgz = 0$$

The exergy expression and exergy equation can be reduced to

$$Ex = [U - T_0S + p_0V]_A - [U - T_0S + p_0V]_0 = [U - TS + pV]_A - [U - TS + pV]_0$$

$$= -\Delta G_A \text{ (the Gibbs free energy)}$$

$$\frac{dEx}{dt} = \frac{d(-\Delta G)}{dt} = -\dot{W} - \dot{E}x_D \text{ or } -\frac{d(-\Delta G)}{dt} = \dot{W} + \dot{E}x_D$$

That is,

$$-\Delta G_A = \int_A^{\Delta G=0} \dot{W}dt + \int_A^{\Delta G=0} \dot{E}x_D dt$$

Which, in the limit of $\int_{A}^{\Delta G=0} \dot{E}x_D dt = 0$ becomes,

$$W = \int_{A}^{\Delta G=0} \dot{W} dt = -\Delta G$$

the same relation as (109).

We now consider an open chemical system with chemical reactants input at $T_A = T_0$ and $p_A = p_0$. This can be an idealized fuel-cell system operating at steady-state operation, $\partial/\partial t = 0$. The system is subject to the additional assumptions

$$\dot{Q}_j = 0 \,(\text{thermally isolated})$$

$$\frac{1}{2}mv^2 = 0$$

$$mgz = 0$$

The general exergy equation for a control volume is

$$\frac{\partial Ex_{cv}}{\partial t} = \sum_j \left(1 - \frac{T_0}{T_j}\right)\dot{Q}_j - \dot{W}_{shaft} + \sum_i \dot{m}_i fe_i - \sum_e \dot{m}_e fe_e - \dot{E}x_D$$

Under the assumptions, the equation and the flow exergy expression reduce to

$$0 = -\dot{W}_{shaft} + \sum_i \dot{m}_i fe_i - \sum_e \dot{m}_e fe_e - \dot{E}x_D$$

and

$$fe = (h - h_0) - T_0(s - s_0) = g - g_0$$

That is,

$$0 = -\dot{W}_{shaft} + \sum_i \dot{m}_i (g - g_0)_i - \sum_e \dot{m}_e (g - g_0)_e - \dot{E}x_D$$

where

$$\sum_i \dot{m}_i (g - g_0)_i = \sum_{i\,of\,R0} \dot{m}_i (g - g_0)_i$$

In which *i of R*0 denotes components of reactant mixture at T_0 and p_0.

The term $\sum_e \dot{m}_e (g - g_0)_e$ can be treated in two options depending on the two choices of the control volume. One is a control volume that includes both the device and the volume exterior to the device exit that is sufficiently large so that exiting product components from the device outlet will be thoroughly mixed, as well as approaching thermal equilibrium with the $T_0 - p_0$ surroundings. Therefore,

$$\sum_e \dot{m}_e fe_e = \sum_{e\,of\,P0} \dot{m}_e fe_e$$

where the *e of P*0 denotes the exit product mixture components at T_0 and p_0.

The second choice is a control volume of the device the exiting product mixture from which is in internal chemical equilibrium but not in thermal equilibrium with the T_0–p_0 surroundings nor be thoroughly mixed. We shall denote the control volume *cv*1, while the control volume exterior to the device *cv*2. In this case

$$\sum_e \dot{m}_e fe_e = \sum_{e\,of\,P} \dot{m}_e fe_e$$

The exergy equation can be written, respectively, for *cv1* and *cv2* as

$$\dot{W}_{shaft} = \left[\sum_{i\,of\,R0} \dot{m}_i (g - g_0)_i - \sum_{e\,of\,P} \dot{m}_e (g - g_0)_e\right] - \dot{E}x_D$$

$$0 = \sum_{e\,of\,P} \dot{m}_e (g - g_0)_e - \sum_{e\,of\,P0} \dot{m}_e (g - g_0)_e - \left(\dot{E}x_D\right)_{irreversibility\ in\ the\ exterior}$$

In sum, the second exergy equation expresses the purely irreversible process outside the device as the exiting components complete their mixing process and thermal (typically cooling down) process. The first exergy equation expresses the electrochemical process driven by the Gibbs function term, $\left[\sum_{i\,of\,R0} \dot{m}_i (g - g_0)_i - \sum_{e\,of\,P} \dot{m}_e (g - g_0)_e\right]$, subject to the irreversibility loss of exergy destruction.

7.5.3 Exergy of Heat and Cold

We now consider an open system subject to a heat input, $\dot{Q}_H$, and a heat output, $\dot{Q}_C$ ($\dot{Q}_C$ is negative). For simplicity, we shall assume the heat output taking place at $T_C = T_0$. That is,

$$\sum_j \left(1 - \frac{T_0}{T_j}\right)\dot{Q}_j = \left(1 - \frac{T_0}{T_H}\right)\dot{Q}_H + \left(1 - \frac{T_0}{T_C}\right)\dot{Q}_C = \left(1 - \frac{T_0}{T_H}\right)\dot{Q}_H$$

The system has no mass exchange with the only exchange to be heat exchange: there is no flow exergy input or output. The exergy equation for a steady-state control volume is

$$0 = \left(1 - \frac{T_0}{T_H}\right)\dot{Q}_H - \dot{W}_{shaft} - \dot{E}x_D$$

or,

$$\dot{W}_{shaft} = \left(1 - \frac{T_0}{T_H}\right)\dot{Q}_H - \dot{E}x_D$$

which is the Carnot–Kelvin formula when the process involves no exergy destruction. Note that

$$\left(1 - \frac{T_0}{T_H}\right)\dot{Q}_H$$

represents the exergy of heat, $\dot{Q}_H$.

Furthermore, an alternative consideration of an open system that receives heat at $T_H = T_0$ leads to

$$\sum_j \left(1 - \frac{T_0}{T_j}\right)\dot{Q}_j = \left(1 - \frac{T_0}{T_H}\right)\dot{Q}_H + \left(1 - \frac{T_0}{T_C}\right)\dot{Q}_C = -\left(1 - \frac{T_0}{T_C}\right)\left(-\dot{Q}_C\right) = \left(\frac{T_0}{T_C} - 1\right)\left(-\dot{Q}_C\right)$$

It follows

$$\dot{W}_{shaft} = \left(\frac{T_0}{T_C} - 1\right)\left(-\dot{Q}_C\right) - \dot{E}x_D$$

That is, $([T_0/T_C] - 1)\left(-\dot{Q}_C\right)$ represents the exergy of cold, $\left(-\dot{Q}_C\right)$. Note that, in this case, the exergy can be greater than the energy, i.e., the cold $\left(-\dot{Q}_C\right)$. This is one example where the second exergy definition does not apply and, strictly speaking, the concept of exergy becomes problematic.

If we overlook the case of the exergy of cold, the above discussion shows that exergy analysis is consistent with the Gibbs function and the Gibbs free energy as well as with the Carnot–Kelvin formula as special cases, respectively. This demonstration encourages the notion that exergy analysis, though not perfect because of definitional contradiction (thus, it is not the final formulation of our analysis), is nonetheless a powerful tool for engineering practice.

7.6 Energy: Exergetic Content of Energy and the Definition of Energy

As long as both pure exergy energies and material exergy are core parts of the theory of exergy, the theory of exergy rests on both the first definition and the second definition with the aforementioned contradiction between them. The only escape out of the dilemma is for the first definition to accept the restriction implied in the second definition, $energy = exergy + anergy$. That is, the "energetic" interpretation of thermodynamics in accordance with MTH. This interpretation restriction is, in fact, the widely held understanding of thermodynamics since Kelvin. In this interpretation, energy is divided into four kinds inclusively

- Pure exergy energy: electrical energy, kinetic energy, and potential energy
- High exergy energy: high temperature heat, electrochemical systems
- Low exergy energy: low-temperature heat
- Zero exergy energy: heat of a body that is in thermodynamic equilibrium with its environment.

The common definition of energy, *energy is the capacity for doing work*, is clearly applicable only to the pure exergy energies, while applicable poorly to high exergy energies and inapplicable to low exergy energy and zero exergy energy. In contrast, relating *exergetic content of an energy system* to be *the capacity for doing work* is applicable to all four cases. I propose the following definition of energy:

> Energy is a conserved quantity that can be neither created nor destroyed; an important characteristic of the quantity is its exergetic content, which is measured in terms of the system capacity for doing work; although energy of a system and its interacting surroundings cannot be destroyed, the system-surroundings capacity for doing work is being lost incessantly.

Thermodynamics had begun as an engineering subject of the study of heat and work (see Fig. 7.1). Joule established the idea that both work and heat are forms of energy, which is universally conserved. Kelvin introduced the general idea of

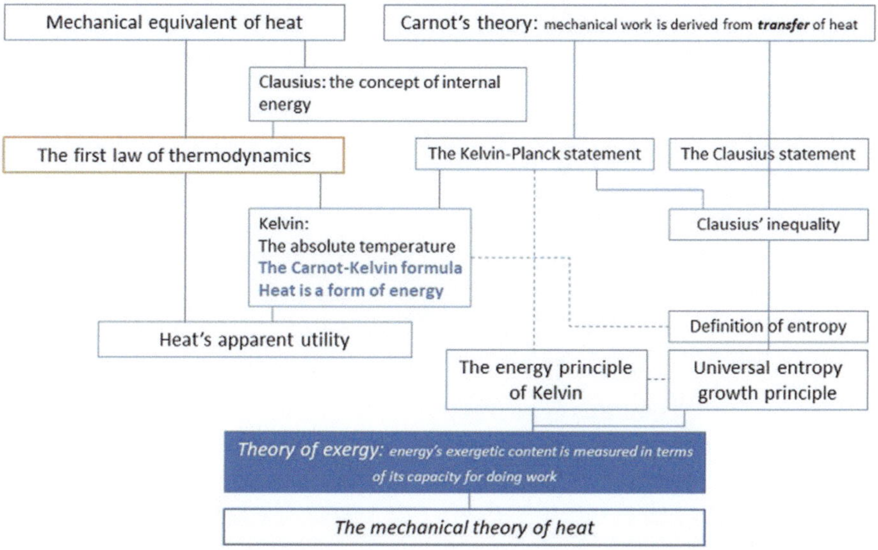

Fig. 7.1 Evolution of thermodynamic thought 3_Mechanical theory of heat (MTH): It identifies the premise of the theory of exergy to be Kelvin's energy principle. Kelvin introduced the general idea of available energy; Gibbs provided the general idea its concrete form, the Gibbs free energy; the formulation of the theory of exergy united stream of heat-energy-related ideas into one general framework completing a full circle. These advances became the core of MTH on how we view everything that happens in nature in terms of energy transformations, either spontaneously or in the reverse direction with a strict limitation

available energy; Gibbs provided the general idea its concrete form, the notion of the Gibbs free energy; finally, the formulation of the theory of exergy transformed/united streams of heat-energy-related ideas into one general framework completing a full circle. Energy and exergy are inseparable twins: one cannot talk about energy intelligently without introducing the concept of exergy.

That energy is a subtle concept was already evident in the necessity of its introduction in terms of five closely related steps/ideas in Table 3.1. Of the five ideas, the most notable is Statement #5, Table 3.1, which is reproduced here

> Energy can be neither created nor destroyed; only the form in which energy exists can be transformed from one form into another.

The meaning of which, the part of the transformation of energy forms, can only be understood in terms of the energy principle: one can place various forms of energy according to their "grade" as determined by exergetic content

Pure exergy energy

↓

Vert high exergy energy

↓

High exergy energy

↓

Moderate exergy energy

↓

Low exergy energy

↓

Heat (waste heat, i.e., zero exergy energy)

What is shown is energies with decreasing grade, from top to bottom. The energy principle or the theory of exergy holds that high-grade energy can be spontaneously transformed into low-grade energy as shown with the arrows. The original principle actually maintains that the tendency of such transformations is universal, but, as we already demonstrated, in Sect. 5.10, the tendency should be interpreted as spontaneous instead of universal. Whereas the first law permits transformation in both directions, from high grade to low grade as well as from low grade to high grade, the energy principle (or the entropy principle) holds such conversions in reverse direction (direction against the arrow) are subject to strict limitation.

The nature of this limitation, other than that reverse transformations are not spontaneous, will be very much a centric topic of our discussion in Chap. 8, as well as the energy doctrine truism: "Everything that happens involves conversion of energy." (Goran Wall, *EXERGETICS* Bucaramanga 2009 [17]).

References

1. Callen HB (1st edition, 1960; 2nd edition, 1985) Thermodynamics and an Introduction to Thermostatistics. Wiley, New York
2. Kondepudi D, Prigogine I (1998) Modern Thermodynamics: From Heat Engines to Dissipative Structures. Wiley, New York
3. Foley A (2008) Entropy, the second law and the concept of "bad" energy! Meeting the Entropy Challenge. AIP Conference Proceedings 1033 (see Ref. [12])
4. Daub EE (1970) Entropy and dissipation. Historical Studies in the Physical Sciences 2:321–354
5. Maxwell JC (1888) Theory of Heat. Dover, 2001 (p. 73)
6. Wang LS (2014) Entropy growth is the manifestation of spontaneity. J Thermodyn 2014 (9 pages; 7 December 2014) 387698

7. Gibbs JW (1961) The Scientific Papers of J. W. Gibbs, Vol. 1: Thermodynamics. Dover (p. 354)
8. Smith C, Wise MN (1989) Energy and Empire: A Biographical Study of Lord Kelvin. Cambridge Univ Press
9. Maxwell JC (1878) Tait's Thermodynamics II. Nature 17 (Feb. 7, 1878):278–280
10. Keenan JH (1941) Thermodynamics, ch. XVII. Wiley, New York (pp. 289–313)
11. Rant Z (1956) Exergie, ein neues Wort fur "Technische Arbeitsfahigkeit" (Exergy, a new word for "technical available work"). Forschung auf dem Gebiete des Ingenieurwesens 22:36–37
12. Meeting the Entropy Challenge (2008) AIP Conference Proceedings 1033. AIP, Melville, New York
13. Bejan A, Tsatsaronics G, Moran M (1996) Thermal Design and Optimization. Wiley, New York
14. Sciubba E, Wall G (2007) A brief commented history of exergy from the beginnings to 2004. Int J Thermodynamics 10(No. 1):1–26
15. Rant Z (1964) Exergie and Anergie. Wissenschaftliche Zeitschrift der TU Dresden 3 (No. 4):1145–1149
16. Baehr H D (1962) Thermodynamik: Eine Einführung in die Grundlagen und ihre technischen Anwendungen
17. Wall G (2009) EXERGETICS Bucaramanga 2009

The Second Law: The Entropy Growth Potential Principle and the Three-Place Relation in *Heat* Phenomena

8

Abstract

Perfection in Gibbsian equilibrium thermodynamics has led to the impression that the Kelvin–Clausius formulation of MTH is flawless. In fact, perfection exists only in equilibrium thermodynamics whereas the Kelvin–Clausius synthesis failed as a project of synthesis of Carnot's theory and MEH to formulate a theory that Carnot aspired to, i.e., engineering thermodynamics. An analysis of MTH's failure is given in this chapter. This analysis arrives at the conclusion that the fundamental error was the equating of correlation with causality, thus, giving heat (and energy) causal agency, which belongs to the second law of thermodynamics in terms of a newly introduced concept, entropy growth potential. Correspondingly, processes that are in the reverse direction to spontaneous processes should be, instead of subject to strict limitation, understood in terms of the triadic relationship.

Keywords

Energy conversion doctrine · The premises of MTH · Energetics · Poincare range · Systems approaching equilibrium with an external heat reservoir · Isolated systems approaching internal equilibrium (pure spontaneity) · Entropy growth in universe for spontaneous changes · Conceptual differentiation of entropy growth in universe (entropy growth potential) · The second law of thermodynamics · Predicative entropic theory of heat (PETH) · Triadic framework · The premises of PETH

L.-S. Wang, *A Treatise of Heat and Energy*, Mechanical Engineering Series,
https://doi.org/10.1007/978-3-030-05746-6_8

8.1 Introduction: The Energy Conversion Doctrine Truism

It is a universally accepted truism that "everything that happens involves conversion of energy," [1, 2] and that energy conversion processes are divided into two kinds: one kind in the direction of spontaneous changes from energy of high-grade form to energy of lower grade form eventually to heat, and the other in the reverse direction including the conversion of heat into mechanical energy; whereas the spontaneous or natural changes are limitless, the reverse-direction changes are subject to strict limitation.

How did we arrive at this understanding on energy? There were three critical steps. First, by 1850, Joule had succeeded to demonstrate equivalence correlation between consumed mechanical work and produced heat [3] and, additionally proposed equivalence between "consumed" heat and produced mechanical work. The equivalence principle (or, MEH) is not, however, an interconvertibility principle. Carnot, earlier in his private writing, had argued that motive power could not be created by mere destruction, or consumption, of the heat of bodies [4].

What happened in 1850–1854 was the second step: Joule, Clausius, and, finally, Kelvin accepted that heat was a form of energy; with the "doctrine of heat energy," they made the *inference* that the doctrine of heat being a form of energy infers "equivalence of heat and work" to be synonymous with "convertibility of heat and work" (see Sect. 4.8). The MEH was thereby upgraded to the interconvertibility principle: evidence of MEH correlation was accepted as evidence of causation. Any doubt Carnot had had was swept away, as interconvertibility became the defining characteristic of thermodynamics as Fermi put it, "*Thermodynamics is mainly concerned with the transformations of* heat *into mechanical work and the opposite transformations of mechanical* work *into heat*" [5]. The principle has the merit of identifying in heat its apparent utility, which together with the discovery of petroleum as a new source of heat made possible the Second Industrial Revolution.

The third step was the result of Kelvin's energy principle, which expanded the meaning of the interconvertibility principle. We noted in Chap. 4 that the energy principle itself was a self-evident proposition as declared by Kelvin in three General Conclusions (only the first two, are reproduced here),

General Conclusion 1. There is at present in the material world a universal tendency to the dissipation of mechanical energy.

General Conclusion 2. Any *restoration* of mechanical energy, without more than an equivalent of dissipation, is impossible in inanimate material processes, and is probably never effected by means of organized matter, either endowed with vegetable life or subjected to the will of an animated creature.

The truism on energy is based on Kelvin's declaration of universal dissipation of mechanical energy into heat and, once dissipated, the dissipated mechanical energy cannot be restored in accordance with General Conclusion 2. The claim of universality amounts to the proposition that the phenomena of entropy growth are *exhausted* by the dissipation of high-grade energy to low-grade energy, i.e., "everything that happens involves entropy growth" is synonymous with "everything that happens involves conversion of energy."

The restoration process is an example of energy conversion in the reverse direction. While the spontaneous transformation from work to heat is limitless, convertibility from heat to work is subject to strict limitation because, otherwise, an unrestricted reverse transformation would have contradicted the Kelvin–Planck statement that heat in itself cannot be converted into work. This was how Kelvin and Clausius incorporated the lesson from the Carnot heat engine to conclude as shown in Sect. 4.6 that

GS-4.b. Heat cannot be converted 100% into work.

But, are the interconvertibility principle and General Statement-*4.b* true? The inference was what Carnot had difficulty with; he did not have a problem of accepting the doctrine of heat energy and MEH (as his unpublished writing [4] showed) and certainly did not deny the role of heat as falling caloric in the production of work. But, he had difficulty with the idea of heat expenditure alone without the transfer of heat as the cause for the production of work.

Carnot's doubt did not die completely. Many have since questioned the validity of universal interconvertibility with regard to heat as a cause of work [6–12]. Yet, none of these arguments has succeeded in refuting universal interconvertibility. One reason is that most of those who argued against interconvertibility advocated a return to the idea of caloric, i.e., against the doctrine of heat energy. That obviously was a mistake.

The attraction of returning to the idea of caloric, wrong though it is, does provide a hint: what we are looking for is the real "drive" of change or process, if not heat.

For facilitating the following discussion, a summary list of the conventional takeaways of MTH is given as the following *general statements* (GSs):

GS-1. Heat's ontology. Heat is a high-entropy form of energy.

GS-2. Equivalence principle. Equivalence exists between consumed mechanical energy and produced heat, and consumed heat and produced mechanical energy.

GS-2.a. Inter-convertibility principle. Heat energy and mechanical energy are universally convertible to each other.

GS-4 or GC of Kelvin. Energy conversion doctrine. Entropy growth is *exhausted* by the spontaneous dissipation of high-grade energy; everything that happens is the result of energy and its transformation from one form to another; Conversion of energy can be classified into two kinds: natural energy conversion and reverse energy conversion.

GS-4.a. or the amended GC of Kelvin. Spontaneous energy conversion: Natural energy conversions are spontaneous and limitless.

GS-4.b. Reverse energy conversion. Reverse energy conversions are subject to strict limitation and heat cannot be converted 100% into mechanical energy.

MTH was the outcome of the Kelvin–Clausius synthesis project with strong energy bias in its effort in incorporating Carnot's theory. This energy bias served well the expediency that has powered industrialization of the past 150 years. Also, it is remarkable that this energy bias does not negatively impact the theory of equilibrium thermodynamics, which is a perfectly coherent system. The reason for that is that energy bias is never explicit in equilibrium thermodynamics. In contrast, energy bias is explicit and front-and-center in *engineering thermodynamics*.

That is because the key question in engineering thermodynamics is what drives the universe. Interconvertibility, or GS-2.a, and energy conversion doctrine, GS-4, appear to provide answer to that question: there is an apparent utility in high temperature heat; or better yet, the idea that energy (see discussion in Chap. 7, as well as discussion on energetics below) is the driver of the universe. In this simplistic view of energy-the-driver (or apparent utility of heat), there is an elephant in the room clothed in entropy growth. Without a clear conception of the elephant, the "standard approach" (see also paragraph below) often views that entropy growth only hinders that drive by degrading energy: "Entropy and dissipation of energy are as inseparable as Siamese twins in the thought of every student of thermodynamics…" [13].

It was not too long ago that Ilya Prigogine recounted the experience of having to overcome this mindset of the standard approach before he came up with the concept of *dissipative structures* [14] as an alternative to the standard approach,

> Among all those perspectives opened by thermodynamics, the one which was to keep my interest was the study of irreversible phenomena, which made so manifest the "arrow of time". From the very start, I always attributed to these processes a constructive role, in opposition to the standard approach, which only saw in these phenomena degradation and loss of useful work…The fact is that it appeared to me that living things provided us with striking examples of systems which were highly organized and where irreversible phenomena played an essential role. Such intellectual connections, although rather vague at the beginning, contributed to the elaboration, in 1945, of the theorem of minimum entropy production, applicable to non-equilibrium stationary states…From the very beginning, I knew that the minimum entropy production was valid only for the linear branch of irreversible phenomena…thus, the question was: What about the stationary states far from equilibrium…Those problems had confronted us for more than twenty years, between 1947 and 1967, until we finally reached the notion of "dissipative structure". [15]

As shown in Chap. 7 and will be in Chap. 9, the entropy principle was successfully incorporated as the centerpiece in equilibrium thermodynamics, which, as its postulational treatments including Callen's treatment [16] made it clear, is a theory that is free of its inductive development of the pre-1865 MTH thermodynamics. But, the assimilation of the entropy principle in engineering thermodynamics in the form of the theory of exergy (Chap. 7) is not satisfactory—which, the exergy theory, remains completely within the pre-1865 framework of MTH and fails to challenge the MTH premises of GS-2.a and GS-4. A suggestion has been made in a recent paper [17] that it is necessary to refute universal interconvertibility, GS-2.a, by assimilating the entropy principle as the centerpiece of engineering thermodynamics, with a clean slate.

8.1.1 Energy Conversion Doctrine and Energetics

This chapter, correspondingly this essay, aims for claiming for the entropy principle its centerpiece placement in engineering thermodynamics pushing aside energy conversion doctrine. Energy conversion doctrine originated in Kelvin's treatment of the second law in terms of available energy *and* the doctrine survived the later introduced entropy principle because that principle has not been assimilated into engineering thermodynamics to updating Kelvin's handling of the second law. That said, the universal popularity enjoyed by energy conversion doctrine may not be possible without one late nineteenth-century development, the short-lived challenge to mechanism and atomism, energetics or *energeticism*, a metaphysical-scientific doctrine that regards energy as the only reality and to view all changes in the world as movements or transformations of energy.

Energetics, the science of energetics, was mainly promoted by Wilhelm Ostwald and Georg Helm and energeticism, the ontology of reality, was promoted by Ostwald and Ernst Mach. The movement can be dated from the first publication of Helm [18] in 1887 to the year of Ostwald's concessional acceptance of micro (atomic) structure of matter in 1908 (thus, the ontology of the mechanical theory triumphed over Ostwald's assertion of energy to be the only reality), a duration barely of two decades. Yet, a movement of such short duration and involving a small number of, though colorful and prominent, participants had significant impacts.

The impact on thermodynamics was a short-lived one. Energeticists maintained that the entropy was not needed in the formulation of the second law and, furthermore, "for Ostwald, thermodynamics is confined to the reciprocal transformations of heat and mechanical energy; it is just one 'chapter' of the more comprehensive science of energetics…the second law of energetics, which is based on the intensity law, is more general than the second law of thermodynamics since it governs all natural 'happening'…" [19]. There was a noisy controversy between energetics, represented by Ostwald and Helm, and thermodynamics, favored by Boltzmann [20]. As well as, a rich correspondence between Ostwald and Planck between October 1891 and July 24, 1898 [19]. In those exchanges, Boltzmann and Planck's view prevailed over energeticists' and, today, entropy and the statistical-molecular approach are universally accepted with energeticism mere a footnote in the history of the science of energy.

But, this small footnote turns out to have a long life. Any consideration of energeticism and its total impacts must be viewed in terms of three elements: its technical-scientific ambition of absorbing thermodynamics as its subdiscipline, its metaphysical ambition, and its narrative ambition of capturing public discourse.

The technical-scientific ambition of energetics was indeed demolished by Boltzmann and Planck in their interaction with energeticists. Thermodynamics and, especially, the second law emerged as universally valid tools for our conception of nature, in such technical-scientific toolbox one can barely find energetics.

Energeticism, though, is more successful with its metaphysical ambition. Because of a residual resistance to scientific materialism among many, the energetic worldview has retained supporters, notable ones include E. Mach, A.J. Lotka, H.T. Odum. One may surmise that the collapse of its technical-scientific ambition should have invalidated its metaphysical claim. The truth is that even though metaphysics is distasteful to many, the energetic worldview legitimizes the widely popular narrative that the world runs on energy conversion, a legitimization stronger than Kelvin's self-evident General Conclusions can provide. Because of that connection, both the metaphysics and popular narrative are active currently.

Evidence for the latter can be found in bestseller books. One example is the 2017 bestseller, *Energy and Civilization, A History* [21] by Vaclav Smil. In the book, he wrote, "The history of successive civilizations, the largest and most complex organisms in the biosphere, has followed this course. Human dependence on ever higher energy flows can be seen as an inevitable continuation of organismic evolution. Wilhelm Ostwald was the first scientist to expand explicitly the second law of energetics to all and any action…" [21: 2]. Ostwald, Lotka, and Odum's influence lives on for popularizing energy conversion doctrine. This chapter gives a critical assessment of this doctrine and offers a entropy centric doctrine in its place.

8.2 Laws of Balance and the Calculation of Entropy Production

The formulation of a new theory of heat can be facilitated with the modern formalism of the second law, which was presented in Sect. 6.4. It is reproduced here in the context of presenting both the first law and the second law together.

The modern formalism of the Brussels School of thermodynamics [22–24] treats entropy as a balance quantity and its production to be always positive. Therefore, both the treatment of energy (the first law) and that of entropy (the second law) bear a similar pattern.

We first consider the first law as the *law of the balance of energy* for a system

$$dE = \delta_E E + \delta_G E$$

Because energy can be neither generated (nor destroyed), the energy generation term becomes

$$\delta_G E = 0$$

and, the energy exchange exists in two forms, i.e.,

$$\delta_E E = \delta Q - \delta W$$

Substitution of both expressions into $dE = \delta_E E + \delta_G E$ yields the first law in the familiar form,

$$dE = \delta Q - \delta W \tag{23A}$$

Or, when kinetic energy and potential energy are negligible, $dE = dU$,

$$dU = \delta Q - \delta W \tag{23}$$

Consider the second law as the *law of the balance of entropy* for a system [22, 24]

$$dS = \delta_E S + \delta_G S \tag{84B}$$

With this starting point, the modern formalism formulates—instead of the entropy principle, i.e., the principle of the increase of entropy (or, the first entropy principle of the second law)—a new version of the second law, which has been called the *second entropy principle* of the second law,

$$\delta_G \geq 0 \tag{88B}$$

i.e., entropy cannot be destroyed and in a real irreversible process the entropy production is always positive. Entropy flow, J_S, is related to heat flow, J_Q, under the condition of local thermodynamic equilibrium, which is the fundamental assumption of the modern formalism [22:345],

$$\frac{\overrightarrow{J_Q}}{T} = \overrightarrow{J_S}$$

It follows, therefore,

$$\delta_E S = \left[\int_A \frac{-\overrightarrow{J_Q}}{T} \cdot \overrightarrow{dA}\right] dt \approx \frac{\delta Q}{T} \tag{86A}$$

The last equality of Eq. (86A) holds, approximately, if T of the system is uniform. Equations (23A), (84B), (88B), and (86A) are the fundamental equations of thermodynamics; Eq. (88B) determines the preferred direction of any spontaneous natural process.

In the cases of finite changes $A \rightarrow B$, Eqs. (23A) and (84B) become

$$\Delta E \equiv E_B - E_A = Q - W \tag{23B}$$

$$\Delta S = S_B - S_A = \Delta_E S + \Delta_G S \tag{84C}$$

8.2.1 Calculation or Determination of Entropy Production

Entropy production is the primary concept in modern formalism. Entropy production can be calculated *explicitly* in terms of phenomenological relationships, e.g., Fourier's law of conduction, between "thermodynamic forces" and "thermodynamic flows" in the form $\delta_G S = FdX$ in which F is the thermodynamics force, e.g., temperature gradient, and dX is the corresponding thermodynamic flow, e.g., heat flow. See, for instance, examples in [25:121–127] and in [25: Chap. 11]. This explicit procedure is applied to individual problems when phenomenological relationships are known and solutions to the phenomenological "governing equations" are obtained.

Alternatively, when we take laws of balance as the primary concept, entropy production $\delta_G S$ can be *implicitly* determined by substituting "the entropy exchange $\delta_E S$ values from Eq. (86A)" and "the system entropy change dS values" from the application of thermodynamic property relationships into Eq. (84B)

$$dS = \delta_E S + \delta_G S$$

The implicit procedure is especially powerful (no need for solving the phenomenological governing equations), and well suited for making inferences based on theoretical *thought experiments*. It will be extensively used in the following discussion.

8.3 The Entropic Drive Corollary

This section and the next consider the production of mechanical work derived from a system's tendency toward equilibrium. In this section, the tendency toward equilibrium is the result of the system's interaction with its surroundings (a surrounding heat reservoir). In the next section, it is the result of a system approaching internal equilibrium.

Henri Poincaré (1854–1912)

Consider a system and its two end states, A and B. The preferred direction of any spontaneous natural process (for instance, in the direction of $A \rightarrow B$) is determined by the condition of positive entropy production. The entropy law asserts that entropy cannot be destroyed, but makes no claim with regard to how large entropy production must be as long as it remains nonnegative. Significantly, a system change between the same $A \rightarrow B$ can take place under a perfect management for achieving the reversibility limit with vanishing entropy production theoretically—only a change of the nature beyond the limit is physically impossible. By setting the limit to what is physically possible in an irreversible world, the second law suggested possibilities that had been hitherto unrecognized in a reversible dynamical world. That insight was realized long ago by Poincaré

> [These thermodynamic laws] can have only one significance, which is that there is a property common to all possibilities; but in the deterministic hypothesis there is only a single possibility, and the laws no longer have any meaning. In the indeterministic hypothesis, on the other hand, they would have meaning, even if they were taken in an absolute sense; they would appear as a limitation imposed upon freedom (Poincaré, 1913, pp. 122–123 [26]).

The more interesting point in Poincaré's comment on the limitation-of-possibilities is not the limitation imposed on the freedom, but the existence of this freedom. This freedom manifests in thermodynamic drive force that can be managed for a range of event possibilities. The possibility and necessity of exercising management are put in this way by Lotka, "The two fundamental laws of thermodynamics are, of course, insufficient to determine the course of events in a physical system. They tell us that certain things cannot happen, but they do not tell us what does happen" [27]. That is, the mathematical expression of the two laws of thermodynamics is not in the form of *governing equations*, which do determine what happens. For thermodynamic systems, *management* or *action*, in addition to the mathematical expression of the two laws, is what makes non-spontaneous events happen.

Poincaré's "single possibility" is the spontaneous natural event (limit), one reference limit of all possible $A \rightarrow B$ processes. The other limit is the reversible $A \rightarrow B$ event. We consider the system in the context of a T_0-environment-reservoir. Let heat exchange of the spontaneous natural event between the (fuel-and-equipment) system and the reservoir to Q_{spon}, which is defined by convention here to be positive if the system gains heat from the reservoir and correspondingly negative if heat is released to the reservoir. Let heat exchange of the corresponding reversible event between the system and the reservoir to be Q_{rev}. And let heat exchange (between the system and the reservoir) of an arbitrary real process (among infinitely many possible $A \rightarrow B$ processes) to Q. The first and second laws infer the following corollary, the *entropic drive corollary* [28]:

> *An event of a system between two end states is characterized by the system-T_0 reservoir heat exchange of the event. The useful range of possible events* (called the Poincare range) *is defined by the spontaneous limit and the reversible limit, with the (system-T_0 reservoir)*

heat *exchanges of the three events (the spontaneous limit, an arbitrary managed-event, and the reversible limit) obeying the inequality,*

$$Q_{spon} \leq Q < Q_{rev} \tag{122}$$

The difference between the system-T_0 reservoir heat exchanges of the two limits, $Q_{rev} - Q_{spon}$, is the reversible free heat *(the reversible* work*) of the system-reservoir,*

$$\left(W_{useful}\right)_{rev} = Q_{rev} - Q_{spon} = T_0(\Delta_P S)_{universe} \tag{123}$$

the second equality of which infers that the thermodynamic drive force is directly related to the entropy *growth potential, not to energy.*

The proof of Eqs. (122) and (123) is as follows. In a spontaneous event, the system undergoes change without interacting with a *work reservoir* or producing useful work that is directly consumed for a purpose. The system may interact with the heat reservoir mechanically, but such mechanical interaction involves neither entropy flow into or out of the heat reservoir nor the production and consumption of *useful work*. The first law assumes the form

$$U_B - U_A = Q_{spon} - W_{mechanical}$$

We now consider a managed event of the system in the direction of a spontaneous process, defined by the same end states of $A \rightarrow B$, which delivers W_{useful} to a work reservoir. The first law becomes for the present managed event

$$U_B - U_A = Q - \left[W_{mechanical} + W_{useful}\right]$$

In the managed process of the system between the same end states, both the system entropy change, $S(B) - S(A)$, and the system internal energy change, $U(B) - U(A)$, remain the same as the that of the spontaneous event. The mechanical work interaction with the heat reservoir also remains the same since the system volume change $V(B) - V(A)$ is the same as well. It follows, therefore, that the *positive* useful work in a managed process is

$$W_{useful} = Q - Q_{spon} \tag{124}$$

Since in the useful range only a positive W_{useful} is considered, Q is expected to satisfy the inequality

$$Q_{spon} \leq Q \tag{125}$$

(It does not mean that a negative W_{useful}, *in this case,* is not possible physically. It only means that such events serve no useful purpose, thus, of no interest in our present consideration.) If the managed event is a perfectly managed reversible event

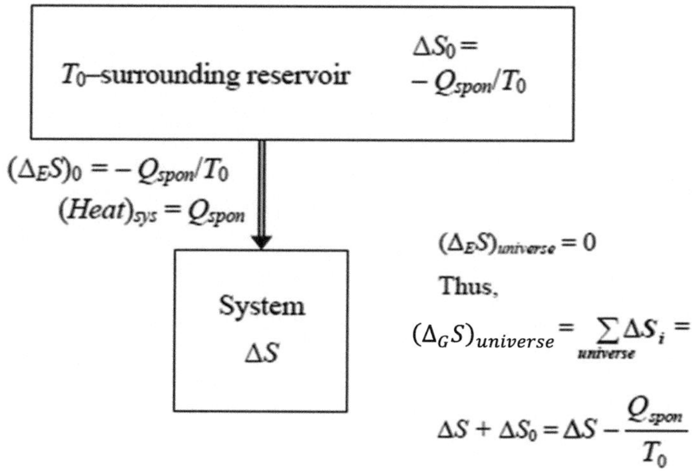

Fig. 8.1 The "universe" of a spontaneous event

$$\left(W_{useful}\right)_{rev} = Q_{rev} - Q_{spon} \tag{124A}$$

This value, $Q_{rev} - Q_{spon}$, can be shown to be related to the "entropy generation" of a spontaneous natural event. First, consider the spontaneous event shown schematically in Fig. 8.1, where the system entropy change is $\Delta S = (S_B - S_A)$, the entropy change of the surrounding reservoir is $\Delta S_0 = -Q_{spon}/T_0$. Since the entropy exchange of the universe $(\Delta_E S)_{universe}$ is by definition zero, the entropy "generation" (we are not sure of it being positive yet) of the universe equals the total entropy change of the universe according to Eq. (84C)

entropy generation in the universe = entropy change of the universe

That is,

$$(\Delta S)_{universe} = \Delta S + \left(-Q_{spon}/T_0\right)$$

the sum of the system entropy change ΔS and the entropy change of the reservoir $\Delta S_0 = -Q_{spon}/T_0$. Note that even though the large dissipative process of entropy production is taking place within the transmission medium in Fig. 8.1, the entropy change of the medium is zero in the above expression due to the assumption of the infinitesimal medium mass.

A *spontaneous natural* event must be associated with positive entropy generation. Whether transformation direction $A \rightarrow B$ or $B \rightarrow A$ will take place spontaneously is dependent on which transformation results in entropy increase. Assuming $A \rightarrow B$ to be the case

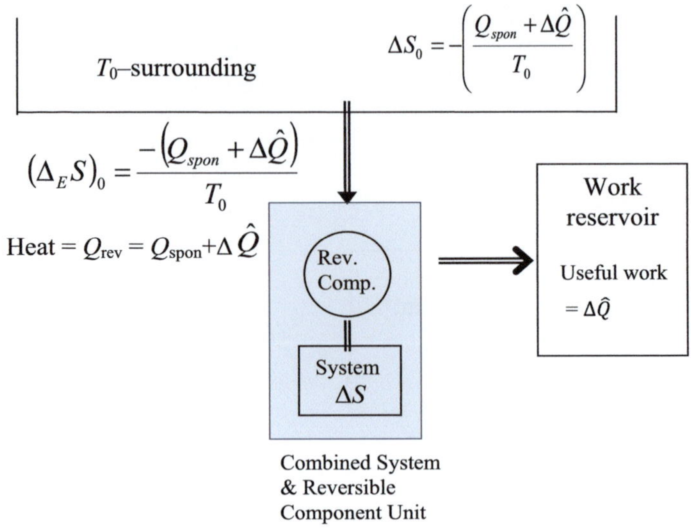

Fig. 8.2 The "universe" of a reversible event, in which the system undergoes managed change between the same end states of the system as shown in Fig. 8.1

$$(\Delta S)_{A\to B} + (-Q_{spon}/T_0)_{A\to B} \geq 0$$

This total entropy growth or generation of the spontaneous transformation, which is positive definite, is denoted by the shorthand notation, $(\Delta_G S)_{universe}$,

$$(\Delta S) + (-Q_{spon}/T_0) \equiv (\Delta_G \boldsymbol{S})_{\boldsymbol{universe}} \tag{126}$$

Consider a **corresponding reversible event** of the system reservoir (which now includes a reversible component and a work reservoir) with the system undergoing the same $A \to B$. The system entropy change, $S(B) - S(A)$, remains the same, and there is no entropy change in the reversible component and in the work reservoir. The entropy change in the new reversible universe can be different from that in the original universe, therefore, only if entropy change in the surrounding heat reservoir

$$(\Delta S_0)_{rev} = (-Q_{rev}/T_0)$$

as shown in Fig. 8.2, differs from that in Fig. 8.1

$$(\Delta S_0)_{spon} = (-Q_{spon}/T_0)$$

in the original universe of Fig. 8.1.

Therefore, the objective of the *reversible component* is to enable the extraction of an additional heat $\Delta\hat{Q}$ from the surrounding reservoir

$$Q_{rev} - Q_{spon} \equiv \Delta\hat{Q} \tag{127}$$

so that the heat reservoir entropy change decreases by a necessary amount (if Q_{spon} is negative in the original spontaneous case, Q_{rev} becomes a smaller negative or even a positive; if it is originally positive, it becomes a greater positive number). The limiting criterion of decrease in the reservoir entropy is to achieve the condition of zero entropy generation for the reversible universe

$$0 = (\Delta_G S)_{reversible-universe} = (\Delta S) + (-Q_{rev}/T_0)$$

Correspondingly,

$$Q_{rev} = T_0 \Delta S$$

Note that the reversible limit may include different reversible processes, all of which, however, correspond to the same Q_{rev} according to equation Eq. (127A). In this sense, therefore, we are speaking of a single reversible $A \rightarrow B$ limit.

With the substitution of $Q_{rev} = T_0\Delta S$ into Eq. (127), the *reversibly extracted additional heat*, $\Delta\hat{Q}$, in view of Eq. (126) becomes

$$\begin{aligned} \Delta\hat{Q} &= T_0\Delta S - Q_{spon} \\ &= T_0\left(\Delta S + \frac{-Q_{spon}}{T_0}\right) = T_0(\Delta_G S)_{universe} \end{aligned} \tag{127A}$$

In reference to the spontaneous event, the reversible $A \rightarrow B$ event extracts an additional heat, the value of which is directly related to spontaneous $A \rightarrow B$ entropy generation. The entropy law assertion of positive entropy generation thus infers $\Delta\hat{Q} \geq 0$ or

$$Q_{spon} \leq Q_{rev}$$

and $\Delta\hat{Q}$ is the maximum possible extracted heat

$$Q - Q_{spon} \leq \Delta\hat{Q} = Q_{rev} - Q_{spon}$$

which is an inference resulting from the maximum work theorem in accordance with Carnot's principle

$$W_{useful} \leq (W_{useful})_{rev} \tag{128}$$

That is,

$$Q \leq Q_{rev} \tag{128A}$$

In sum, the first part of the corollary, Eq. (122), results from Eqs. (125), and (128A); the second part, Eq. (123), results from Eqs. (124A), (127), and (127A)

$$\left(W_{useful}\right)_{rev} = \Delta\hat{Q} = T_0(\Delta_G S)_{universe}$$

Note that whereas in Eq. (126) entropy growth, $(\Delta_G S)_{universe}$, stands originally for entropy generation as a manifestation of dissipative processes in the universe, here in Eq. (123) the same term, quantitatively, represents the production of reversible work as a manifestation of the constructive role of "entropic" events in the universe. Rather than entropy *growth* as it is originally called, the term in the context in Eq. (123) should be more properly called by the name of *entropy growth potential*, and thus,

$$(\Delta_{\boldsymbol{G}} S)_{universe} = (\Delta_{\boldsymbol{P}} S)_{universe} \tag{129A}$$

That is, the above equation becomes

$$\left(W_{useful}\right)_{rev} = \Delta\hat{Q} = T_0(\Delta_P S)_{universe} \tag{123}$$

8.4 Entropic Drive Corollary for Isolated Systems: Pure Spontaneity

We now consider isolated composite systems with a tendency towards internal equilibrium. The principle of the increase of entropy, Eq. (74), directly applies, and the only change in entropy in the universe is that of the isolated systems

$$(\Delta S)_{iso-sys} = \Delta S \geq 0 \tag{74}$$

where ΔS is the system entropy change when the system reaches internal equilibrium. The approach toward equilibrium for isolated systems involves no interaction with a heat reservoir.

In view of Eq. (123), an intriguing question is whether the concept of entropy growth potential is applicable to isolated systems

$$\left[(\Delta_{\boldsymbol{G}} S)_{universe} =\right](\Delta S)_{iso-sys} = (\Delta_{\boldsymbol{P}} S)_{universe} \tag{129B}$$

In that case, since a spontaneous change in an isolated system does not involve interaction with a reservoir, the system entropy growth potential can extract heat from heat reservoir of any temperature, i.e., reversible work becomes, as it was first shown in [29]

$$W_{rev} = T_{reservoir}\Delta S \tag{130}$$

While $(\Delta_P S)_{univ}$ in Eq. (123) depends on T_0, ΔS in Eq. (130) is independent of any reservoir temperature. Therefore, if Eq. (129B) applies, Eq. (130) would be applicable for any available reservoir (see examples below) at arbitrary temperature as long as the isolated system returns to its natural final end state.

Here, in this section, two examples as thought experiments are investigated for establishing the validity of Eq. (130).

The first example is the free expansion process. The process was quoted by Planck to argue against GS-4.b (see Sect. 8.7). The same point was made by Fermi in an expanded form in the echo of Planck's observation

> An essential part of Lord Kelvin's postulate is that the transformation of the heat into work be the *only* final result of the process. Indeed, it is not impossible to transform into work heat taken from a source all at one temperature provided some other change in the state of the system is present at the end of the process. Consider, for example, the [spontaneity-driven] isothermal expansion of an ideal gas that is kept in thermal contact with a source of heat at the temperature T [see Problem 6.3]. Since the energy of the gas depends only on the temperature, and the temperature does not change during the process, we must have $\Delta U = 0$. From the first law, Eq. (15), we obtain, then, L [i.e., W]= Q. That is, the work, L, performed by the expanding gas is equal to the heat Q which it absorbs from the source. There is thus a complete transformation of heat, Q, into work L. This, however, is not a contradiction of Kelvin's postulate, since the transformation of Q into L is not the only final result of the process. At the end of the process, the gas occupies a volume larger than it did at the beginning [5: in a Footnote on p. 30].

A perfect example of what Fermi and Planck had in mind is the spontaneous *Joule free expansion* and its transformation into a corresponding reversible event.

The following is a description of the "continuous free expansion," a version of *free expansion* given in [16:73, 99]. Consider a composite system consisting of two compartments of equal volume (the two compartments are separated by a piston with perfect seal). The first compartment is filled with an ideal gas at T_1 and the second is evacuated at vacuum (Fig. 6.1). If the ideal gas is permitted to push against the piston and expand into the evacuated compartment region, thereby increasing its volume from V_1 (V_A) to $2V_1$ (V_B). If the exterior walls of the composite system are rigid and adiabatic, thus no heat exchange and no work during the quasi-static free expansion occurs. It follows

$$U_B - U_A = Q - W = 0$$

Therefore, the temperature of the ideal gas remains constant at T_1, and the corresponding system (ideal gas) entropy change is

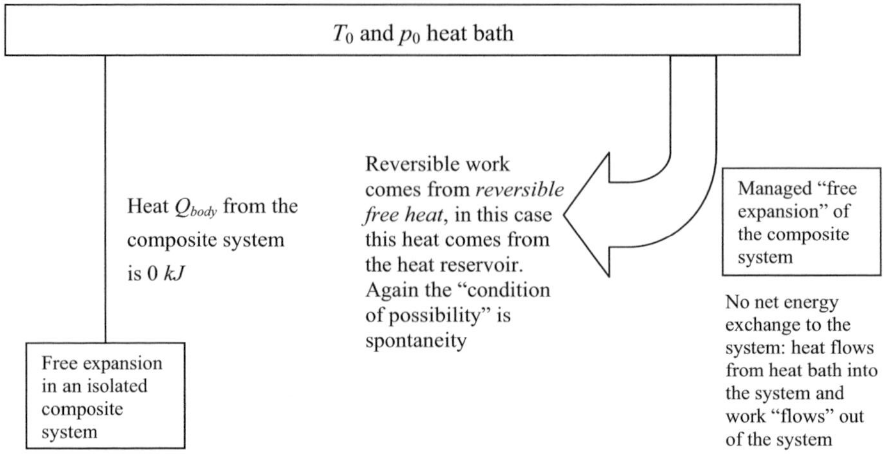

Fig. 8.3 Spontaneous event of a free expansion composite system on left; reversible mechanism (piston-cam-weight) managed reversible event on right. The figure depicts the case of the composite system temperature, T_1 to be equal to the heat bath temperature T_0

$$\Delta S \equiv S_B - S_A = Nc_V ln\frac{T_B}{T_A} + NRln\frac{V_B}{V_A} = NR\ln 2$$

Imagine that in the reversible event as described by Fermi the piston is now connected to a cam-weight (work reservoir) mechanism that balances the force exerted by the gas on the piston—and the walls of the system are now diathermic and the whole composite system is submerged in a heat reservoir/bath at T_1 (see Fig. 6.2 and Fig. 8.3). Repeat the consideration of gaseous expansion to double the initial volume. The near balance of net force on the piston results in slow expansion of gas along the same quasi-static path of the original continuous free expansion. Note that the process in the present case is reversible due to the nearly balanced net force and the slowness of expansion and correspondingly the slow heat extraction from the T_1 bath, the amount of which is

$$Q_{Rev} = \int_{V_1}^{2V_1} TdS = T_1 \Delta S = NRT_1 \ln 2$$

A simple calculation of the resulting isothermal gaseous expansion shows

$$W_{Rev} = \int_{V_1}^{2V_1} pdV = NRT_1 \int_{V_1}^{2V_1} \frac{dV}{V} = NRT_1 \ln 2 = \Delta \hat{Q}$$

The schematic of the heat balance is shown as Fig. 8.3. Reversible work is *reversible free* heat, this heat comes from the heat bath/reservoir.

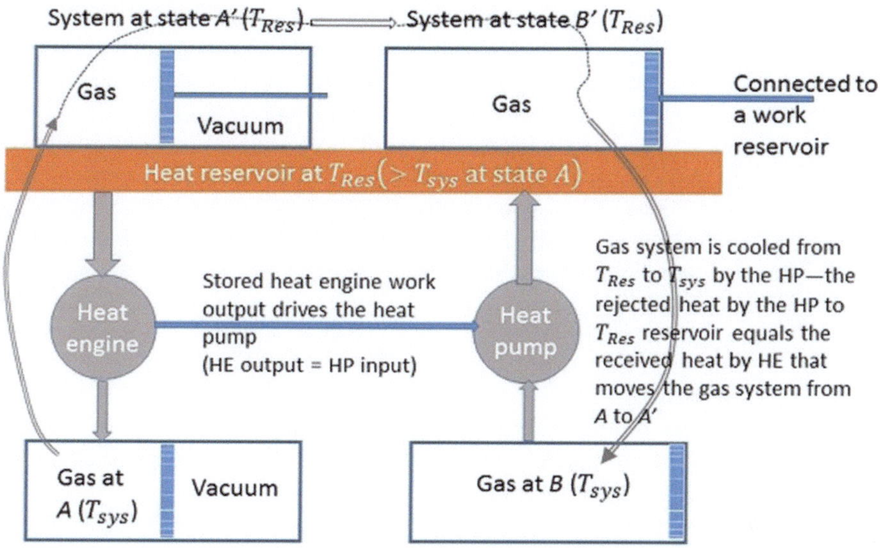

Fig. 8.4 Reversible mechanism managed expansion of gas depicting the case of the composite system temperature, T_1 ($\mathbf{T_{sys}}$) to be different from the heat bath temperature T_0 (T_{Res})

A moment of reflection shows that if the temperature of the reservoir $T_{Reservoir}$ is different from the gas temperature T, reversible work will be given by the same expression $T_{Res}\Delta S$. With a heat reservoir of T_{Res} higher than system T_{Sys}, a Carnot engine operates between the system and the T_{Res} reservoir with the system serving as a heat sink changing from A to A′ at T_{Res} (Fig. 8.4). This step produces a work output from the heat engine and absorbs an amount of heat from the T_{Res} reservoir. This is then followed by isothermal expansion of the system from A' to B' yielding $T_{res}\Delta S = NRT_{res}ln(V_B/V_A) = NRT_{res}ln2$ work as described in top of Fig. 8.4 (the same process shown in Fig. 8.3 with the T_0 of the bath replaced by the T_{res} of the reservoir). The system is then cooled by a heat pump driven with the stored power output of the heat engine returning from B' to B and the cooling operation rejects heat to the reservoir. Note that the required power input of the heat pump equals the heat engine power output and correspondingly the rejected heat by the heat pump equals the absorbed heat by the heat engine. Hence, the net outcome the whole operation is the extraction of heat of the amount

$$T_{res}\Delta S$$

from the heat reservoir transforming it 100% into work. Unlike the example of a Carnot heat engine, there is no intrinsic advantage in a cold heat reservoir in the case of isolated systems. Instead, a colder heat reservoir will proportionally yield a smaller amount of work.

The second example is the case of "temperature equalization spontaneity" within two parts of an isolated thermal composite system: A composite system consists of gaseous subsystem$^{(1)}$ and subsystem$^{(2)}$ (same gas in both) of equal mole numbers ($N^{(1)} = N^{(2)} = N$), initially at p and $(T_{\text{initial}})^{(1)}$, and p and $(T_{\text{initial}})^{(2)}$, respectively. Assume $(Nc_p)^{(1)} = (Nc_p)^{(2)} = 10\ kJ/K$ (the total heat capacity of the composite system is 20 kJ/K), $(T_{\text{initial}})^{(1)} = 450\ K$ and $(T_{\text{initial}})^{(2)} = 200\ K$, and a special heat reservoir at $T_0 = 300\ K$. The spontaneous event is now a cooling event of the composite system from an average temperature of 325 to 300 K (in both subsystems). The corresponding heat exchange is

$$Q_{spon} = -\left[(Nc_p)^{(1)}(450 - 300) + (Nc_p)^{(2)}(200 - 300)\right] = -500\,\text{kJ}$$

We note that this is an arbitrarily constructed spontaneous event, not the normal spontaneous event of an isolated system that interacts with no heat reservoir (as considered again in example below).

The reversible event is an isentropic event of the composite system with a "silent" or "non-participating" heat reservoir. That is,

$$Q_{rev} = 0\,\text{kJ}$$

Thus,

$$\Delta\hat{Q} = 0 - (-500) = 500\,\text{kJ}$$

The isentropic event is brought about by a Carnot heat engine going through a sequence of micro-cycles: During each micro-cycle the Carnot engine extracts $\frac{T^{(1)}}{T^{(2)}}\delta Q^{(2)}$ from subsystem 1 and rejects $\delta Q^{(2)}$ amount of heat to subsystem 2. $\delta Q^{(2)}$ is related to the temperature change of subsystem 2, $\delta Q^{(2)} = Nc_p dT^{(2)}$, and changes in $T^{(1)}$ and $T^{(2)}$ are correlated according to $T^{(1)} \times T^{(2)} = 450 \times 200$. It can be readily shown, noting that for an isentropic event $T^{(1)} \cdot T^{(2)} = 300^2$

$$W_{Rev} = \int \delta W = \int \delta Q^{(1)} - \delta Q^{(2)} = \int_{T^{(2)}=200}^{300} 10dT^{(2)}\left(\frac{T^{(1)}}{T^{(2)}} - 1\right) = \Delta\hat{Q} = 500\text{kJ}$$

The schematic heat relations are shown in Fig. 8.5.

This second example also represents a case of converting heat 100% into work. In this case, heat appears to be from the source system of the thermal composite body. But, the above treatment, though correct, is not satisfactory since it neither conforms to neither Eq. (123) nor Eq. (130). It is misleading to view this to be a case of transformation of change in the form of energy of the source system into work. Instead, it should be viewed like the free expansion example to be a case of

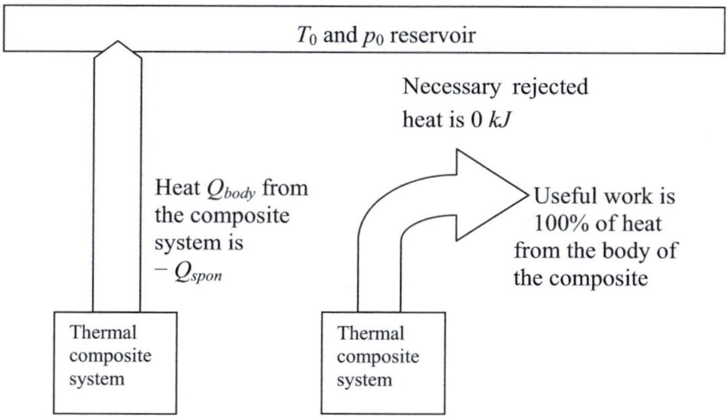

Fig. 8.5 Thermal composite system. The system undergoes a spontaneous event on left; the system undergoes a reversible event on right as managed by a Carnot micro-cycle machine

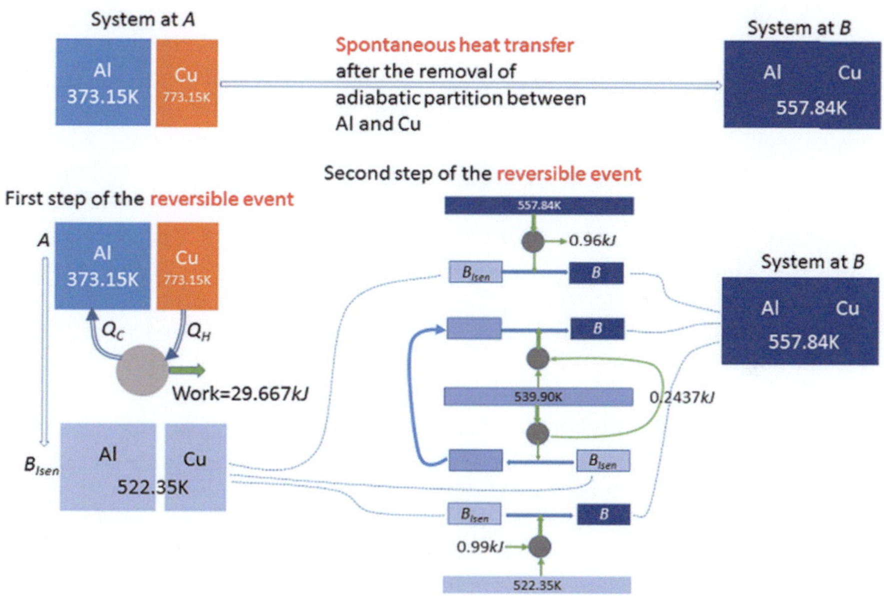

Fig. 8.6 Schematics of an isolated thermal system with pure spontaneity: Top depicts the spontaneous event and bottom depicts three reversible events with the thermal system in interaction with three different heat reservoirs

transformation of pure spontaneity in the system into work by extracting heat from the reservoir. This is explained in the example below as described in Fig. 8.6, where the problem is treated in conformity to Eq. (130).

Example

This example of an isolated composite thermal system (Fig. 8.6) has been discussed in Chaps. 3 and 5: blocks X and Y [X is aluminum $\left(c_{pX} = 0.900 \text{ kJ/kg} \cdot \text{K}\right)$ with m_X= 0.5 kg, Y is copper $\left(c_{pY} = 0.386 \text{ kJ/kg} \cdot \text{K}\right)$ with m_Y= 1 kg] initially at 100 and 500 °C, respectively (state ***A***), are brought together undergoing a **spontaneous** heat transfer process to a final state ***B*** of 557.84 *K* (see Problem 3.8) with corresponding entropy change 0.0549 kJ/K (see Problem 5.4).

Now consider a Carnot heat engine operating between X and Y producing work while the system undergoes a **reversible** change to an isentropic temperature at internal thermal equilibrium with corresponding reversible work output (lower left part of Fig. 8.6). This is essentially the reversible event considered in Fig. 8.5. But, instead of using a special heat reservoir for the spontaneous event as it was done in Fig. 8.5, here the spontaneous event is under isolated condition (see the top of Fig. 8.6), involving no reservoir. For the comparison of the two events, one needs in the reversible event to bring the system back from state $\boldsymbol{B_{Isen}}$ (522.35 K, after the first isentropic step) to the final end state ***B***—calling it the *second step* of the reversible events. A moment of reflection shows three possible reversible events (of an infinite number of possibilities) associated with three different heat reservoir temperatures (see Fig. 8.6).

Show: (a) For the first step of the reversible event, the isentropic temperature to be 522.35 K and determine the corresponding reversible work. (b) Describe the second step of the reversible event for a heat reservoir at 557.85 K and the corresponding additional work output of the second step (confirming that the overall work output is equal to the prediction of [129]). (c) Describe the second step of the reversible event for a heat reservoir at 522.35 K and the corresponding heat pump work input of the second step (confirming that the overall work output is equal to the prediction of [129]). (d) Consider a second step of the reversible event for a heat reservoir at a particular given temperature so that according to (129), the overall work output of the reversible event (for both steps) is the same as the isentropic work. That is, the second step requires no (nor produces) additional work. Determine the reservoir temperature.

Solution

(a)

$$0 = m_X c_{pX} ln\frac{T_{Isen}}{373.15} + m_Y c_{pY} ln\frac{T_{Isen}}{773.15} = 0.5 \times 0.9 \times ln\frac{T_{Isen}}{373.15}$$
$$+ 1 \times 0.386 \times ln\frac{T_{Isen}}{773.15} T_{Isen} = 522.3521 \text{ K}$$

Note that this process involves no heat loss, thus work output equals a decrease in the internal energy of the composite system according to the first law:

$$W_{rev} = U_{ini} - U_{EndOfStep} = m_X c_{pX}(373.15 - 522.35) + m_Y c_{pY}(773.15 - 522.35)$$
$$= 29.667\,\text{kJ}$$

(b) A Carnot engine receives heat $(557.84/T)\delta Q$ from the reservoir and rejects $\delta Q = \left(m_X c_{pX} + m_Y c_{pY}\right)dT$ to the system, the difference of which equals the work output

$$W_{2ndStep} = 0.836 \int_{522.35}^{557.84} \left(\tfrac{557.84}{T} - 1\right)dT = 0.9857kJ$$
$$W_{rev} = T_{res}\Delta S = 30.6527kJ \leftrightarrow W_{Isen} + W_{2ndStep} = 29.667 + 0.9857$$
$$= 30.6527\text{kJ CHECK!}$$

(c) A Carnot heat pump extracts heat $(522.35/T)\delta Q$ from the reservoir and supplies $\delta Q = \left(m_X c_{pX} + m_Y c_{pY}\right)dT$ to the system, the difference of which equals the required HP work input

$$W_{2ndStep} = 0.836 \int_{522.35}^{557.84} \left(1 - \tfrac{522.35}{T}\right)dT = 0.9643\,\text{kJ}$$
$$W_{rev} = T_{res}\Delta S = 28.7027kJ \leftrightarrow W_{Isen} + W_{2ndStep} = 29.667 - 0.9643$$
$$= 28.7027kJ\text{ CHECK!}$$

(d) Equation (129) can be used for determining the reservoir temperature

$$T_{res} = W_{Isen}/\Delta S = 29.667/0.054949 = 539.901\text{ K}$$

Referring to Fig. 8.6, for a heat reservoir at this intermediate temperature the system can be brought to the reservoir with the operation of a Carnot heat engine—then afterward be brought to 557.84 K with the operation of a Carnot heat pump. Using the procedures of (b) and (c), respectively, it can be shown that the heat work engine output (0.2437 kJ, exactly equals the heat pump work input (0.2437 kJ). Thus, the second step involves no network with the reservoir at this particular temperature.

The above example/thought experiment serves to show that the capacity of the isolated thermal composite to produce reversible work derives from the same source as the isolated free expansion composite system: a source of *pure spontaneous tendency found* in *all* isolated composite systems. Even if a Carnot heat engine is used to operate isentropically by absorbing heat from the hot part of the system and rejecting heat to the cold part, that such operation is possible is derived from the *spontaneity* of the system. This can be seen clearly in particular in Case (d) above: Compare the spontaneous event and the reversible event. As the end states of the reversible event are the same as the spontaneous isolated event, it is clear that the system in the reversible event gives out no heat and, thus, the produced work is derived from heat from the heat reservoir at 539.901 K. The erstwhile isentropic process produced work is *effectively* the "extracted heat" that the system is able to extract from the reservoir during the second part of the reversible event.

Though this is a subtler example than that shown in Fig. 8.3, it demonstrates the same case of conversion of extracted heat 100% into work driven purely by entropy growth potential. (From a practical point of view, there is no reason to employ a heat reservoir with temperatures lower than 539.901 K (in this case) if the purpose is solely work production. Heat reservoirs are used here for the purpose of theoretically comparing the spontaneous event to the reversible events.)

These thought experiments, including that of Sect. 5.10, confirm that Eq. (130) is applicable to isolated systems. The significance of that is that whereas the theory of exergy at its current stage is formulated only for systems tending to equilibrium with a surrounding reservoir. The extension of exergy analysis to isolated systems tending to internal equilibrium will require further formulation involving complications associated with the treatment of composite systems. The present approach, Eqs. (123) and (130), applies to both cases of external and internal equilibriums based on the single conceptual set of entropy growth/entropy growth potential (see next section). The parsimony of the present approach shows heuristic, as well as conceptual, value in entropy growth potential discussed below.

8.5 The Entropy Growth Potential Principle

We have come to the vantage point to identify the ingredient, *entropy growth potential*, for replacing *universal interconvertibility*.

Why did the MTH need the *universal interconvertibility* principle? Recall again that thermodynamics was the study of heat and heat engines. While the Carnot–Kelvin formula provided the specific idea of maximum possible mechanical work to be derived from heat under one particular idealized condition, the principle provided the general idea of where mechanical work is to be derived in heat engines from heat under general conditions, i.e., it provided the "conceptual" framework for heat's apparent utility. The theory of exergy then extended that into energy-as-the-driver, without, however, challenging the principle itself.

8.5.1 Conceptual Differentiation of Entropy Growth and Entropy Growth Potential

It is correct to view that expended work causes the production of heat, but the real reason that expended work causes the production of heat is an inference from universal entropy growth. To think in terms of expended heat to be the cause for the production of work as a matter of energetic relation without explicit use of the idea of universal entropy growth amounts to give heat the sole role of causing work in the dyadic framework of heat's apparent utility. It is a role which heat cannot fulfill [10]. Because of this incorrect assignment of a role for heat, the universal inter-convertibility principle deprives the second law of its constructive function. The second law of the MTH as the law of universal entropy growth is consequently incomplete.

For making the second law into a complete law, it is necessary to conceptually differentiate entropy growth and entropy growth potential [17]. The role, one that heat cannot fulfill, can be fulfilled perfectly by entropy growth potential (EGP), $(\Delta_P S)_{universe}$, defined

$$\left[(\Delta_G S)_{universe}\right]_{spon} = (\Delta_P S)_{universe} \tag{129A}$$

or,

$$\left\{\left[(\Delta_G S)_{universe}\right]_{spon} = \right\}(\Delta S)_{iso-sys} = (\Delta_P S)_{universe} \tag{129B}$$

As it was shown in Sects. 8.3 and 8.4 that in association with the system going from its initial state to its final state in interaction with a reservoir, or as an isolated system, there are infinite number of possible events. These infinite numbers of possible events are bookended by the spontaneous event and the reversible event called the Poincare range. Every event in a Poincare range shares the same EGP, $(\Delta_P S)_{universe}$. But, the entropy growth of each specific event, $\left[(\Delta_G S)_{universe}\right]_{event}$, is different ranging from

$$\left[(\Delta_G S)_{universe}\right]_{spon} = (\Delta_P S)_{universe}$$

for the spontaneous event to

$$\left[(\Delta_G S)_{universe}\right]_{rev} = 0$$

for the reversible event. With, for a specific event in the Poincare range, its entropy growth falls between the two limits

$$0 \leq \left[(\Delta_G S)_{universe}\right]_{event} \leq \left[(\Delta_G S)_{universe}\right]_{spon} \tag{131}$$

The fact that every event shares the same EGP reminds us that the specific EGP is the driver of every event in the given Poincare range. That is,

> Entropy *growth potential (EGP) is the driver of every event in a given Poincare range; in the mechanical theory of* heat *(the MTH, i.e., standard thermodynamics), the "consumption" of heat and the "consumption" of energy are treated as proxies of EGP, which is the real driver The correct identification of the real driver leads to* (in Sects. 8.6 and 8.7), *in departure from the MTH, how we classify "energy conversion" processes.*

By extension, *general* EGPs are drivers of *all* events in nature, which will be called the *principle of entropy growth potential.*

The erstwhile spontaneous entropy *growth*, $(\Delta_G S)_{universe}$, serves as the drive for reversible work, $(\Delta_P S)_{universe}$ in Eqs. (123) and (130). The very same quantity, initially indicating dissipation, represents instead constructive drive potential (*entropy growth potential*) giving rise to the maximum reversible work if a perfect control mechanism is provided, to valuable work if a real-world heat engine operational mechanism is provide, or to moderate amounts of kinetic and potential mechanical energy in mechanism-free natural convective fluid bodies. In sum, the full expression of the second law entails a conceptual differentiation. By adding to *universal* entropy *growth principle* (the first part of the law) the *principle of entropy growth potential* (the second part of the law that is differentiated from the first part), we have the complete second law of thermodynamics.

The second law can now fulfill its essential function (see the next section and Fig. 8.7). This new second part of the second law serves as the new cornerstone in a new theory of heat as outlined below.

8.6 The Predicative Entropic Theory of Heat

A case can be made that Carnot's theory (and consequently thermodynamics) is a predicative theory of heat. An ontological theory of heat attempts to answer, in the most general possible terms, the question, "What is heat?" A predicative theory of heat, instead, attempts to answer the question, "What is it to say something about heat?" An important question for a predicative theory is its category [30]. A brief introduction to category is given here.

8.6.1 Peirce's Reduction Thesis and Carnot's Theory as a Triadic Relational Theory of Heat

Charles Sanders Peirce presented a paper entitled, "On a new list of categories," to the *American Academy of Arts and Sciences* [30] in which he outlined a *theory of predication* involving three universal categories. The claim of the universal categories of three-ness (monads, dyads, and triads) arose from the mathematical idea about the *reducibility* of *n-adic* relations. According to the reduction thesis

a. Triads are necessary because genuinely triadic relations cannot be completely analyzed in terms of monadic and dyadic predicates, and
b. Triads are sufficient because there are no genuinely tetradic or larger polyadic relations; all higher arity *n-adic* relations can be analyzed in terms of triadic and lower arity relations.

In recent years, Burch [31] has offered proof of the reduction thesis.

We may use the reduction thesis as the organizing principle of evolution in thermodynamic thought. It began with the caloric theory as a one-place relation theory that heat is conserved as presented in Chap. 2. From that beginning a more comprehensive *dynamical theory,* i.e., *mechanical theory of* heat [MTH], was formulated on the basis of the MEH as the interconvertibility principle, which becomes the standard theory of the present time. The interconvertibility principle is a statement of two-place relation.

Carnot's theory has been interpreted as the conversion of high-temperature heat into mechanical energy and that there is limitation in the amount of high-temperature heat to be converted into mechanical energy. Carnot himself, though, emphasized the importance of low-temperature heat sink, "Heat alone is not sufficient to give birth to the impelling power: it is necessary that there should also be cold; without it, the heat [at high temperature] would be useless" [4]. Accordingly, we may consider an alternative way of interpreting the Carnot heat engine.

Again, the setup is the transfer of heat energy from a heat body at T_A to a body at low temperature T_B in association with the production of mechanical energy; the three grades of energy involved are again mechanical energy, W_{rev}, high-temperature heat energy at T_A, Q_A, and low temperature heat energy at T_B. But, instead of considering the conversion of heat energy from a heat body at T_A to mechanical energy, we consider a new way of understanding the Carnot heat engine beginning with considering two events in association with the setup

1. The spontaneous event of the pure heat transfer process, and
2. The reversible event for the production of mechanical energy

Since the same amount of heat energy exits from the T_A heat body in both events, we pose the question not as how much of the heat energy from a T_A heat body is converted to mechanical energy, but what is the relation between the difference in heat energies exchanged with the T_B heat body of the two events and the amount of mechanical energy in the reversible event.

In answering this question, it is useful to consider entropy growth in the spontaneous event. In this event, the same Q_A amount of heat energy that leaves T_A heat body enters the T_B heat body. Therefore, the entropy growth in the universe which in this case equals the entropy change in the universe is

$$(\Delta_G S)_{univ} = -\frac{Q_A}{T_A} + \frac{Q_A}{T_B}$$

What happens to heat energies from the T_B heat body during the two events are

1. The spontaneous event Q_A is added to the T_B heat body.
2. The reversible event, according to standard Carnot heat engine treatment,

$$Q_A \cdot \frac{T_B}{T_A} \text{ is added to the } T_B \text{ heat body}$$

We note now that the mechanical energy (W_{rev}), $Q_A\left(1 - \frac{T_B}{T_A}\right)$, equals the difference between the amounts of heat energy added to the T_B heat body for the two events. Also, it equals

$$W_{rev} = T_B \cdot \left(-\frac{Q_A}{T_A} + \frac{Q_A}{T_B}\right) = T_B \cdot (\Delta_G S)_{univ}$$

Correspondingly,

$$W_{rev} = T_B \cdot (\Delta_P S)_{univ}$$

The natural interpretation of the new way of considering the Carnot heat engine is that entropy growth potential is the driver for all events defined by the pair of initial and final states, which enables the *extraction* of heat energy from the T_B heat body converting it into mechanical energy of the same amount. In this framework, we need no more to consider the causal agency of heat but instead identify "heat-and-cold", i.e., entropy growth potential, to be the driver. Carnot's theory is, therefore, identified to be a rudimentary predicative theory of heat of triadic relation.

8.6.2 The Predicative Entropic Theory of Heat (PETH)

With the formulation of the entropy growth potential principle, Carnot's theory of heat is updated as a predicative entropic theory of heat (PETH) representing the three-place relationship in general *heat* phenomena, as shown in Fig. 8.7 and Sect. 8.7.

The new principle and the new theory give us the tool for the task of correcting errors in the conventional takeaways of MTH as summarized in Sect. 8.1. First of all, the residual objection to the definition of heat in Sect. 5.6 can be overcome once the new principle strips away heat's causal agency. For each of the rest of MTH's

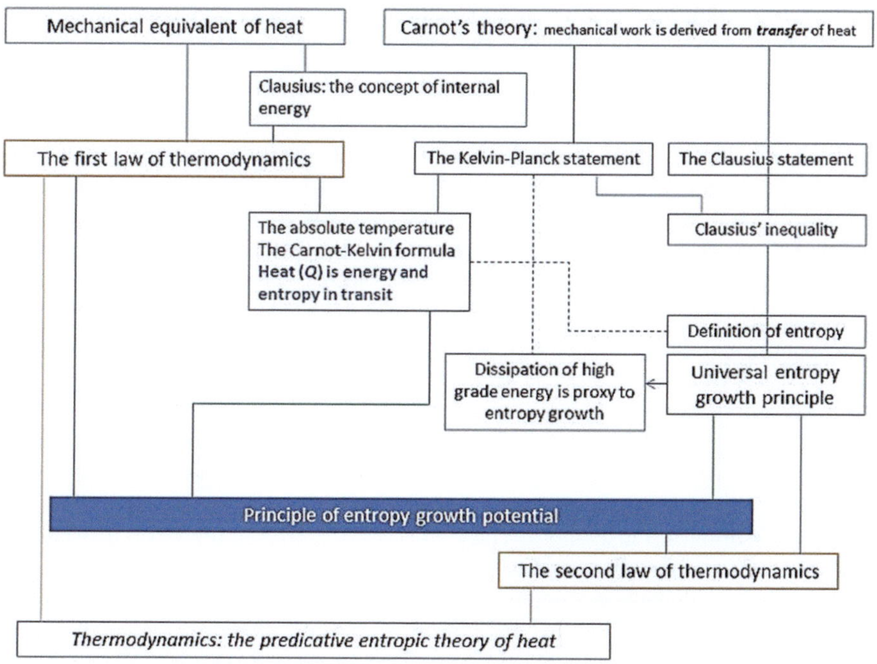

Fig. 8.7 The predicative entropic theory of heat: by replacing universal interconvertibility with the principle of entropy growth potential, engineering thermodynamics is liberated from MTH by successfully placing heat in its categories, both ontologically and predicatively

errors and what PETH's new understanding will be in each case, it is useful to go over the *general statements*, which appeared in Sect. 8.1, one by one.

On Heat's ontology. Unlike others who have written on refutation of interconvertibility principle, PETH, by no means, entertains a return to the caloric framework in returning to Carnot's thinking. Carnot's theory is not a caloric theory of heat. The paramount difference between the caloric theory of heat and Carnot's theory is that Carnot's theory is not an ontological theory of heat. It is a relational or predicative theory of heat and, as a predicative theory, it can incorporate either the ontology of the caloric theory of heat or the ontology of the mechanical theory of heat. Carnot embraced the caloric theory in his 1824 publication, but in his private writing [4] indicated predisposition in accepting the ontology of the mechanical theory. As the update of Carnot's theory, PETH accepts GS-1

GS-1. **Heat's ontology**. Heat is a high-entropy form of energy.

On equivalence principle. The PETH also accepts GS-2. Everyone does. The only clarification is that GS-2 should be revised by replacing "consumption" of heat with "extraction" of heat

GS-2. Equivalence principle. Equivalence exists between consumed mechanical energy and produced heat, and extracted heat and produced mechanical energy.

On interconvertibility. From his own statement of "Considering…heat is not a substance, but a dynamical form of mechanical effect, we perceive that there must be an equivalence between mechanical work and heat, as between cause and effect…," [32], pp. 174–200], as well as his biographers [33], Thomson thought that universal interconvertibility followed necessarily from the ontological doctrine that heat is a form of energy. That inference was never established on solid ground. From the very beginning, a referee of Joule's 1850 paper pointed out, correctly, that universal interconvertibility did not follow from the MEH (cited in [34]). Equating equivalence correlation with causality by Joule, Thomson, and Clausius was a classic error. It is an error that has been made by many in different contexts because, while causation may exist behind correlation, the details of causation or true sources of causation are multifold and a simple presumption of equating of two can be wrong. The above worded GS-2 clarifies that equivalence between extracted heat and produced mechanical energy does not mean that heat causes mechanical energy. Causation lies elsewhere. The PETH, by identifying universal causation in EGP, rejects the interconvertibility principle.

On energy conversion doctrine. Energy conversion doctrine is based on the universality claim of the energy principle—or, entropy growth is exhausted by the spontaneous dissipation of high-grade energy. The energy principle, however, was proposed as a self-evident principle, without a solid ground (Uffink characterized it as "the unargued statements of Kelvin" [35]). The principle's claim of universality being false was pointed by Planck as cited in Sect. 5.10 as well as elaborated in the Section. PETH rejects "GC of Kelvin" and replaces energy conversion doctrine with amended GS-4

Amended GS-4. Entropy growth potential (EGP) is the universal driver of changes in nature; changes in nature can be classified into two kinds: natural energy conversion or pure spontaneous entropy growth, and processes involving changes in reverse direction.

Clarification of changes in nature will be discussed below dealing with the next two General Statements.

On spontaneous energy conversion. Conversion of mechanical work into heat is spontaneous and, in such spontaneous energy conversion processes, entropy growth is realized and entropy growth potential is consumed. We can understand such a process in a dyadic framework. Spontaneity, i.e., EGP, of such processes is in fact the driver of all processes (see below) except to note that spontaneity can also exist in *pure* EGP form (see Sects. 5.10 and 8.4) in which EGP is derived from natural tendency in initial state → final state with *no* change in form of system energy. General Statement GS-4.a remains unchanged

GS-4.a. Spontaneous energy conversion. Natural energy conversions and pure EGP driven changes are spontaneous and limitless.

We now come to the last General Statement, which deserves special attention. The next section devotes to that purpose as well as to further comment on the triadic framework of PETH.

8.7 The Triadic Framework: All Reversible Processes Are Heat Extraction Processes

With the exception of energy conversion in natural convection processes, conversion of heat into mechanical work is not natural but man-made. These man-made processes are not spontaneous being in the reverse direction to spontaneous process. There is the widespread notion that for all reverse processes, 100% conversion of heat into mechanical energy is not possible, e.g., consideration of Carnot heat engine in Sect. 4.6 concludes that the maximum amount of heat out of Q_A that can be physically converted into work is $Q_A(T_A - T_B)/T_A$. However, an alternative way of considering the Carnot heat engine in Sect. 8.6.1 shows that the EGP in association with the hot and cold heat bodies, $(\Delta_P S)_{univ}$, enables the *extraction* of heat energy from the T_B heat body converting it into mechanical energy of the same amount. That is, the "extracted heat" is converted 100% into work.

The possibility of heat being 100% converted into work is not merely a game of how we count "heat." In Sects. 5.10 and 8.4, it is demonstrated by thought experiments that for every isolated composite system, its EGP, which is a pure EGP, enables the *extraction* of heat energy from a corresponding heat reservoir converting it 100% into work. In all those cases, no other form of heat is involved in the thought experiment processes. The claim, *heat cannot be converted 100% into mechanical energy*, is, therefore, demonstrably false. This was, of course, pointed out by Fermi (in Sect. 8.4) and, earlier, by Planck

> We often find the second law stated as follows: The change of mechanical work into heat may be complete, but, on the contrary, that of heat into work must needs be incomplete, since, whenever a certain quantity of heat is transformed into work, another quantity of heat must undergo a corresponding and compensating change; e.g. transference from higher to lower temperature. This is quite correct in certain very special cases, but it by no means expresses the essential feature of the process, as a simple example will show…If a perfect gas be allowed to expand, doing external work, and be prevented from cooling by connecting it with a heat reservoir of higher temperature, the temperature of the gas, and at the same time its internal energy, remains unchanged, and it may be said that the amount of heat given out by the reservoir is completely changed into work without an exchange of energy taking place anywhere. Not the least objection can be made to this. [36: 81–82]

General Statement GS-4.b has to be rejected. The issue, however, is more than a matter of the falsity that heat cannot be converted 100% into mechanical energy. It is about the concept of reverse energy conversion: whether such processes should be characterized as energy conversion.

Spontaneous processes are dyadic natural energy conversion processes in which the consumption of EGP and the full realization of entropy growth occur at the same time. In an engineering conversion of heat into mechanical work, however, entropy growth and entropy growth potential must be differentiated (see Sect. 8.5): In a reversible conversion, only entropy growth potential in the hot-body/cold-body pair is consumed with no entropy growth, while in a real engineering conversion entropy growth potential is consumed with partial entropy growth (see Eq. [131]). In both, entropy growth is delayed or partially delayed, and its eventual realization occurs when all work is dissipated or when all orders created by work collapse at some point in the future. In all cases, the ultimate ending is always the same: all entropy growth potentials turn into dissipated heat eventually (heat, i.e., dissipated heat, is the ultimate end rather than being a cause by any means).

Conversion of heat into work cannot be characterized as dyadic energy conversion processes. This is the fundamental error in the dyadic framework of heat's apparent utility. Instead, General Statement GS-4.b is amended into the Amended GS-4.b

> **Amended GS-4.b. Reversible-like process**. The EGP associated with natural energy conversions and/or pure EGP can be managed to produce *reversible-like* and/or reversible processes, which involves changes in reverse direction, in a triadic framework; there is, in fact, no *pure* reverse energy conversion process.

The idea of reverse energy conversion of heat into work was what Carnot and Kelvin aspired to understand. That is, to understand reversibility. There is, however, no *pure* reverse energy conversion; a change in reverse direction such as heat to work is always accompanied by EGP, which is often in the form of natural energy conversion in a spontaneous direction in a triadic framework.

Triadic framework, or triadic relation, is the essence of reversibility or reversible-like processes (see Chap. 6). Thinking in terms of energy conversion tends to encourage dyadic thinking, which only characterizes perfectly spontaneous/nonreversible processes (Chap. 6). For understanding reversibility, we need to go not only beyond thinking dyadically (which is absolutely necessary) but also reexamining its standard definition—*a reversible process is a process whose direction can be "reversed" by inducing infinitesimal changes to some property of the system* via *its surroundings, with no increase in entropy. Throughout the entire reversible process, the system is in* **thermodynamic equilibrium** *with its surroundings.* This definition, correct though it is as a descriptive criterion of naturalistic processes, does not provide the prescriptive operational requirement that is necessary for bringing about reversibility. A reversible process cannot be captured as a naturalistic process. For achieving reversibility and real efficiency, we need to think in terms of EGP, the triadic relation of reversibility, and extraction of heat instead of the consumption of heat.

These are new concepts in the PETH, its premises are summarized in Table 8.1 as corrections to the MTH premises.

Correspondingly, the energy conversion doctrine truism is replaced with the EGP-centric doctrine

Table 8.1 MTH premises versus PETH premises

MTH premises		PETH premises	
GS-1	Heat's ontology ✓✓	GS-1	Heat's ontology
GS-2	Equivalence principle (MEH) ✓✓	GS-2	Equivalence principle (MEH)
GS-2a	Interconvertibility principle		
		GS-3	Entropy growth potential principle ✓✓
GS-4	Energy conversion doctrine ⇨	Amended GS-4	EGP-centric doctrine ✓
GS-4a	Spontaneous energy conversion ✓	GS-4a	Spontaneous energy conversion
GS-4b	Reverse energy conversion ⇨	Amended GS-4b	Reversible-like processes in triadic framework ✓

Everything that changes is driven by entropy growth potential, and these changes are divided into two kinds: one kind of natural spontaneous changes of high-grade form to low-grade form eventually to heat and the other of reversible-like processes including the "conversion" of heat into mechanical energy; whereas the natural changes are dyadic and non-reversible, the reverse-like changes are triadic and made to happen as the result of efficacious causation for achieving efficiency.

8.7.1 Definition of Waste Heat

The useful range of managing EGP is subject to the limits, as in Eq. (122), the lower and the upper limits of system-reservoir heat exchange corresponding to the spontaneous event and the reversible event, respectively. The existence of the Poincare range, defined by the two limits, suggests two alternative heat concepts. For a managed event of system-reservoir with heat exchange Q, the management performance of the event can be characterized by (see Fig. 8.8)

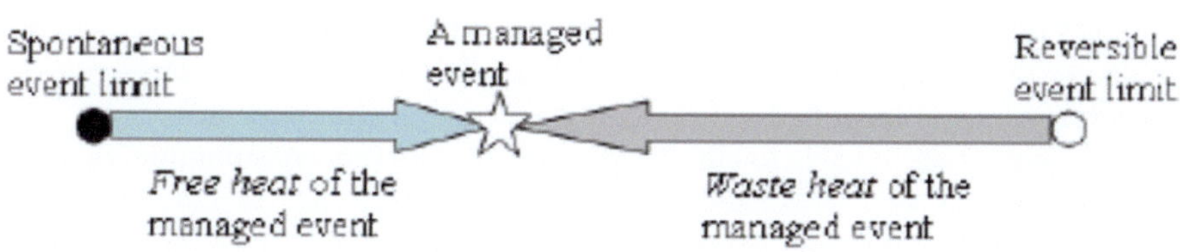

Fig. 8.8 Meaning of waste heat in terms of the two reference-end-events of Poincare range

Waste heat, H_{waste}, is the measure of Q in reference to reversible heat extracted, the upper limit,

$$H_{waste} = Q_{rev} - Q$$

and correspondingly,
Free heat is the measure of Q in reference to spontaneous heat extracted, the lower limit,

$$Free\ heat = Q - Q_{spon}$$

Both concepts offer useful perspectives for the reversible limit. Reversibility means either achieving the elimination of waste heat or the maximization of free heat, i.e., the reversible free heat. Waste heat versus free heat is the alternative of the "glass is half empty" versus the "glass is half full", and both are the manifestation of the degree of success in the management of the EGP. Again, it is worthwhile to repeat that waste heat in a process is neither directly measured by its discarded heat, e.g., heat discarded by a Carnot cycle is not waste heat; nor, in an adiabatic process which discards no heat, necessarily means that it involves no waste heat, e.g., in the case of free expansion the adiabatic process incurs waste heat (see Problem 8.4).

8.7.2 Kinds of EGP's: Stock EGP and Natural (Ongoing) EGP

In the context of their management, EGP can be distinguished into two kinds:

1. Entropic drive of the latent or stock kind (which may be referred to as the *stock spontaneity* or *stock EGP*).

   1.a. The usual EGP that we call energy systems
   1.b. Pure EGP, the special kind of stock EGP that have potential for producing work without energy degradation as considered in Sects. 5.10 and 8.4.

2. Entropic drive of the ongoing kind (*ongoing spontaneity*, i.e., *natural EGP*).

The most common examples of spontaneity of the former kind are fossil fuels. Examples of the latter kind are sun and wind, etc. The usefulness of sun and wind is not due to their materiality but to their existence as phenomena of exergy transfer processes, which is identified with ongoing entropy growth processes.

While stock energy leads to entropy growth only after its consumption, ongoing EGP phenomena are associated with entropy growth as *ongoing* processes regardless whether they are "consumed" or not. Entropically speaking, the "consumption" of the latter is of entirely different matter from the consumption of stock energy and, thus, we should accord the application of natural EGP with different measurements, e.g., whether energy derived from natural EGP should be excluded from "energy input" in *coefficient of performance*, *COP = Heat supplied or removed/ Energy input*. That is, for many applications, especially low-temperature application, a creative way of counting energy can promote the use of natural EGP while reserving the use of stock EGP only for managing *generalized* triadic processes. Treatment of these kinds of problems will be made elsewhere.

8.7.3 Additional Examples of Heat Extraction

The triadic framework of PETH is applied to a number of classical examples to reinforce the idea that all reversible and reversible-like processes are heat extraction processes.

Let us consider now the cold case of the classical Carnot cycle (Fig. 8.9b). Imagine the cold body undergoing first a spontaneous heating process (Fig. 8.9a) receiving from the reservoir an amount of heat, which equals to that the system receives from the Carnot engine in Fig. 8.9b. This amount of heat is designated as Q_{Spon}, which relates to the system entropy change $Q_{Spon} = T\Delta S$. The spontaneous entropy production, $(\Delta_P S)_{Spon}$, is $\Delta S + \left(-Q_{Spon}\right)/T_0$

The corresponding reversible change, Fig. 8.9b, in which, the system-and-its-reversible-component (RC) unit receives from the reservoir an amount of heat $Q_{Rev} = T_0\Delta S$ (Eq. 110A), and Q_{Rev} is greater than Q_{Spon} by $\Delta \cdot \hat{Q} = T_0\Delta S - T\Delta S$. This greater amount of heat exchange results in vanishing entropy production

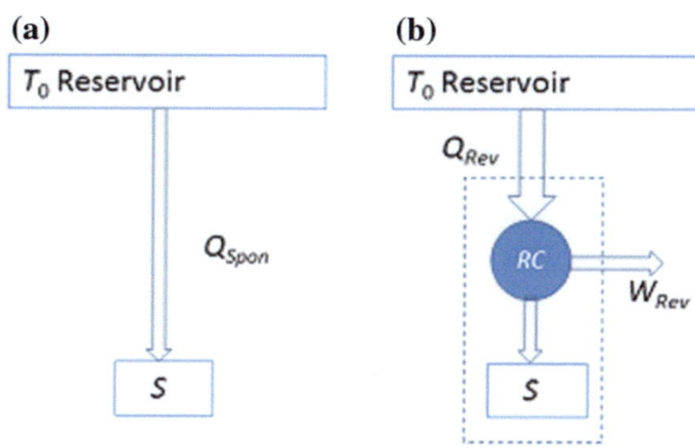

Fig. 8.9 Spontaneous heating (**a**) and reversible heating (**b**) of a cold body, S

$$(\Delta_P S)_{Rev} = \Delta S + \frac{-Q_{Rev}}{T_0} = \Delta S + \frac{-T_0 \Delta S}{T_0} = 0$$

as well as defines the reversible work

$$W_{Rev}\left[\Delta\hat{Q}\right] = T_0\Delta S - T\Delta S = T_0\Delta S\left(1 - \frac{T}{T_0}\right) = Q_H\left(1 - \frac{T_L}{T_H}\right)$$

8.7.4 Reversible Free Heat $\Delta\hat{Q}$ and Free Energy ΔF

Note that the additional extracted heat $\Delta\hat{Q}$, or $Q_{rev} - Q_{spon}$, is defined in terms of two distinctive events between the same end states A and B (see the entropic drive corollary). Another closely related concept, the concept of system *free energy* ΔF_A, is defined in terms of the system state (e.g., A) and a universal reference state 0

$$\textit{Free energy of a system at state } A \equiv \Delta F_A$$

Reversible free heat of process $A \rightarrow B$ is thus equal to the difference between free energy at A, ΔF_A, and free energy at B, ΔF_B,

$$\Delta\hat{Q} = \Delta F_A - \Delta F_B \equiv (\Delta F)_{AB}$$

Reversible free heat equals free energy if ΔF_B=0. That is, free energy at a system at state A is the reversible free heat of the system in terms of the end states of A and the final end state at a universal reference state 0.

An example of cooling combustion gas. The unifying treatment can be applied to an energy source in the form of combustion gas undergoing a constant-pressure quasi-static *cooling* from an initial state at flame temperature to a final state at thermal equilibrium with a surrounding heat reservoir. In this case, $Q_{Spon} = \Delta H = -(H_{ini} - H_0) \equiv -HV$, where HV denotes the combustion gas *heating value*, a positive number. Since $\Delta H = \int_{T_{initial}}^{T_0} TdS = -\int_{T_0}^{T_{initial}} TdS$, this enthalpy change is related to entropy change

$$\Delta H = -T_{MEAN}(S_{initial} - S_0)$$

in terms of a *mean gas temperature*

$$T_{MEAN} \equiv \frac{\int_{T_0}^{T_{initial}} TdS}{S_{initial} - S_0} = \frac{HV}{S_{initial} - S_0} \qquad (132)$$

and $Q_{rev} = -T_0(S_{initial} - S_0) = \frac{-T_0}{T_{MEAN}} HV$. It follows, therefore,

$$Q_{Rev} - Q_{Spon} = -\frac{T_0}{T_{MEAN}} HV - (-HV)$$

And

$$(\Delta F)_{comb\_gas} = \Delta\hat{Q} = Q_{Rev} - Q_{Spon} = HV\left(1 - \frac{T_0}{T_{MEAN}}\right) \tag{133}$$

Equation (133) will be referred to as the generalized Carnot–Kelvin formula. This simple formula, which assumes the same form of the Carnot–Kelvin formula, Eq. (54A), is one of the new results of free energy derived here. Note that in Eq. (133) $\Delta F_B = 0$, thus

$$(\Delta F)_{AB} = \Delta F_A$$

The same notation applies to Eqs. (134) and (134A) below.

8.7.5 Chemical Composite Systems: Gibbs Free Energy

Changes to chemical or physical composite systems at the same temperature with its surrounding ($T = T_0$) are driven by affinities of various kinds. An exothermic chemical reacting system can be represented schematically by Fig. 8.10, in which *CCS* is a chemical composite system. Here, Q_{Spon} and Q_{Rev} are not directly related as in cases of thermal heat sources. They do not always, therefore, have the same sign. Instead, the spontaneous heat exchange, $Q_{Spon} = \Delta H$, which is a negative number for exothermic reactions, and the reversible heat exchange, $Q_{Rev} = T_0\Delta S = T\Delta S$, may or may not share the same sign dependent on the property of the chemical mixture.

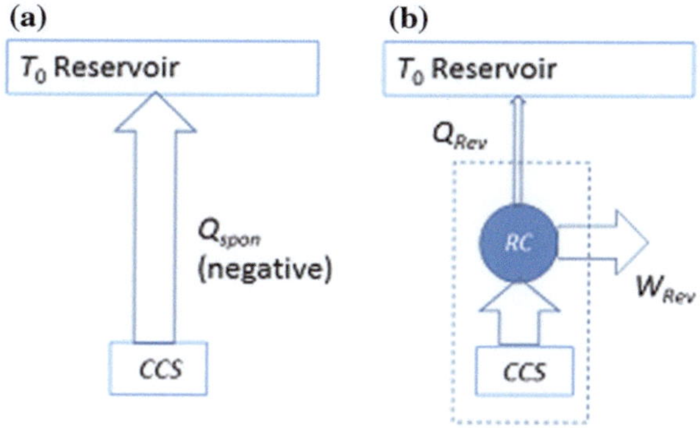

Fig. 8.10 Spontaneous exothermic chemical reaction **a** in a chemical composite system (CCS), and **b**, its corresponding reversible event

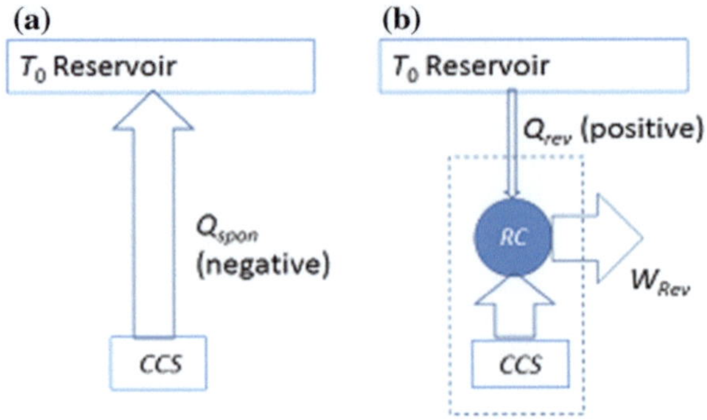

Fig. 8.11 Spontaneous exothermic chemical reaction in a CCS (**a**), and its corresponding reversible event (**b**). In this case, the system entropy change, ΔS, is positive corresponding to a positive Q_{rev}. Thus, $W_{rev} > |Q_{spon}|$

First consider the case of Q_{Rev} having the same sign as Q_{Spon}. As always the inequality $Q_{Rev} > Q_{Spon}$ applies, i.e., Q_{Rev} is a smaller negative number than Q_{Spon} with a corresponding ΔS being a small negative number as shown in Fig. 8.10b.

The same remark that was made for the isolated composite systems applies here. The true "cost" of work production is not the energy in terms of the enthalpy of the spontaneous heat dissipation, but the spontaneity that corresponds to this enthalpy of combustion. Spontaneity or free heat is the true measure of the value of the reactant mixture, not its enthalpy of combustion per se. Or, in the case of a positive ΔS, the reversible heat exchange $Q_{Rev} = T\Delta S$ of a small heat absorption is shown in Fig. 8.11b.

Free energy for both the above case of negative ΔS (Fig. 8.10b) and the case of positive ΔS here (Fig. 8.11b) equals

$$\Delta F = Q_{Rev} - Q_{Spon} = T\Delta S - \Delta H = -\Delta G \tag{134}$$

Free energy for the case of positive ΔS as shown in Fig. 8.11b (shown as arrow W) is even greater than the fuel heating value $-\Delta H$ (shown as arrow $|Q_{Spon}|$).

An endothermic chemical reacting system can be represented schematically by Fig. 8.12, in which the CCS is again a chemical composite system. Both Q_{Spon} here in Fig. 8.12 and Q_{Spon} in the case of exothermic reaction (Figs. 8.10 and 8.11) are not due to a "preexisting" temperature difference between the CCS and the reservoir (Fig. 8.9) but to the "emerging" temperature difference that results from a chemical reaction which is ultimately driven by chemical affinity. Again, the same expression for free energy applies

$$\Delta F = \Delta Q = -\Delta G \tag{134A}$$

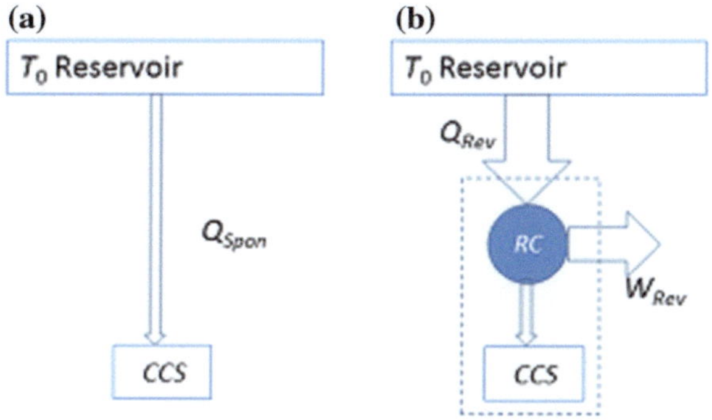

Fig. 8.12 Spontaneous endothermic chemical reaction (**a**) and its corresponding reversible event (**b**)

These above examples reinforce the idea that, in the triadic framework, all reversible processes are heat extraction processes. The notion of heat extraction cannot be overstressed for the comprehension of reversible-like heat phenomena.

8.8 Entropy Growth Potential and Reversibility's Triadic Framework

The energy conversion doctrine was a giant step toward understanding heat and energy. However, the triumph of the Kelvin–Clausius synthesis (MTH) was tainted with misplacing heat in its relational category. The accomplishment of the MTH can be preserved, as shown in Table 8.1, as long as it is understood that spontaneous change in forms of energy is proxy to entropy growth, which is the real driver of nature as manifested by the existence of pure entropy growth involving no change in energy form. The entropy growth potential principle, which supplants interconvertibility principle and under which the energy principle is subsumed, corrects the relational error of the MTH and serves (by supplanting the interconvertibility principle) as the cornerstone for a new theory of heat, the predicative entropic theory of heat (Fig. 8.7), as the logical extension of the mechanical theory of heat.

Rather than dividing energy conversion processes into those in spontaneous direction and those in reverse direction, in the PETH, changes in nature are divided into spontaneous changes and reversible-like changes. The essence of reversible-like processes is that they are characterized in triadic relations. Emden famously wrote "In the huge manufactory of natural processes, the principle of entropy occupies the position of manager, for it dictates the manner and method of the whole business, whilst the principle of energy [conservation] merely does the bookkeeping, balancing credits and debits." [37] One way to think in the triadic framework is to think in terms of the triad of EGP, heat from heat reservoir, and mechanical work (Fig. 8.13). Another way to think in the *generalized* triadic framework is the triad of manager powered by stock energy (i.e., stock EGP), natural EGP, and useful outcome (e.g., low temperature heat for buildings). In all

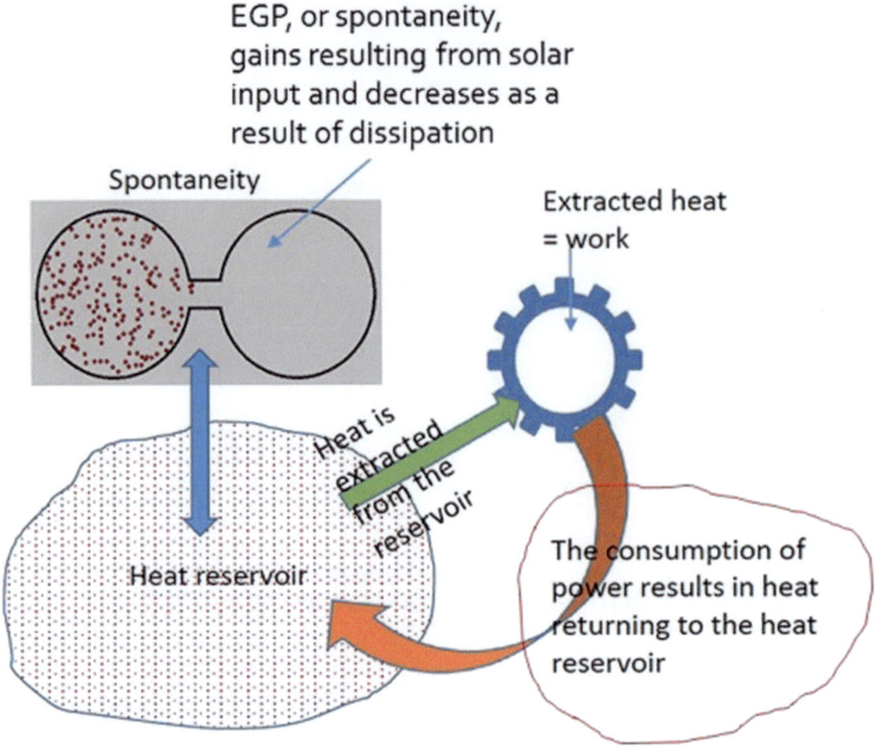

Fig. 8.13 Production of work results from the extraction of heat powered by EGP. Note that all reversible-like processes have their entropy growth delayed. However, eventually all work produced will be dissipated or the orders created by work will be destroyed so that ultimately the delayed entropy growth will be realized resulting in heat as the ultimate end results

cases in the triadic framework, efficacious management through entropic thinking is key to success.

The predicative entropic theory embraces the ontological position of the mechanical theory rejecting both the ontology of the caloric theory and the ontology of energeticism. But, it departs from the mechanical theory of heat in placing the second law as its centerpiece with the implication that entropy growth potential is the driver of all change with energy or available energy to be its proxy. This is a radically different understanding of the second law, which is well expressed by Joshua Floyd [38]:

> The point of re-examining the second law in the context of the sustainability of evolutionary systems is not to say everything is OK with the industrial age—far from it. The point is to show…that the tendency for living systems to become more complex is not somehow in defiance of the second law. In fact, it is more meaningful to characterise the phenomenon described by the second law as a fundamental *enabler* of all activity, and hence of all life.

> …Clearly, human techno-economic systems have the capacity to emerge in harmful forms. But if we are to find viable pathways beyond our present global ecological, techno-economic, social and cultural problems, our efforts to create these better futures will benefit from the most comprehensive and thorough understandings of current realities and future possibilities that we can construct. Better knowledge of how our basic physical foundations operate is an essential part of this. [38:1042)]

The predicative entropic theory can be an essential part of the knowledge by clearly envisioning possibilities in the Poincare range and in the role of natural EGP.

Problems

8.1 An ideal gas of 0.1 kmol at the initial state of 298.15 K and 303.9 kPa occupies one chamber of a composite system. The other chamber (of double volume of the first) contains a vacuum. The two chambers are separated by a frictionless piston. The piston is now connected to a mechanism that balances the force exerted by the gas on the piston and is equipped with work storage capacity—and the whole composite system is submerged in a heat reservoir/bath at 298.15 K. The ideal gas undergoes an isothermal expansion from its initial volume V_1 to its final volume 3 V_1. Determine V_1, the final pressure, and the EGP of the composite system. Also, determine the entropy change of the ideal gas expansion process, heat absorbed by the ideal gas during isothermal expansion, and work produced by the ideal gas during isothermal expansion. What is the total entropy change of the composite system and the bath?

$$0.816\,\mathrm{m}^3;\, 101.3kPa;\, EGP = 0.9134\,\mathrm{kJ/K}$$

$$0.9134\,\mathrm{kJ/K}$$

$$Q = T_0\Delta S = 272.3\,\mathrm{kJ}$$

$$W = T_0\Delta S = 298.15 \cdot 0.913 = 272.3\,\mathrm{kJ}$$

$$(\Delta S)_{univ} = 0.9134 + \frac{-Q}{298.15} = 0$$

8.2 Consider the case of a cold body initially at T, which is lower than T_0 of the surrounding reservoir. Heat flows in this case out of the reservoir into the body. Show that the total entropy production corresponding to the spontaneous process of the cold body approaching thermal equilibrium with the reservoir is

$$(\Delta_P S)_{spon\tan eous} = -\frac{C_p(T_0 - T)}{T_0} + C_p \ln\frac{T_0}{T} \geq 0$$

(The above expression may be written as—by introducing $\frac{T-T_0}{T_0} \equiv x - \left[\ln\frac{1}{1+x} - x\right]$. It can be readily shown that $\left[\ln\frac{1}{1+x} - x\right] \geq 0$ when $-1 \leq x \leq 0$.)

8.3 Consider the spontaneous process as discussed in Prob. 8.2. Prescribe in terms of theoretical operating steps how one can convert the spontaneous change into a reversible change between the same end states and show that the reversible work obtained is equal to

$$T_0(\Delta_P S)_{univ.}.$$

8. 4 (a) Give the thermodynamic definition of *waste* heat.

(b) Consider a Carnot heat engine operating between a heat reservoir at $T_H = 600\,\text{K}$ (which is the system driver of the process) and a heat sink at $T_C = 300\text{K}$, and absorbing the amount of heat Q_H from the heat reservoir. What is the value of heat rejection Q_C by the Carnot heat engine to the heat sink in terms of Q_H? What is the waste heat in this case (again in terms of Q_H)?

(c) Consider a free expansion piston cylinder (which is the system-driver of the process), which is made of two chambers of equal volume V each—one of which is filled with an ideal gas and the other is evacuated (i.e., in vacuum state). The piston cylinder is perfectly insulated and its piston is allowed to expand in a "free expansion" process so that at the end of the process the gas-filled chamber volume is doubled (the evaluated chamber volume $\rightarrow 0$). What is the change in the temperature of the gas? What is the waste heat for this process?

8.5 The Carnot–Kelvin formula $W_{rev}(Kelvin) = Q_A - Q_B = Q_A\left(1 - \frac{T_B}{T_A}\right)$ suggests the energetic idea of *theoretical necessity* in "discarding a fraction of the heat" in *all* energy conversion processes, which has been the source of widespread confusion. Whereas the Wang formula

$$W_{rev}(Wang) = T_0(\Delta_P S)_{spon} = Q_{rev} - Q_{spon}$$

(where T_0 is the temperature of the heat sink [i.e., $T_0 = T_B$], and $(\Delta_P S)_{spon}$ is the entropy growth if Q_A amount of heat is transmitted spontaneously from the heat source at T_A to the heat sink at T_B) surmises the efficiency to be dependent on effective "extracting heat", thus suggesting instead the issue to be not a matter of loss of heat but efficient use of entropy growth potential for absorbing heat.

(a) Show the expression of $(\Delta_P S)_{spon}$ in terms of T_A, T_B, and Q_A, and that the substitution of the obtained expression into the Wang formula reduces the formula to the Carnot–Kelvin formula.

(b) There are two alternative explanations of the efficiency of a Carnot cycle:

(i) Higher reversible work with lower T_B is due to stronger entropy growth potential, thus, a greater amount of heat extracted.
(ii) Or, higher reversible work with lower T_B is due to smaller heat to be discarded to the sink.

Although both are factually correct for the case at hand, explain why the first of the two is a preferred explanation.

8.6 Imagine Rumford's cannon experiment: by immersing a cannon barrel in a water tank of 100 gallons, with water initially at 30 °C and the barrel was bored with a specially blunted boring tool until water starts to boil. The measured work input is 110,694 kJ. Determine the entropy growth at the end of the experiment.
Now consider a long time afterward at the completion of the boring when the water in the tank is cooled down to the surrounding temperature of 30 °C. What will be the eventual entropy growth? Explain that the eventual entropy growth is the result of the boring process and the cooling process in water (determine the value of entropy growth due to the cooling process). Going back to the initial setup of a "system" as the work reservoir of 110,694 kJ and a heat reservoir at 30 °C, we consider the process as a spontaneous dissipation of mechanical energy of 110,694 kJ into heat. Without the details of the actual processes described above, what can you tell in terms of the entropy growth and the entropy growth potential of this *mechanical energy* → *heat* example?

8.7 *Thermal spontaneity* example: The above cooling part of Prob. 8.6 can be analyzed in a similar manner to Prob. 8.2 and Prob. 8.3. Given a system of 100 gallons of water at 100 °C in the surroundings (heat reservoir) at 30 °C. What are the entropy growth and the entropy growth potential.
Spontaneous event: What is the entropy growth in a spontaneous cooling event?
Reversible event (*Heat* → *mechanical energy* example): Conversion, more precisely the extraction, of heat into mechanical energy is possible only with the existence of thermal spontaneity under the control of reversible-like management. Describe such a reversible event and explain why such an event entails no entropy growth.
Explain why both events are driven by entropy growth potential.

8.8 Show in detail steps the derivation of the generalized Carnot–Kelvin formula.

8.9 An isolated composite system consists of gaseous subsystem$^{(1)}$ and subsystem$^{(2)}$ (same gas in both) of equal mole numbers ($N^{(1)} = N^{(2)} = N$), initially at p and T_1, and p and T_2, respectively. They are insulated from each other but connected pneumatically with a valve in closed position initially. After the valve opens, the composite system eventually reaches thermal equilibrium internally. Assuming constant c_V and c_p, the final equilibrium temperature, T_m, will be

$$T_m = \frac{T_1 + T_2}{2}$$

Show that the total entropy change of the composite system is

$$\Delta S = Nc_p \ln \frac{T_m^2}{T_1 T_2} = Nc_p \ln \frac{(T_1 + T_2)^2}{4T_1 T_2}.$$

Explain why the composite system entropy change equals the process entropy production $\Delta S = (\Delta_P S)_{spon}$.

8.10 Consider the second example in Sect. 8.4 as shown in Fig. 8.5: Assume Nc_p = 10 kJ/K (the total heat capacity of the composite system is 20 kJ/K), T_1 = 450 K and T_2 = 200 K, the temperature of the reservoir T_0 = 325 K. Calculate $(\Delta_P S)_{spon}$. Prescribe in terms of theoretical operating steps how one can convert the spontaneous change into a reversible change between the same end states and show that the reversible work obtained is equal to

$$T_0(\Delta_P S)_{spon}.$$

Hint: The theoretical operating steps will be a two-stage operation—the first stage, 1 → 2S, is an *isentropic* stage of a Carnot engine operating between the two subsystems. Because it is an isentropic stage

$$T^{(1)} = \frac{450 \times 200}{T^{(2)}} \text{ and } T^{(1)} = T^{(2)} = T_{2S} = \sqrt{450 \times 200} = 300\text{K}$$

Correspondingly, the Carnot engine produces

$$W_{1 \to 2S} = \int dW = \int_{200\,\text{K}}^{300\,\text{K}} Nc_p \left[\frac{T^{(1)}}{T^{(2)}} - 1 \right] dT^{(2)}$$

The second stage, 2S → 2, is a heating of the composite system/thermal mass toward 325 K.

$$(\Delta_P S)_{spon} - 1.6009\,\text{kJ/K}$$

$$W_{1\to 2S} = 500\,\text{kJ}$$

$$W_{2S\to 2} = 20.278\,\text{kJ}$$

$$Q_{rev} = \int_{300}^{325} 20 \times \frac{325}{T_{system}} dT_{system} = 520.278\,\text{kJ}$$

8.11 Problem 8.10 can be considered alternatively in terms of a reservoir T_0 = 300 K. The spontaneous event is now a cooling event of the composite system from an average temperature of 325–300 K (in both subsystems). The reversible event is now an isentropic event of the composite system with a "silent" or "non-participating" reservoir. Determine Q_{spon}. Show that a simple system of heat capacity 20 kJ/K initially at 325 K undergoing cooling to 300 K will also experience the same Q_{spon}. Now consider the transformation of both systems to their corresponding reversible events. Determine their respective Q_{rev} and the corresponding reversible works.

$$-\,500\,\text{kJ}$$

$$(Q_{rev})_{first} = 0$$

$$(Q_{rev})_{second} = -4803\,\text{kJ}$$

$$(W_{rev})_{first} = 500\,\text{kJ}$$

$$(W_{rev})_{second} = -4803 - (-500) = 19.7\,\text{kJ}$$

(NOTE: All the Q_{spon} heat of the first system is converted into work since the heat is associated with large spontaneity, while only a small fraction of Q_{spon} of the second system is converted into work since it is associated with a much smaller spontaneity. The production of work is principally a matter of spontaneity.)

8.12 Two blocks A and B are initially at 100 and 500 °C, respectively. They are brought together and isolated from the surroundings. They are allowed to reach a final state of internal thermal equilibrium by an isentropic process brought about with a Carnot heat engine operating between the two blocks. Determine the final equilibrium temperature of the blocks and the useful work produced by the Carnot heat engine, which manages the entropic drive force of the isolated composite system approaching internal thermal equilibrium isentropically. Block A is aluminum $[c_p = 0.900\,\text{kJ/kg}\cdot\text{K}]$ with m_A= 0.5 kg and block B is copper $[c_p = 0.386\,\text{kJ/kg}\cdot\text{K}]$ with m_A= 1.0 kg.

$$T_0 = 522.35\,\text{K},\ 29.67\,\text{kJ}$$

(Note: The difference between W [29.67 kJ] here and the product of T_0 and entropy change in Prob. 3.8 [=28.7 kJ] is the small amount of work derived from the reversible cooling of the whole two blocks from 557.85 to 522.35 K)

References

1. Atkins P (2010) The Laws of Thermodynamics. Oxford Univ Press
2. Wall G (2009) EXERGETICS Bucaramanga 2009
3. Joule JP (1850) On the mechanical equivalent of heat. Phil Trans R Soc 140:61–82
4. Carnot S (1878) Reflexions sur la Puissance Motrice du Feu, Carnot H (ed) (Fragments of writings published posthumously by H. Carnot). Gauthier-Villars, Paris, France
5. Fermi E (1956) Thermodynamics. Dover (p. 48)
6. Mach E (1896) Principles of the Theory of Heat. Reidel, Dordrecht; Verlag von J. A. Barth, Leipzig
7. Callendar HL (1911) The caloric theory of heat and Carnot's principle. Proc Phys Soc London 23:153–189
8. Falk G (1985) Entropy, a resurrection of caloric—a look at the history of thermodynamics. Eur J Phys 6:108–115
9. Truesdell C (1980) The Tragicomical History of Thermodynamics: 1822–1854. Springer, New York
10. Job G, Lankau T (2003) How harmful is the first law? Ann NY Acad Sci 988:171–181
11. Fuchs H (2010) The Dynamics of Heat (Springer-Verlag, New York)
12. Mares JJ et al. (2008) Phenomenological approach to the caloric theory of heat. Thermochimica Acta 474:16–24
13. Daub EE (1970) Entropy and dissipation. Historical Studies in the Physical Sciences 2:321–354
14. Prigogine I (1969) Structure, Dissipation and Life. Theoretical Physics and Biology. Versailles (1967), North-Holland Publ. Company, Amsterdam. It is in this communication that the term "structure dissipative" was used for the first time.
15. Prigogine I (1977) Biographical Prepared for the occasion of 1977 Nobel Prize in Chemistry
16. Callen HB (1st edition, 1960; 2nd edition, 1985) Thermodynamics and an Introduction to Thermostatistics. Wiley, New York
17. Wang LS (2017) The second law: From Carnot to Thomson-Clausius, to the theory of exergy, and to the entropy-growth potential principle. Entropy 2017, 19, 57
18. Helm G (1887) Die Lehre von der Energie, histoisch-kritisch entwickelt. A Felix, Leipzig
19. Deltete RJ (2012) Planck, Ostwald, and the second law of thermodynamics. The Journal of the International Society for the History of Philosophy of Science 2(No. 1):121–146
20. Deltete RJ (1999) Helm and Boltzmann: Energetics at the Lubeck Naturforscherversammlung. Synthese 119(No.1/2):45–68
21. Smil V (2017) Energy and Civilization, A History. MIT Press, Cambridge, MA
22. Kondepudi D, Prigogine I (1998) Modern Thermodynamics: From Heat Engines to Dissipative Structures. Wiley, New York
23. Prigogine I (1947) Etude Thermodynamique des Phénomènes Irréversibles. Dunod, Paris
24. Coveney PV (1988) The second law of thermodynamics: entropy, irreversibility and dynamics. Nature 333(No.2): 409–415. In the article, Coveney wrote, "From an epistemological viewpoint, the contributions of Prigogine's Brussels School are unquestionably of signal importance. The myth of a completely timeless, deterministic Universe is henceforth replaced by a world in which static affairs are enlarged to embrace the probabilistic kinetics of process; in which reversibility and irreversibility are accorded equal objectivity; and in which

the notions of 'being' and 'becoming' are unified within a single conceptual framework." (p. 414)
25. Kondepudi D (2008) Introduction to Modern Thermodynamics. Wiley, Hoboken, New Jersey
26. Poincaré H (1913) Science and Hypothesis. The Science Press, Lancaster, PA (pp. 122–123)
27. Lotka AJ (1922) Natural selection as a physical principle. Proceedings of National Academy of Science (PNAS) 8 (No. 6):151–154
28. Wang LS (2013) Exergy or the entropic drive. Int J Exergy 12 (No.4):491–521
29. Wang LS (2014) Entropy growth is the manifestation of spontaneity. J Thermodyn 2014 (9 pages; 7 December 2014) 387698
30. Peirce CS (1868) On a new list of categories. Proceedings of the American Academy of arts and Sciences 7: 287–298 (presented May 14, 1867)
31. Burch R (1991) A Peircean Reduction Thesis: Foundations of Topological Logic. Texas Univ. Press
32. Thomson W (Lord Kelvin) (1911) Mathematical and Physical Papers of William Thomson 1:1–571. Cambridge Univ Press
33. Smith C, Wise MN (1989) Energy and Empire: A Biographical Study of Lord Kelvin. Cambridge Univ Press
34. Young J (2015) Heat, work and subtle fluids: a commentary on Joule (1850) 'On the mechanical equivalent of heat'. Phil Trans R Soc A 373 (Issue 2039): 20140348
35. Uffink J (2001) Bluff your way in the Second Law of Thermodynamics. Stud Hist Phil Sci Part B: 2001 Studies in Hist and Philo of Modern Physics 32(No.2):305–395
36. Planck M (1969) Treatise on Thermodynamics, 3rd edition. Dover, New York
37. Emden R (1938) Why do we have winter heating? Nature 141:908–909
38. Floyd J (2007) Thermodynamics, entropy and disorder in futures studies. Futures 39 (2007):1029–1044

Applications to Special States of Thermodynamic Equilibrium: Gibbsian Thermodynamics for Physical and Chemical Applications

9

Abstract

Gibbs demonstrated in 1878 that the entropy principle (introduced 1865) can be a foundation for the general theory of thermodynamic equilibrium as he noted, "…the general increase of entropy which accompanies the changes occurring in any isolated material system [suggests] that when the entropy of the system has reached a maximum, the system will be in a state of equilibrium." Such extreme principle has already been discussed in Chap. 7. This chapter provides a more complete but condensed treatment of Gibbsian equilibrium thermodynamics. Though this chapter is not the core of this book, it is noted that its material is easy to grasp due to the mathematical elegance of the content and every engineer (target reader of the book) can benefit from its mastery.

Keywords

Fundamental functions of state (e.g., Gibbs U-V-S surface) · Open systems · Multicomponent closed systems · Gibbsian thermodynamics · The Euler equation · Maxwell relations · Basic tools for determining thermodynamic functions · Chemical reaction and combustion · Enthalpy of formation · Chemical equilibrium

L.-S. Wang, *A Treatise of Heat and Energy*, Mechanical Engineering Series,
https://doi.org/10.1007/978-3-030-05746-6_9

9.1 The Fundamental Functions of State and the Fundamental Differentials

J. W. Gibbs (1839–1903)

We present the formulation of Gibbsian equilibrium thermodynamics and discuss its applications to physical and chemical systems. These are not only important for the thermodynamic treatment of physical and chemical systems but also provide useful tools of property data required in the engineering treatment of thermodynamic systems.

Equation (64) in Chapter 5 is reproduced here

$$dU = TdS - pdV \tag{64}$$

It suggests that the set of *U*-*S*-*V* is a unique one. Their functional relationship may be written as

$$f(U, S, V) = 0 \tag{141}$$

Equation (64) is known under the name of *fundamental differential* or *principal exact differential.* Correspondingly, Eq. (141) may be referred to as a *fundamental thermodynamic function of state* in the sense that its differentiation results in Eq. (64).

While a general thermodynamic functional relationship contains information on the nature of equilibrium states of a system in terms of some of its equilibrium properties, a fundamental thermodynamic function captures the *complete* thermodynamic information about the system. Callen [1:28] wrote "if the fundamental relation [fundamental function] of a particular system is known all conceivable thermodynamic information about the system is ascertainable from it." The reader is reminded again that this statement is correct in so far as all conceivable information about the system in and by itself not with regard to how the system interacts with its surroundings in the interconnected universe. This chapter treats the problems of the former kind while Chap. 10 will comment on the matter of the latter.

Beginning with the energy representation of the fundamental function, Eq. (141)

$$U = U(S, V) \tag{141A}$$

we compute the first differential

$$dU = \left(\frac{\partial U}{\partial S}\right)_V dS + \left(\frac{\partial U}{\partial V}\right)_S dV \tag{142}$$

Comparing Eq. (142) with Eq. (64)

$$T \equiv \left(\frac{\partial U}{\partial S}\right)_V = T(S, V) \tag{143}$$

$$-p \equiv \left(\frac{\partial U}{\partial V}\right)_S = -p(S, V) \tag{144}$$

One notes that Eqs. (143) and (144) are two functions of states, while Eq. (141A) is a fundamental function of state. If the concrete expression of EQ. (141A), i.e., a fundamental equation of state, is known, the concrete expressions of Eqs. (143) and (144) individually can be readily obtained by taking the differentiation of the fundamental equation of state.

Conversely, in the opposite direction, it requires the use of both equations of state to construct the single fundamental equation of state, as we shall show in the following example by making use of the entropy representation of the fundamental function.

Consider the inverted case of the entropy representation of Eq. (141A)

$$S = S(U, V) \tag{145}$$

It follows that, by comparing Eq. (145) with Eq. (65),

$$dS = \frac{1}{T} dU + \frac{p}{T} dV \tag{65}$$

the corresponding functions of states in the entropy representation are

$$\frac{1}{T} = \left(\frac{\partial S}{\partial U}\right)_V = \frac{1}{T}(U, V) \tag{146}$$

$$\frac{p}{T} = \left(\frac{\partial S}{\partial V}\right)_U = \frac{p}{T}(U, V) \tag{147}$$

9.1.1 Equations of State for Ideal Gases and the Ideal Gas Fundamental Equation of State

Consider the case of ideal gases. The thermal equations of state, Eq. (3) in Chapter 1, is reproduced here

$$\frac{p}{T} = \frac{NR}{V} \tag{3}$$

The caloric equation of state, Eq. (34B), as reproduced, is

$$U = U_0 + Nc_V(T - T_0)$$

If it is assumed, arbitrarily, $U_0 = Nc_V T_0$, it becomes

$$U = Nc_V(T)T \text{ or } \frac{1}{T} = \frac{Nc_V}{U} \tag{34C}$$

Substitution of Eqs. (4) and (34C) into (65) yields

$$dS = \frac{Nc_V}{U}dU + \frac{NR}{V}dV \tag{148}$$

Equation (148) can be integrated to obtain the fundamental equation of state for ideal gases

$$S = S_0 + Nc_V \ln\frac{U}{U_0} + NR\ln\frac{V}{V_0} \tag{149}$$

As a fundamental equation of state, Eq. (149), all conceivable thermodynamic information about an ideal gas can be derived from it. First, the thermal equation of state, Eq. (4), and the caloric equation of state Eq. (34C), result immediately from the differentiation of Eq. (149), in view of Eqs. (146) and (147),

$$\frac{1}{T} = \left(\frac{\partial S}{\partial U}\right)_V = \frac{Nc_V}{U} \tag{34C}$$

and

$$\frac{p}{T} = \left(\frac{\partial S}{\partial V}\right)_U = \frac{NR}{V} \tag{3}$$

The differentiation of Eq. (34C), in turn, yields the specific heat c_V,

$$c_V \equiv \frac{1}{N}\left(\frac{\partial U}{\partial T}\right)_V = c_V(T)$$

The differentiations of Eq. (4), $V = V(T,p) = \frac{NRT}{p}$, in turn, yield the volume expansivity, β, and the isothermal compressibility, κ_T,

$$\beta \equiv \frac{1}{V}\left(\frac{\partial V}{\partial T}\right)_p = \frac{1}{T} \tag{150}$$

$$\kappa_T \equiv \frac{-1}{V}\left(\frac{\partial V}{\partial p}\right)_T = \frac{1}{p} \tag{151}$$

A general consideration of these properties will be used in Sect. 9.5.2.

9.2 Open Systems

The first law equation has been expressed in terms of closed systems. It can be generalized into the expression of energy conservation for open systems. It will be sufficient to consider one stream through which matter flows into or out of an open system (a control volume system) enclosed within a control volume boundary surface *C*. We shall first suppose that *C* is a rigid surface and the system within is filled with a homogeneous substance whose state is uniform.

The opening of boundary surface *C* is assumed to be a cylindrical channel as shown in Fig. 9.1 (reproduced from Kestin-A Course in Thermodynamics, Fig. 13.8, p. 582), in which a general deformable *C* is shown. During a period of time *dt*, it is assumed that a mass *dm*, whose properties will be described by symbols with the subscript ξ, is forced into the system or discharged from it. The internal energy of the system changes correspondingly from U to $U + dU$.

In order to write down the First Law for a closed system, it is necessary to draw a boundary including *dm* throughout the period *dt*. Consider the case of matter flowing into the system. One can then imagine a fictitious cylinder piston which is at the beginning of the period, state *a*, at the position defining the cylindrical

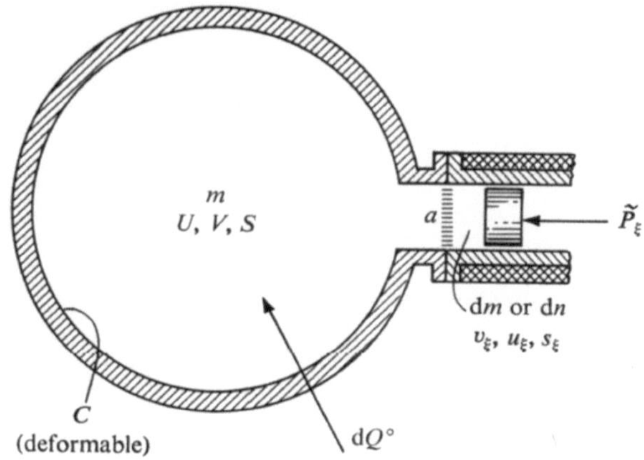

Fig. 9.1 Application of the first law equation to an open system (reproduced from Kestin-A Course in Thermodynamics, Fig. 13.8, p. 582)

volume of V_ξ – where $V_\xi = v_\xi(dm)$ [v_ξ is the mass-specific volume] just outside the opening of C. Imagine the fictitious piston pushes the substance mass dm slowly and quasi-statically (i.e., internal reversibly) into C during the period. At the end of the period, state b, the piston is positioned at the opening of C. At the same time, the system exchanged a quantity of heat δQ (absorbed a quantity of heat if δQ is positive) with the surroundings and performed a quantity of useful work δW_{useful}. The internal energy at state a is thus, $U + u_\xi dm$, and at state b it is $U + dU$, where dU denotes a change in internal energy within C during $a \to b$. The corresponding change in the volume of the imaginary closed system defined by the fictitious cylinder piston and C is equal to the negative of $V_\xi = v_\xi(dm)$. That is, a compression work $pdV_\xi = pv_\xi dm$ is done on the system. Hence the first law, Eq. (22) in Chapter 3, can be written

$$(U + dU) - (U + u_\xi dm) = \delta Q - \left(\delta W_{useful} - pv_\xi dm\right)$$

or

$$dU = (\delta Q + [u_\xi + pv_\xi]dm_{in}) - \delta W_{useful} \tag{152}$$

Note that $u + pv = h$. It can be readily generalized to

$$dU = (\delta Q + h_\xi dm_{in}) - \left(\delta W_{useful} + h_\xi dm_{out}\right) \tag{152A}$$

or in the case of multiple inlets and outlets

$$dU = \left(\delta Q + \sum_{in\_i} (h_\xi)_i dm_i\right) - \left(\delta W_{useful} + \sum_{out\_j} (h_\xi)_j dm_j\right) \tag{152B}$$

9.3 Open Systems with Semi-permeable Membrane Opening, and Multicomponent Closed Systems

Gibbs introduced the idea of *chemical potential* in his magnum opus entitled *On the Equilibrium of Heterogeneous Substances* [2, (1875) and (1878)]. The heterogeneous substance system may be either an open system with a semi-permeable membrane opening or a closed system of multicomponents.

Consider the former case of a mixing substance system in C (of uniform state at U-V-S-p-T-m [or N]), while a homogeneous substance ξ of mass dm in the cylinder inlet (or outlet), with a semi-permeable membrane partition separating C and the cylinder [3:582–583]. This implies that the pressure force of the cylinder piston equals the partial pressure p_ξ of the ξ component of the mixing substance. It will be assumed that the temperature of the homogeneous substance ξ equals that of the mixing substance. In this manner, adding or extracting mass dm will be reversible, and this will enable us to write the second law as well as the first law for the change in the open system of mixing substance.

We shall consider no useful work W_{useful} here but admit a deforming boundary C as shown in Fig. 9.1 with work due to deformable C (which is absent in Eq. (152)),

$$\delta W = pdV$$

Note that p is the total pressure of the mixing substance, not the partial pressure p_ξ of its ξ-component, while $h_\xi = u_\xi - p_\xi v_\xi$. With these stipulations, Eq. (152) becomes

$$dU = (\delta Q + h_\xi dm) - pdV \tag{153}$$

Since the process is reversible, and the system entropy exchange assumes $\delta_E S = s_\xi dm + \frac{\delta Q}{T}$, Eq. (84B) in Chapter 6 becomes

$$dS = \delta_E S = s_\xi dm + \frac{\delta Q}{T}$$

Substitution of which into Eq. (153) yields

$$dU = TdS - pdV + (h_\xi - Ts_\xi)dm \tag{154}$$

The coefficient of dm is known as the *chemical potential*, which was first introduced by Gibbs. We shall opt for the molar representation dN replacing the mass dm. Denote the mole-based chemical potential by μ. Equation (154) is then written as

$$dU = TdS - pdV + \mu dN \tag{154A}$$

for a system interacting with a single component.

Equation (154A) can be generalized either for open systems interacting with n components each with its corresponding membrane partition (with no bulk mass flow in or out other than diffusion through membranes) or for closed systems (with no mass exchange but with possible Q or/and W exchange) of multiple components with chemical reaction (see below)

$$dU = TdS - pdV + \sum_{i=1}^{n} \mu_i dN_i \tag{154B}$$

This is the fundamental differential for multicomponent systems, which is also known under the name of *Gibbs equation.* Note that

$$\mu_i = h_i - Ts_i$$

and the corresponding fundamental function of state

$$U = U(S, V, N_1, \ldots N_n) \tag{155}$$

Equations (154B) and (155) form the starting point of Gibbsian thermodynamics.

Callen noted "The postulatory formulation of thermodynamics features states, rather than processes, as fundamental constructs. Statements about Carnot cycles and about the impossibility of perpetual motion of various kinds do not appear in the postulates, but state functions, energy, and entropy become the fundamental concepts. An enormous simplification in the mathematics is obtained, for processes then enter simply as differentials of the state functions" [1:viii (1960)]. While I disagree with this statement in the general formulation of thermodynamics concerning the central issues of heat and work, Callen's comment does apply to Gibbsian thermodynamics, which addresses physical/chemical properties and chemical and phase equilibrium problems and in which the questions of heat and work (especially the useful work) and the concept of reversibility are absent. It is useful to note that, in this regard, Eqs. (154B) and (155) are universally valid, since in the equations the nature of processes is immaterial.[1]

[1]It is useful to remind the readers that among the following set of equations:

$$dU = \delta Q - \delta W$$
$$dU = TdS - \delta W$$
$$dU = \delta Q - pdV$$
$$dU = TdS - pdV \text{ or } dU = TdS - pdV + \textstyle\sum_{i=1}^{n} \mu_i dN_i,$$

only the first and the fourth are always valid for all processes, while the second and third are valid for processes that meet internal reversibility condition (see Sect. 6.5).

9.4 Formal Structure of Gibbsian Thermodynamics

The formal structure of Gibbsian thermodynamics is once again expressed in terms of the fundamental function of state

$$U = U(S, V, N_1, \ldots N_n) \tag{155}$$

with the corresponding fundamental differential or exact differential

$$dU = TdS - pdV + \sum_{i=1}^{n} \mu_i dN_i \tag{154B}$$

Equations (155) and (154B) imply the following functions of state:

$$T \equiv \left(\frac{\partial U}{\partial S}\right)_{V,N_1,\ldots,N_n} = T(S, V, N_1, \ldots, N_n) = T(s, v) \tag{156}$$

$$-p \equiv \left(\frac{\partial U}{\partial V}\right)_{S,N_1,\ldots,N_n} = -p(S, V, N_1, \ldots, N_n) = -p(s, v) \tag{157}$$

$$\mu_j \equiv \left(\frac{\partial U}{\partial N_j}\right)_{S,V,N_{i\neq j}} = \mu_j(S, V, N_1, \ldots, N_n) = \mu(s, v) \tag{158}$$

The second equalities in Eqs. (156)–(158) apply for single-component simple systems, for which temperature, pressure, and chemical potential can be written as functions of molar entropy and molar volume. For such (single-component simple) systems, the fundamental function and the fundamental differential may be written in terms of molar variables of s and v as

$$u = u(s, v) \tag{141B}$$

$$du = Tds - pdv \tag{64A}$$

9.4.1 The Euler Equation

The fundamental function, Eq. (155), is a homogeneous first-order function. An important consequence of this mathematical property is that, for any λ,

$$U(\lambda S, \lambda V, \lambda N_1, \ldots, \lambda N_n) = \lambda U(S, V, N_1, \ldots, N_n)$$

Differentiating which with respect to λ

$$\frac{\partial U(\lambda S,\ldots)}{\partial(\lambda S)}\frac{\partial(\lambda S)}{\partial\lambda}+\cdots+\frac{\partial U(\ldots,\lambda N_i,\ldots)}{\partial(\lambda N_i)}\frac{\partial(\lambda N_i)}{\partial\lambda}+\cdots=\frac{d(\lambda U)}{d\lambda}$$
$$=U(S,V,N_1,\ldots,N_n)$$

or

$$\frac{\partial U(\lambda S,\ldots)}{\partial(\lambda S)}S+\cdots+\frac{\partial U(\ldots,\lambda N_i,\ldots)}{\partial(\lambda N_i)}N_i+\cdots=U(S,V,N_1,\ldots,N_n)$$

The equation is true for any λ and in particular for $\lambda = 1$, in which case, it takes the form in view of Eqs. (156) to (158)

$$TS-pV+\sum_{i=1}^{n}\mu_i N_i=U \tag{159}$$

This is known as the Euler equation, which is useful as we shall see in the following.

9.4.2 Alternative Fundamental Functions and Fundamental Differentials

The fundamental equation of state, Eq. (149) for ideal gases is

$$S=S_0+Nc_V\ln\frac{U}{U_0}+NR\ln\frac{V}{V_0}$$

which may be rewritten as

$$U=U_0\left(\frac{V}{V_0}\right)^{-(k-1)}exp\left(\frac{S-S_0}{Nc_V}\right)$$

The information contained in the resulting fundamental function, U as function of S and V, is likewise all inclusive.

It is sometimes useful to choose a different set of independent variables, and the question arises whether the resulting U-function (of the new set of independent variables, e.g., S-p or T-V instead of the original S-V) retains the same usefulness as a fundamental function. In these cases for the purpose of preserving the all-inclusive information the dependent variable, U can be transformed by a

Legendre transformation [1:137–149] into a new variable, i.e., an alternative fundamental functional variable, ψ.

Consider the case of the transformation of U-S-V into ψ-S-p, in which ψ is the new dependent variable

$$\psi \equiv U - \left(\frac{\partial U}{\partial V}\right)_{S,N_i} V = U - (-p)V = U + pV \tag{100}$$

which is identified as the enthalpy H.

Consider the case of the transformation of U-S-V into ψ-T-V, one gets

$$\psi \equiv U - \left(\frac{\partial U}{\partial S}\right)_{V,N_i} S = U - TS \tag{100}$$

which is the Helmholtz function A_H (we already used the term in Sect. 7.1).

Consider next the case of the transformation of A_H-T-V into ψ-T-p, one gets

$$\psi(= G) = A_H - \left(\frac{\partial A_H}{\partial V}\right)_{T,N_i} \cdot V = A_H + pV = U - TS + pV \tag{101}$$

which will be called the Gibbs function, G.

The complete set of new dependent variables is summarized in Table 9.1 (for simplicity this part of treatment [and in Sects. 9.5 and 9.6] is limited to single-component simple systems only).

One notes in view of the Euler equation that the Gibbs function is

$$G = U + pV - TS = \sum_{i=1}^{n} \mu_i N_i \tag{160A}$$

which is consistent, in view of $\mu_i = h_i - Ts_i = g_i$, with

Table 9.1 Thermodynamic potentials and their independent variables

Independent variables, X and Y	Dependent thermodynamic functional variable ψ derived from Legendre transformation	Name of the alternative dependent variable ψ	Fundamental function $\psi = \psi(X, Y)$
S and V	U		U-S-V
S and p	$U + pV$	Enthalpy H	H-S-p
T and V	$U - TS$	Helmholtz function A_H	A_H-T-V
T and p	$H - TS = A_H + pV$ $= U + pV - TS$	Gibbs function G	G-T-p

$$G = g_1N_1, \ldots, g_nN_n \tag{160B}$$

The corresponding fundamental differentials, or principal exact differentials, to (154B) are

$$dH = TdS + Vdp + \sum_{i=1}^{n} \mu_i dN_i \tag{161}$$

$$dA_H = -SdT - pdV + \sum_{i=1}^{n} \mu_i dN_i \tag{162}$$

$$dG = -SdT + Vdp + \sum_{i=1}^{n} \mu_i dN_i \tag{163}$$

It is important to note that the set of alternative fundamental functional *variables*, U-H-A_H-G, in Table 9.1/Table 7.1 is not identical with the set of thermodynamic *potentials*, entropy S, the Helmholtz function A_H, and the Gibbs function G as shown in Table 7.2 even though the corresponding fundamental functions (functional *relationships*) are the same. The alternative fundamental functional variables and their independent variables are used for the purpose of retaining the *complete* thermodynamic information about the system (as we shall see below); the alternative thermodynamic potentials are used for determining equilibrium states based on the maximization or minimization of the potentials (Sects. 7.1 and 9.6). This subtle difference is rarely made clear in the thermodynamic literature.

9.4.3 The Maxwell Relations

Consider function $\psi = \psi(X, Y)$ in Table 9.1 and its differential

$$d\psi = \left(\frac{\partial\psi}{\partial X}\right)_Y + \left(\frac{\partial\psi}{\partial Y}\right)_X$$

The theory of multivariable functions gives us the following relation:

$$\left(\frac{\partial^2\psi}{\partial Y\partial X}\right) \equiv \left[\frac{\partial}{\partial Y}\left(\frac{\partial\psi}{\partial X}\right)_Y\right]_X = \left[\frac{\partial}{\partial X}\left(\frac{\partial\psi}{\partial Y}\right)_X\right]_Y \equiv \left(\frac{\partial^2\psi}{\partial X\partial Y}\right) \tag{164}$$

In addition, we have the following second and third relations for the set any three variables x-y-z (not necessarily the sets of fundamental functional ones of ψ-X-Y):

$$\left(\frac{\partial x}{\partial y}\right)_z = \frac{1}{\left(\frac{\partial y}{\partial x}\right)_z} \tag{165}$$

$$\left(\frac{\partial x}{\partial y}\right)_z \left(\frac{\partial y}{\partial z}\right)_x \left(\frac{\partial z}{\partial x}\right)_y = -1 \tag{166}$$

Application of Eq. (164) above to Eqs. (154B), (161)–(163) yields the following Maxwell relation:

$$\left(\frac{\partial T}{\partial V}\right)_S = -\left(\frac{\partial p}{\partial S}\right)_V \tag{167}$$

$$\left(\frac{\partial T}{\partial p}\right)_S = -\left(\frac{\partial V}{\partial S}\right)_p \tag{168}$$

$$\left(\frac{\partial S}{\partial V}\right)_T = -\left(\frac{\partial p}{\partial T}\right)_V \tag{169}$$

$$-\left(\frac{\partial S}{\partial p}\right)_T = \left(\frac{\partial V}{\partial T}\right)_p \tag{170}$$

The Maxwell relations, Eqs. (167) to (170), will be very useful in the determination of thermodynamic properties as shown in the following Sect. 9.5.

9.5 Determination of Thermodynamic Properties Based on Measurable Data

The set of measurable thermodynamic properties is as follows: the properties of mechanics: mass m, mole number N, volume V, force, pressure p; the properties of thermodynamic systems that are introduced by thermodynamic theory: temperature T, volume expansivity β, isothermal compressibility κ_T, molar heat capacity c_p. The case of c_V is an interesting one; c_V is measurable for gaseous substances, but not so for neither liquids nor solids because it is impossible to keep the volume for a system constant during such measurement.

9.5.1 Basic Tools

It can now be demonstrated that all the other thermodynamic properties (thermodynamic state variables) can be determined in terms of these measurable properties.

The basic "tools" are the set of fundamental differentials, Eqs. (154B), (161)–(163), and the set of the Maxwell relations, Eqs. (167)–(170). The former equations are written in the following forms by considering, for instance, U as a function of any pair of independent variables x and y:

$$\left(\frac{\partial U}{\partial x}\right)_y = T\left(\frac{\partial S}{\partial x}\right)_y - p\left(\frac{\partial V}{\partial x}\right)_y \tag{171}$$

The rest of the three equations can be similarly written as

$$\left(\frac{\partial H}{\partial x}\right)_y = T\left(\frac{\partial S}{\partial x}\right)_y + V\left(\frac{\partial p}{\partial x}\right)_y \tag{172}$$

$$\left(\frac{\partial A_H}{\partial x}\right)_y = -S\left(\frac{\partial T}{\partial x}\right)_y - p\left(\frac{\partial V}{\partial x}\right)_y \tag{173}$$

$$\left(\frac{\partial G}{\partial x}\right)_y = -S\left(\frac{\partial T}{\partial x}\right)_y + V\left(\frac{\partial p}{\partial x}\right)_y \tag{174}$$

Equations (171)–(174) and Eqs. (167)–(170) enable the determination of thermodynamic properties as shown in the following.

9.5.2 Examples

Before the demonstration by examples, recall the following definitions of molar heat capacity and other derivative forms of properties:

$$c_p = \frac{1}{N}\left(\frac{\partial H}{\partial T}\right)_p = \left(\frac{\partial h}{\partial T}\right)_p = T\left(\frac{\partial S}{\partial T}\right)_p$$
$$c_V = \frac{1}{N}\left(\frac{\partial U}{\partial T}\right)_V = \left(\frac{\partial u}{\partial T}\right)_V = T\left(\frac{\partial S}{\partial T}\right)_V$$

$$\beta \equiv \frac{1}{V}\left(\frac{\partial V}{\partial T}\right)_p = -\frac{1}{V}\left(\frac{\partial S}{\partial p}\right)_T$$

in view of (170)

$$\kappa_T \equiv \frac{-1}{V}\left(\frac{\partial V}{\partial p}\right)_T$$

One of the most common examples of property relationships is the construction of $V = V(x, y)$ and $\psi = \psi(x, y)$, where x and y can be any pair of independent variables. The most common and useful pairs are (T, p), (T, V), or (p, V). Consider the case $V = V(T, p)$

$$dV = \left(\frac{\partial V}{\partial T}\right)_p dT + \left(\frac{\partial V}{\partial p}\right)_T dp = V(\beta dT - \kappa_T dp)$$

If a reference point (T_0, p_0) is chosen and data of $\beta(T, p)$ and $\kappa_T(T, p)$ are available for a given substance, the above equation can be used to construct the equation of state for the substance (including cases of liquid or solid).

Consider the example $U = U(T, V)$

$$dU(T, V) = \left(\frac{\partial U}{\partial T}\right)_V dT + \left(\frac{\partial U}{\partial V}\right)_T dV$$

The coefficient of the first term of RHS is

$$\left(\frac{\partial U}{\partial T}\right)_V = Nc_V$$

The coefficient of the second term of RHS, by using Eq. (171), becomes

$$\left(\frac{\partial U}{\partial V}\right)_T = T\left(\frac{\partial S}{\partial V}\right)_T - p\left(\frac{\partial V}{\partial V}\right)_T = T\left(\frac{\partial p}{\partial T}\right)_V - p$$

where the second equality is due to Eq. (169). It follows by using Eq. (166) and the definitions of volume expansivity and isothermal compressibility

$$\left(\frac{\partial U}{\partial V}\right)_T = T\left(\frac{\partial p}{\partial T}\right)_V - p = T\frac{\left(\frac{\partial V}{\partial T}\right)_p}{-\left(\frac{\partial V}{\partial p}\right)_T} - p = \frac{T\beta}{\kappa_T} - p$$

The differential dU then becomes

$$dU = Nc_V dT + \left(\frac{T\beta}{\kappa_T} - p\right)dV \tag{175}$$

Equation (175) is the general expression for dU valid for any substance. Considering the case of ideal gases with $\beta = \frac{1}{T}$ and $\kappa_T = \frac{1}{p}$, the second term of RHS drops out in Eq. (175), and it reduces to

$$dU = Nc_V dT + \left(\frac{T\beta}{\kappa_T} - p\right)dV = Nc_V dT \tag{175A}$$

which is in agreement with Joule's law, Eq. (34) in Chapter 3.

Now consider the case of an "idealized" solid or liquid under the assumption

$$\beta \to 0(= \mathbf{O}[\varepsilon]) \text{ and } \kappa_T \to 0\left(= \mathbf{O}\left[\varepsilon^2\right]\right)$$

where ε is an infinitesimal number; the above symbol means that the thermal expansion coefficient is an infinitesimal number and isothermal compressibility is an even smaller number of the order of magnitude of the square of β. Equation (145) becomes for this case

$$dU = Nc_V dT + \left(\frac{T\beta}{\kappa_T} - p\right)dV = Nc_V dT + T\frac{dV(= \mathbf{O}[\varepsilon])}{\frac{\kappa_T}{\beta}(= \mathbf{O}[\varepsilon])} \tag{175B}$$

The last term is an indeterminate fraction, being ${}^{\mathrm{O}[\varepsilon]}/_{\mathrm{O}[\varepsilon]}$, or $\mathrm{O}[\infty]\cdot\mathrm{O}[\varepsilon]$. Equation (175B) is, therefore, a useless expression of U.

It is, therefore, necessary to consider $U = U(T,p)$

$$dU(T,p) = \left(\frac{\partial U}{\partial T}\right)_p dT + \left(\frac{\partial U}{\partial p}\right)_T dp$$

Correspondingly,

$$\left(\frac{\partial U}{\partial T}\right)_p = T\left(\frac{\partial S}{\partial T}\right)_p - p\left(\frac{\partial V}{\partial T}\right)_p = Nc_p - pV\beta$$

$$\left(\frac{\partial U}{\partial p}\right)_T = T\left(\frac{\partial S}{\partial p}\right)_T - p\left(\frac{\partial V}{\partial p}\right)_T = -T\left(\frac{\partial V}{\partial T}\right)_p + pV\left[\frac{-1}{V}\left(\frac{\partial V}{\partial p}\right)_T\right] = V(p\kappa_T - T\beta)$$

It follows

$$dU = N\left(c_p - pv\beta\right)dT + V(p\kappa_T - T\beta)dV \tag{176}$$

In the limit of ideal gases, Eq. (176) reduces to

$$dU = N\left(c_p - R\right)dT + V(1 - 1)dV = Nc_V dT \tag{176A}$$

which is in agreement with the ideal gas result Eq. (175A). In the limit of an idealized solid and liquid as incompressible substances, i.e.,

$$\beta \to 0 \, (= \mathbf{O}[\varepsilon]) \text{ and } \kappa_T \to 0 (= \mathbf{O}[\varepsilon^2])$$

Equation (176) reduces to

$$dU = N(c_p - pv\beta)dT + V(p\kappa_T - T\beta)dV = Nc_p dT + 0 = Nc_p dT \tag{176B}$$

We have thus the interesting result that the proportionality constant of dU to dT for ideal gases is c_V, while the proportionality constant of dU to dT for "incompressible" substances is c_p. The indeterminate term in (175B) is a finite value accounting for the difference between the value of c_V and c_p, as shown in the following.

The following two dS expressions are left as Problem 8.2 of this chapter:

$$dS = \frac{Nc_V}{T} dT + \frac{\beta}{\kappa_T} dV$$
$$dS = \frac{Nc_p}{T} dT - V\beta dp$$

Note again, for incompressible substances, $\beta \to 0$ $(= \mathbf{O}[\varepsilon])$ and $\kappa_T \to 0$ $(= \mathbf{O}[\varepsilon^2])$, the first dS expression is useless involving indeterminate. For such substances, the second dS yields

$$dS = \frac{Nc_p}{T} dT,$$

Correspondingly,

$$\Delta S = Nc_p ln \frac{T_2}{T_1} \tag{71}$$

which was given as Eq. (71) in Sect. 5.3.4 without proof.

We now take the differentiation of the first dS expression with respect to T holding p constant,

$$\left(\frac{\partial S}{\partial T}\right)_p = \frac{Nc_V}{T}\left(\frac{\partial T}{\partial T}\right)_p + \frac{\beta}{\kappa_T}\left(\frac{\partial V}{\partial T}\right)_p = \frac{Nc_V}{T} + \frac{\beta}{\kappa_T} V\beta$$

Note that the LHS term is equal to $\frac{Nc_p}{T}$. It follows

$$Nc_p = Nc_V + \frac{T(Nv)\beta^2}{\kappa_T}$$

or

$$c_p - c_V = \frac{Tv\beta^2}{\kappa_T} \tag{177}$$

The two molar heat capacities will be equal if and only if the volume expansivity β becomes zero (changing from positive to negative). The only such case occurs for water at 1 atm and 4 °C. This unique water property is the reason for an important phenomenon for the survival of freshwater fishes in lakes that become frozen with ice sheets formed at the top in the wintertime.

9.5.3 Why the Whole of Fresh Water Lakes Do Not Freeze in Winter?

Water at a temperature above 4 °C turns over convectively if its temperature increases with depth due to the cold water being heavier at the top while the warm water being lighter at the bottom. Thus, a uniform temperature condition prevails quickly throughout the lake—in a state of *convective equilibrium*

$$\left(\frac{\partial T}{\partial p}\right)_S = \left(\frac{Tv\beta}{c_p}\right) = \mathbf{O}[\varepsilon]$$

according to the second dS expression in the above. Once the lake water temperature reaches 4 °C, however, the opposite density stratification—a stably stratified lake water condition (in a state of *stratified equilibrium* or *conductive equilibrium*)—occurs as a result of its unique *negative* volume expansivity: no convective turnover any more, and heat transfer takes place by conduction, which is a very slow process. Therefore, the lake freezes over to form ice at the top layer by layer. Fish can thus find a livable habitat under the ice top with the water underneath remaining at temperature 4 °C.

9.5.4 Convective Equilibrium of Atmospheric Air at Hydrostatic Equilibrium

A body of water of a lake at convective equilibrium (unless it is extraordinarily deep) is isothermal. The body of atmospheric air at convective equilibrium is not isothermal because of its height, which leads to variation in pressure and the resulting compression heating for downward-moving air or expansion cooling for up-current air. These effects may be treated as adiabatic, more accurately isentropic, processes according to the second dS expression for ideal gases

$$\left(\frac{\partial T}{\partial p}\right)_S = \left(\frac{Tv\beta}{c_p}\right) = \frac{v}{c_p} = \frac{k-1}{k}\frac{T}{p}$$

That is,

$$\frac{dT}{T} = \frac{k-1}{k}\frac{dp}{p}$$

Since the atmosphere is in hydrostatic equilibrium and an increase in height is accompanied by a decrease in pressure

$$dp = -\rho g dh = -\frac{gM}{R}\frac{p}{T}dh$$

Substitution of this into the previous equation gives

$$\frac{dT}{dh} = -\frac{k-1}{k}\frac{gM}{R}$$

Assuming $k_{air} = 1.4$; $g = 9.807$ m/s^2; $M_{air} = 28.97$; R =8.314 kJ/kmol-K, we obtain

$$\frac{dT}{dh} = -9.76 \times 10^{-3}\,\text{K/m} = -9.76\,\text{K/km}$$

This value is somewhat larger than the observed value of lapse-rate (linear decrease of temperature with altitude) of -6.5 K/km. The difference is owing to having neglected the effect of condensation of water vapor in the expanding mass of air. This effect can be factored into the prediction by equating the predicted lapse rate with the observed value

$$\frac{k_{eff}-1}{k_{eff}}\frac{gM}{R} = -6.5 \times 10^{-3}\,\text{K/m}$$

to determine an effective k of the air that incorporates the condensation of water vapor. It is found $k_{eff} = 1.235$.

9.6 Thermal Equilibrium and Mechanical Equilibrium

The state of system equilibrium is characterized by the vanishing of entropy production, which in the case of composite systems becomes the extremization (maximization or minimization) of a corresponding thermodynamic potential (Table 7.2). For the case of an isolated composite system, the appropriate thermodynamic potential is the entropy, or T_0S, which reaches a maximum at equilibrium. The idea of the maximum entropy principle was first suggested by Gibbs [2:354 (1878)]

> It is an inference naturally suggested by the general increase of entropy which accompanies the changes in any isolated material system that when the entropy of the system has reached a maximum, the system will be in a state of equilibrium.

9.6.1 Thermal Equilibrium

The suggestion can be made specific by considering an isolated thermal composite system consisting of a gaseous subsystem$^{(1)}$ and subsystem$^{(2)}$ with $c_V^{(1)}$ and $c_V^{(2)}$, and $N^{(1)}$ and $N^{(2)}$. The two subsystems are initially at $V^{(1)}$ and $T_i^{(1)} \equiv T_1$, and $V^{(2)}$ and $T_i^{(2)} \equiv T_{2,}$ respectively. They are insulated from each other initially. $V^{(1)}$ and $V^{(2)}$ will remain unchanged. The adiabatic wall separated the two subsystems is then replaced with a diathermic wall at the initial end state of i, and the composite system eventually reaches thermal equilibrium internally, $T_f^{(1)} = T_f^{(2)}$, at the final (equilibrium) end state of f. Assuming constant c_V (and c_p), the energy balance requires

$$U_i^{(1)} + U_i^{(2)} = U_f^{(1)} + U_f^{(2)}$$

or

$$\begin{aligned} 0 &= \left(U_f^{(1)} - U_i^{(1)}\right) + \left(U_f^{(2)} - U_i^{(2)}\right) \\ &= N^{(1)} c_V^{(1)} \left(T_f^{(1)} - T_1\right) + N^{(2)} c_V^{(2)} \left(T_f^{(2)} - T_i^{(2)}\right) \end{aligned}$$

The entropy of the composite system is

$$S = S^{(1)}\left(U^{(1)}, V^{(1)}\right) + S^{(2)}\left(U^{(2)}, V^{(2)}\right)$$

for which $V^{(1)}$ = constant and $V^{(2)}$ = constant, and $U^{(1)} + U^{(2)}$ = constant.

Let $\delta Q = dU^{(1)} = -dU^{(2)}$ be the amount of heat transferred from subsystem$^{(2)}$ to subsystem$^{(1)}$ in a given differential dt

$$dS = \frac{\partial S}{\partial Q}\delta Q = \frac{\partial S}{\partial U^{(1)}} dU^{(1)} = \frac{\partial S^{(1)}}{\partial U^{(1)}} dU^{(1)} + \frac{\partial S^{(2)}}{\partial U^{(2)}} \frac{\partial U^{(2)}}{\partial U^{(1)}} dU^{(1)}$$

or

$$dS = \left(\frac{1}{T^{(1)}} + \frac{-1}{T^{(2)}}\right) dU^{(1)}$$

The condition of equilibrium demands that dS vanishes at equilibrium

$$0 = (dS)_{equili} = \left(\frac{1}{T^{(1)}} + \frac{-1}{T^{(2)}}\right)_{equili} dU^{(1)}$$

for arbitrary values of $dU^{(2)}$, whence,

$$\frac{1}{T_f^{(1)}} = \frac{1}{T_f^{(2)}}$$

A numerical example. Let the two subsystems be two ideal gases with the same mole number $N^{(1)} = 2$ and $N^{(2)} = 3$ kmol and with $c_V^{(1)} = \left(3/2\right)R$ and $c_V^{(2)} = \left(5/2\right)R$. The initial temperatures of the two subsystems are $T^{(1)} = T_1 = 250$ K and $T^{(2)} = T_2 = 350$ K. The adiabatic wall separated the two subsystems is then replaced with a diathermic wall. Consider a quasi-static process, i.e., after a given amount of heat (e.g., $\Delta U^{(1)}$) transferred from subsystem$^{(2)}$ to subsystem$^{(1)}$ though the diathermic wall, the original adiabatic wall is placed back as separation wall so that the two systems individually will "relax" into thermal equilibrium internally within each system. The "replacement with diathermic wall and the placing-back of adiabatic wall" quasi-static step is then repeated a sufficient number of times until the two systems reach thermal equilibrium at the final equilibrium state.

The entropy functions of the two subsystems can be obtained by integrating

$$dS^{(1)} = N^{(1)}\frac{3}{2}R\frac{dT^{(1)}}{T^{(1)}} + N^{(1)}R\frac{dV^{(1)}}{V^{(1)}} = \frac{6}{2}R\frac{dT^{(1)}}{T^{(1)}}$$
$$dS^{(2)} = N^{(2)}\frac{5}{2}R\frac{dT^{(2)}}{T^{(2)}} + N^{(2)}R\frac{dV^{(2)}}{V^{(2)}} = \frac{15}{2}R\frac{dT^{(2)}}{T^{(2)}}$$

It follows that

$$S = \frac{6}{2}R\ln\frac{T^{(1)}}{250} + S^{(1)}(250) + \frac{15}{2}R\ln\frac{T^{(2)}}{350} + S^{(2)}(350)$$

From the first law consideration,

$$U^{(1)} + U^{(2)} = U_1 + U_2$$

i.e.,

$$T^{(2)} = T_2 - \frac{N^{(1)}}{N^{(2)}}\frac{c_V^{(1)}}{c_V^{(2)}}\left(T^{(1)} - T_1\right) = 350 - \frac{6}{15}\left(T^{(1)} - 250\right) = 450 - \frac{2}{5}T^{(1)}$$

Finally, we find

$$S = \frac{6}{2}R\ln\frac{T^{(1)}}{250} + S^{(1)}(250) + \frac{15}{2}R\ln\frac{450 - \frac{2}{5}T^{(1)}}{350} + S^{(2)}(350)$$

Note that $U^{(1)} - U^{(1)}(250\ \text{K}) = 3R(T^{(1)} - 250)$. A quasi-static trajectory can be constructed in terms of $[S - S^{(1)}(250\ \text{K}) - S^{(2)}(350\ \text{K})]$ versus $[U^{(1)} - U^{(1)}(250\ \text{K})]$ or $U^{(1)}$. See Table 9.2 and Fig. 9.2.

Maximization of entropy coincides with equalization in the subsystem temperatures. This is, of course, expected. The consideration of chemical equilibrium in Sect. 9.9 will follow the same line of reasoning. Yet the result there will not be so intuitively obvious.

9.6.2 Mechanical Equilibrium

In fact, even simple mechanical equilibrium introduces some surprise issue that is not readily so obvious to our intuition.

First, consider the case without surprise that an isolated composite system consisting of two simple subsystems separated by a movable diathermal piston/wall that is impervious to the flow of matter. The process of piston motion and heat flow across the diathermal wall is subjected to the closure conditions

$$U^{(1)} + U^{(2)} = CONSTANT$$
$$V^{(1)} + V^{(2)} = CONSTANT$$

Table 9.2 Quasi-static trajectory toward internal thermal equilibrium

$T^{(1)}$(K)	$T^{(2)}$	$U^{(1)} - U^{(1)}(250\ \text{K})$	$S - S^{(1)}(250\ \text{K}) - S^{(2)}(350\ \text{K})$	
250	350	0	0	
260	346	249.4	0.2615	
270	342	498.8	0.4778	
280	338	748.3	0.6512	
290	334	997.7	0.7842	
300	330	1247.1	0.8785	
310	326	1496.5	0.9359	
320	322	1745.9	0.9579	
321	321.60	1770.9	0.9582	
321.43	**321.43**	1781.6	**0.9583**	←Max S at equilibrium

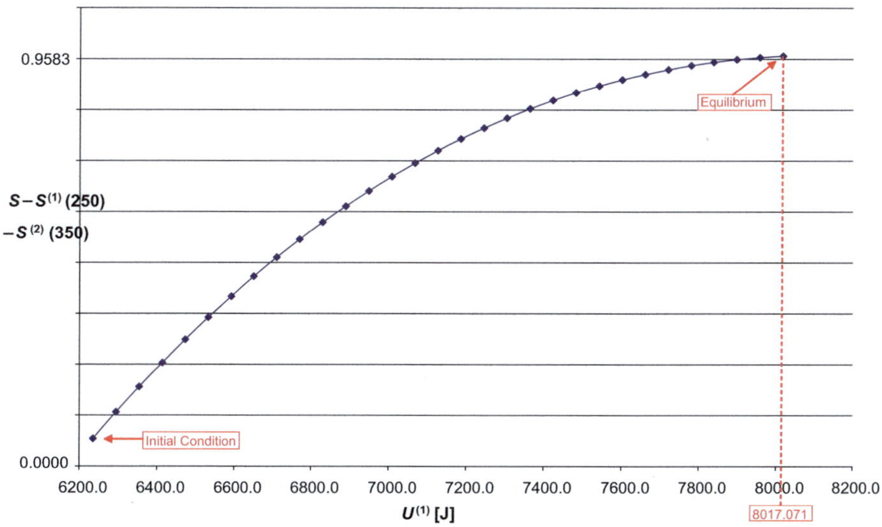

Fig. 9.2 Quasi-static trajectory toward internal mechanical equilibrium: total entropy versus internal energy of the first subsystem

The entropy maximization principle requires that at equilibrium

$$0 = dS = \frac{\partial S^{(1)}}{\partial U^{(1)}} dU^{(1)} + \frac{\partial S^{(1)}}{\partial V^{(1)}} dV^{(1)} + \frac{\partial S^{(2)}}{\partial U^{(2)}} dU^{(2)} + \frac{\partial S^{(2)}}{\partial V^{(2)}} dV^{(2)}$$

By the closure conditions,

$$\begin{aligned} dU^{(2)} &= -dU^{(1)} \\ dV^{(2)} &= -dV^{(1)} \end{aligned}$$

whence

$$0 = \left(\frac{1}{T^{(1)}} - \frac{1}{T^{(2)}}\right)_{equili} dU^{(1)} + \left(\frac{p^{(1)}}{T^{(1)}} - \frac{p^{(2)}}{T^{(2)}}\right)_{equili} dV^{(1)}$$

As this equation must hold independently for arbitrary values of $dU^{(1)}$ and $dV^{(1)}$, we must have

$$\left(\frac{1}{T^{(1)}} - \frac{1}{T^{(2)}}\right)_{equili} = 0$$
$$\left(\frac{p^{(1)}}{T^{(1)}} - \frac{p^{(2)}}{T^{(2)}}\right)_{equili} = 0$$

That is, at internal thermodynamic equilibrium mechanically and thermally, the equality of temperature and pressure prevails

$$T^{(1)} = T^{(2)}$$
$$p^{(1)} = p^{(2)}$$

Consider a slightly revised problem by making the movable but impervious to matter piston adiabatic. The problem is thus a mechanical equilibrium problem without realizing at the same time internal thermal equilibrium. Callen [1:5 and 53–54] calls this a unique *indeterminate* problem. Note that this is a mechanical, or pressure driven, problem of the kind of which the Joule free expansion is a specific case. The defining character of both is the absence of a work reservoir. Like free expansion, therefore, the quasi-static work (assuming $p^{(1)} > p^{(2)}$),

$$\delta W = \left(p^{(1)} - p^{(2)}\right) dV^{(1)}$$

does not apply simply because there is no δW to be stored away in a work reservoir. Work, as represented on RHS of the above expression in this mechanical equilibrium problem, cannot be a purely mechanical process. It necessarily involves frictional entropic processes. Without a full entropic consideration—explicitly or tacitly—one will be (as Callen was) perplexed by the problem. The detail of the consideration will be left as an exercise in Problem 9.6, where it will be shown that the problem becomes determinate with additional assumption.

A hint of the required additional assumption lies in the Joule free expansion problem: The free expansion takes place in a composite system composing one gaseous subsystem and one subsystem of vacuum devoid of matter. It involves the same kind of issue that the gaseous expansion work against vacuum must be entropically dissipated just as a general pressure-driven process does. This thus gives rise to the question of the *distribution* of the dissipative process of the eddy kinetic energy into thermal energy. However, in free expansion, no such question is being asked not because of the absence of the issue, but because all dissipation into thermal energy is by default—or tacitly assumed to be—distributed in the gaseous subsystem.

9.7 Gaseous Mixtures and Their Properties

The treatment of mixtures was introduced in Sect. 1.6 in terms of *Dalton's laws*

$$\frac{p_i}{p} = \frac{N_i}{N} = x_i \tag{8}$$

Properties of mixtures were considered in Sect. 5.9

$$U(T) = \sum_{i=1}^{n} N_i u_i = N \sum_{i=1}^{n} x_i u_i(T) \tag{76}$$

$$H(T) = \sum_{i=1}^{n} N_i h_i = N \sum_{i=1}^{n} x_i h_i(T) \tag{77}$$

$$S(T,p) = N \sum_{i=1}^{n} x_i s_i(T,p) + NR \sum_{i} x_i(-\ln x_i) \tag{80A}$$

9.7.1 Specific Gibbs Function of Mixture in Terms of T-p

We are now in a position to consider the Gibbs function of a mixture of ideal gases:

$$\begin{aligned} G(T,p) &= H(T) - TS(T,p) \\ &= \sum_i N_i h_i(T) - T \sum_i N_i F_i(T) + NRT \ln\frac{p}{p_0} - TR \sum_i N_i(-\ln x_i) \end{aligned} \tag{178}$$

where $F_i(T)$, which is a function of T, represents the value of the first two terms on RHS of Eq. (80). One can define another function(s) of T as follows:

$$\varphi_i(T) \equiv \frac{h_i(T)}{RT} - \frac{F_i(T)}{R}$$

With that definition, Eq. (178) can be written as

$$\begin{aligned} G(T,p) &= RT\left[\sum_i N_i \varphi_i(T) + \sum_i N_i \ln\frac{p}{p_0} + \sum_i N_i \ln x_i\right] \\ &= RT \sum_i N_i\left(\varphi_i(T) + \ln\frac{p}{p_0} + \ln x_i\right) \end{aligned} \tag{178A}$$

Since by definition $G = \sum_i N_i g_i$, the molar Gibbs function is

$$\mu_i = g_i(T, p, x_i) = RT\left(\varphi_i(T) + \ln\frac{p}{p_0} + \ln x_i\right) \quad (179)$$

Equations (178A) and (179) will be useful equations of state for ideal gas for the determination of chemical equilibrium of chemical reaction in Sect. 9.9. Before we study the equilibrium question, we shall first consider the enthalpy balance of chemical reaction.

9.8 Combustion Chemical Reactions and Enthalpy Balance

Combustion is exothermic chemical reaction (see Figs. 8.10 and 8.11), the chemical reaction that releases heat during reaction resulting in elevated product temperature unless heat is removed from the product mixture. All hydrocarbon–fuel/air reactions (almost all common fuels are hydrocarbon) are exothermic reactions. When the amount of O_2 or air in the reactant mixture is exactly right—neither deficient nor excessive, i.e., minimum needed for the complete combustion—the mixture is called a *stoichiometric* mixture. The air is called the *stoichiometric* or *theoretical* air (or 100% theoretical air); the combustion is called *theoretical combustion.* For instance, the theoretical combustion of methane/oxygen is

$$CH_4 + 2O_2 \rightarrow CO_2 + 2H_2O$$

Air is modeled as one mole of oxygen and 3.76 mol of nitrogen.

The theoretical combustion of methane and air is

$$CH_4 + 2(O_2 + 3.76N_2) \rightarrow CO_2 + 2H_2O + 7.52N_2$$

Mole numbers of each species in the two reaction equations are balanced. Correspondingly, the mass balance of the reaction is satisfied. The resulting air–fuel ratio, AF, can be readily calculated. For instance, AF for the case of stoichiometric methane/air is

$$AF = \frac{m_{air}}{m_{methane}} = \frac{2(32 + 3.76 \times 28.013)}{16.043} = 17.12 \quad (180)$$

Consider the theoretical combustion of octane (C_8H_{18}) and air

$$C_8H_{18} + 12.5(O_2 + 3.76N_2) \rightarrow 8CO_2 + 9H_2O + 47N_2$$

The air–fuel ratio for this case is

$$AF = \frac{m_{air}}{m_{octane}} = \frac{12.5(32 + 3.76 \times 28.013)}{114.231} = 15.03$$

The air–fuel ratio of the common fuel *gasoline* is 14.6. Octane is sometimes used as a model for gasoline since its AF, 15.03, is close to gasoline's AF, 14.6.

The fuel–air ratio is the inverse of AF

$$FA = \frac{1}{AF} \tag{180A}$$

Another parameter used is the *equivalence ratio*, which is defined as

$$equivalence\_ratio \equiv \frac{FA_{actual}}{FA_{stoichiometric}} \tag{181}$$

A theoretical combustion $(AF)_{stoichiometric}$ corresponds to an equivalence ratio of unity; lean mixtures have an equivalence ratio less than unity with larger AF values; rich mixtures have an equivalence ratio greater than unity with smaller AF values.

A lean fuel–air mixture reaction example (a 140% theoretical air and octane mixture, i.e., 40% excess air) is

$$C_8H_{18} + 17.5(O_2 + 3.76N_2) \rightarrow 8CO_2 + 9H_2O + 5O_2 + 65.8N_2$$

The air–fuel ratio is

$$AF = \frac{17.5(32 + 3.76 \times 28.013)}{114.231} = 21.04$$

i.e., $FA = 0.0475$ and correspondingly,

$$equivalence_{ratio} \equiv \frac{FA_{actual}}{FA_{stoichiometric}} = \frac{0.0475}{1/15.03} = 0.714$$

9.8.1 Enthalpy of Formation

The product mixture is composed of different chemical elements from those in the reactant mixture. Consistently assigned values of enthalpy must be determined before energy balance or enthalpy balance can be carried out for a reaction. That is, the enthalpy value of any species must be based on a *chemically* consistent scheme; the enthalpy thus determined is called enthalpy of formation. Step 1 is to assign the enthalpy *of formation* of all the stable elements (carbon atom C; molecular hydrogen H_2; molecular nitrogen N_2; molecular oxygen O_2) at standard temperature and pressure to be zero. Step 2 is an experimental program of determining the enthalpy of formation of species that is directly related to the stable elements by simple reactions. Take the example

$$\mathrm{H_2} + {}^1\!/\!_2\,\mathrm{O_2} \rightarrow \mathrm{H_2O} + Q_{to\,be\,removed}$$

The $Q_{to\ be\ removed}$ is found experimentally to be 141,780 kJ/kg or (2.016) (141780) = 285,830 kJ/kmol for the case of liquid water in the product mixture. The energy balance implies that

$$H_{\mathrm{H_2}} + H_{\frac{1}{2}\mathrm{O_2}} = H_{\mathrm{H_2O}(l)} + 285,830$$

Since the *enthalpy of formation* of both hydrogen, $H_{\mathrm{H_2}}$, and oxygen, $H_{\frac{1}{2}\mathrm{O_2}}$, are zero, the enthalpy of formation of one kmol of water (liquid)—designated now by the notation $\left(\bar{h}_f^0\right)_{\mathrm{H_2}o}$—satisfies the enthalpy balance

$$0 = \left(\bar{h}_f^0\right)_{\mathrm{H_2O}(l)} + 285,830$$

Therefore,

$$\left(\bar{h}_f^0\right)_{\mathrm{H_2O}(l)} = -285,830\,\mathrm{kJ/kmol}$$

The corresponding value for water in vapor phase in the product is

$$\left(\bar{h}_f^0\right)_{\mathrm{H_2O}(g)} = -241,820\,\mathrm{kJ/kmol}$$

Consider another elementary reaction

$$\mathrm{C} + \mathrm{O_2} \rightarrow \mathrm{CO_2} + Q_{to\,be\,removed}$$

The $Q_{to\ be\ removed}$ is found experimentally in this case to be 393,520 kJ/kmol. Therefore,

$$\left(\bar{h}_f^0\right)_{\mathrm{CO_2}} = -393,520\,\mathrm{kJ/kmol}$$

Let us consider an example of Step 3 of indirect determination of the enthalpy of formation with the following (which is also a simple reaction):

$$CO + {}^{1}\!/_{2}\, O_2 \rightarrow CO_2 + Q_{to\,be\,removed}$$

The $Q_{to\ be\ removed}$ is found experimentally in this case to be 393,520 kJ/kmol. The enthalpy balance is

$$\left(\bar{h}_f^0\right)_{CO} + 0 = \left(\bar{h}_f^0\right)_{CO_2} + 282,990$$

That is,

$$\left(\bar{h}_f^0\right)_{CO} = -393520 + 282990 = -110,530\,\text{kJ/kmol}$$

Another example,

$$CH_4 + 2O_2 \rightarrow CO_2 + 2H_2O + Q_{to\,be\,removed}$$

The $Q_{to\ be\ removed}$ is found experimentally in this case to be 890,868 kJ/kmol. The enthalpy balance is

$$\left(\bar{h}_f^0\right)_{CH_4} + 0 = \left(\bar{h}_f^0\right)_{CO_2} + 2\left(\bar{h}_f^0\right)_{H_2O} + 890,330$$

That is,

$$\left(\bar{h}_f^0\right)_{CH_4} = -393520 + 2(-285830) + 890330 = -74,850\,\text{kJ/kmol}$$

Through such a research program, the *enthalpy-of-formation* values for most hydrocarbons have been compiled and can be found in Table 9.3 including *n*-Octane (liquid).

$$\left(\bar{h}_f^0\right)_{C_8H_{18}(l)} = -249,950\,\text{kJ/kmol}$$

Table 9.3 Enthalpy of formation, Gibbs function of formation and absolute entropy at 25 °C and 1 atm

Substance	Formula	$\bar{h}_f^o$ (kJ/kmol)	$\bar{g}_f^o$ (kJ/kmol)	$\bar{s}^o$ (kJ/kmol-K)
Carbon	$C(s)$	0	0	5.74
Hydrogen	$H_2(g)$	0	0	130.68
Nitrogen	$N_2(g)$	0	0	191.61
Oxygen	$O_2(g)$	0	0	205.04
Carbon monoxide	$CO(g)$	−110,530	−137,150	197.65
Carbon dioxide	$CO_2(g)$	−393,520	−394,360	213.80
Water vapor	$H_2O(g)$	−241,820	−228,690	188.83
Water	$H_2O(l)$	−285,830	−237,180	69.92
Hydrogen peroxide	$H_2O_2(g)$	−136,310	−105,600	232.63
Ammonia	$NH_3(g)$	−46,190	−16,590	192.33
Methane	$CH_4(g)$	−74,850	−50,790	186.16
Acetylene	$C_2H_2(g)$	+226,730	+209,170	200.85
Ethylene	$C_2H_4(g)$	+52,280	+68,120	219.83
Ethane	$C_2H_6(g)$	−84,680	−32,890	229.49
Propylene	$C_3H_6(g)$	+20,410	+62,720	266.94
Propane	$C_3H_8(g)$	−103,850	−23,490	269.91
n-Butane	$C_4H_{10}(g)$	−126,150	−15,710	310.12
n-Octane	$C_8H_{18}(g)$	−208,450	+16,530	466.73
n-Octane	$C_8H_{18}(l)$	−249,950	+6,610	360.79
n-Dodecane	$C_{12}H_{26}(g)$	−291,010	+50,150	622.83
Benzene	$C_6H_6(g)$	+82,930	+129,660	269.20
Methyl alcohol	$CH_3OH(g)$	−200,670	−162,000	239.70
Methyl alcohol	$CH_3OH(l)$	−238,660	−166,360	126.80
Ethyl alcohol	$C_2H_5OH(g)$	−235,310	−168,570	282.59
Ethyl alcohol	$C_2H_5OH(l)$	−277,690	−174,890	160.70
Oxygen	$O(g)$	+249,190	+231,770	161.06
Hydrogen	$H(g)$	+218,000	+203,290	114.72
Nitrogen	$N(g)$	+472,650	+455,510	153.30
Hydroxyl	$OH(g)$	+39,460	+34,280	183.70

Source From JANAF, *Thermochemical Tables* (Midland, MI: Dow Chemical Co., 1971); *Selected Values of Chemical Thermodynamic Properties*, NBS Technical Note 270-3, 1968; and *API Research Project 44* (Carnegle Press, 1953)

9.8.2 Fuel Heating Value (HV: HHV and LHV)

With the *enthalpy of formation* values compiled, the *heating values* of a reaction can be calculated (see Table 9.4). For instance, take the following reaction:

$$C_8H_{18}(l) + 17.5(O_2 + 3.76N_2) \rightarrow 8CO_2 + 9H_2O(l) + 5O_2 + 65.8\,N_2$$

Table 9.4 Properties of some common fuels and hydrocarbons

Fuel (phase)	Formula	Molar mass (kg/kmol)	Density[a] (kg/L)	Enthalpy of vaporization[b] (kJ/kg)	Specific heat[a] c_p (kJ/kg K)	Higher heating value[c] (kJ/kg)	Lower heating value[c] (kJ/kg)
Carbon (*s*)	C	12.011	2	–	0.708	32,800	32,800
Hydrogen (*g*)	H_2	2.016	–	–	14.4	141,800	120,000
Carbon monoxide (*g*)	CO	28.013	–	–	1.05	10,100	10,100
Methane (*g*)	CH_4	16.043	–	509	2.20	55,530	50,050
Methanol (*l*)	CH_4O	32.042	0.790	1168	2.53	22,660	19,920
Acetylene (*g*)	C_2H_2	26.038	–	–	1.69	49,970	48,280
Ethane (*g*)	C_2H_6	30.070	–	172	1.75	51,900	47,520
Ethanol (*l*)	C_2H_6O	46.069	0.790	919	2.44	29,670	26,810
Propane (*l*)	C_3H_8	44 097	0.500	335	2.77	50,330	46,340
Butane (*l*)	C_4H_{10}	58.123	0.579	362	2.42	49,150	45,370
1-Pentene (*l*)	C_5H_{10}	70.134	0.641	363	2.20	47,760	44,630
Isopentane (*l*)	C_5H_{12}	72.150	0.626	–	2.32	48,570	44,910
Benzene (*l*)	C_6H_6	78.114	0.877	433	1.72	41,800	40,100
Hexene (*l*)	C_6H_{12}	84.161	0.673	392	1.84	47,500	44,400
Hexane (*l*)	C_6H_{14}	86.177	0.660	366	2.27	48,310	44,740
Toluene (*l*)	C_7H_8	92.141	0.867	412	1.71	42,400	40,500
Heptane (*l*)	C_7H_{16}	100.204	0 684	365	2.24	48,100	44,600
Octane (*l*)	C_8H_{18}	114.231	0.703	363	2.23	47,890	44,430
Decane (*l*)	$C_{10}H_{22}$	142.285	0.730	361	2.21	47,640	44,240
Gasoline (*l*)	$C_nH_{1.87n}$	100–110	0.72–0.78	350	2.4	47,300	44,000
Light diesel (*l*)	$C_nH_{1.8n}$	170	0.78–0.84	270	2.2	46,100	43,200
Heavy diesel (*l*)	$C_nH_{1.7n}$	200	0.82–0.88	230	1.9	45,500	42,800
Natural gas (*g*)	$C_nH_{3.8n}N_{0.1n}$	18	–	–	2	50,000	45,000

[a]At 1 atm and 20 °C
[b]At 25 °C for liquid fuels, and 1 atm normal boiling temperature for gaseous fuels
[c]At 25 °C. Multiply by molar mass to obtain heating values in kJ/kmol

The enthalpy balance of the reaction is

$$\left(\bar{h}_f^0\right)_{C_8H_{18}(l)} + 0 = 8\left(\bar{h}_f^0\right)_{CO_2} + 9\left(\bar{h}_f^0\right)_{H_2O(l)} + 0 + HHV$$

Here, the fuel heating value is the higher heating value (HHV) because water in the product is in the liquid form. The balance in numerical terms

$$-249,950 = 8 \times (-393,520) + 9 \times (-285,830) + HHV$$

gives the *n*-Octane (liquid) HHV

$$HHV = 5,470,680\,\text{kJ/kmol} = 47,891\,\text{kJ/kg}$$

Repeat the same fuel–air mixture now with water in the product in the vapor form. The heating value, in this case, is called the lower heating value (LHV),

$$C_8H_{18}(l) + 17.5(O_2 + 3.76N_2) \rightarrow 8CO_2 + 9H_2O(g) + 5O_2 + 65.8\,N_2$$

The enthalpy balance of the reaction is

$$\left(\bar{h}_f^0\right)_{C_8H_{18}(l)} + 0 = 8\left(\bar{h}_f^0\right)_{CO_2} + 9\left(\bar{h}_f^0\right)_{H_2O(g)} + 0 + LHV$$

Correspondingly,

$$-249,950 = 8 \times (-393,520) + 9 \times (-241,820) + LHV$$

i.e.,

$$\begin{aligned} LHV &= 5,074,590\,\text{kJ/kmol}\_C_8H_{18}(l) \\ &= 44,424\,\text{kJ/kg}\_C_8H_{18}(l) \end{aligned}$$

The corresponding LHV based on one kg of working fluid of a stoichiometric octane–air mixture is

$$LHV = 44,424 \times \frac{F/A}{1 + \frac{F}{A}} = 2771.3\,\text{kJ/kg}\_\textit{workingfluid}$$

9.9 Chemical Equilibrium (Gaseous Reaction Product Composition)

The product mixtures considered in Sect. 9.8 exist at the standard temperature, at which condition there is negligible dissociation in the product. The situation is different when the product is maintained at high elevated temperatures in that dissociation may not be negligible. In that case, the composition of the product mixture is determined in accordance with chemical equilibrium criterion on the basis of minimum Gibbs function under the constraint of constant T and p.

It is necessary to construct for a given reaction with a given composition of reactant mixture the constant T and p quasi-static process in terms of G as a function of *extent of reaction*, ξ. Take a simple reaction and assume the stoichiometric (theoretical) fuel–air reactant mixture, the *balancing equation* (the state of *micro-reversibility* between reactant and product in an equilibrium mixture) of the reaction is

$$CO_2 \rightleftarrows CO + {}^1\!/_2\, O_2$$

This is a balancing equation for the dissociation reaction; this is how reaction equations are given in the EQUILIBRIUM CONSTANT TABLE (Table 9.5); all reactions in the table are dissociation reactions with the natural logarithms of the equilibrium constants shown as negative values. Since in most cases, we study *recombination* reactions, we shall rewrite the balancing equation

$$CO_2 + {}^1\!/_2\, O_2 \rightleftarrows CO_2$$

(Note that the natural logarithm of the corresponding equilibrium constant will be simply the positive value of that of the original dissociation reaction in the table.) Besides the equilibrium constant, the principal use of the balancing equation is to find the *stoichiometric coefficient* of each component in the reaction, v_i. That is, an arbitrary reaction of $A + B \rightarrow C + D$, in terms of stoichiometric coefficients is

$$0 \rightleftarrows v_A A + v_B B + v_C C + v_D D$$

In this case, therefore,

$$v_{CO} = -1$$
$$v_{\frac{1}{2}O_2} = -{}^1\!/_2$$
$$v_{CO_2} = +1$$

Table 9.5 Natural logarithms of the equilibrium constant K_p

The equilibrium constant K_p for the reaction $v_A A + v_B B \rightleftharpoons v_C C + v_D D$ is defined as $K_p \equiv \frac{P_C^{v_C} P_D^{v_D}}{P_A^{v_A} P_B^{v_B}}$

Temp. (K)	$H_2 \rightleftharpoons 2H$	$O_2 \rightleftharpoons 2O$	$N_2 \rightleftharpoons 2N$	$H_2O \rightleftharpoons H_2 + 1/2\,O_2$	$H_2O \rightleftharpoons 1/2\,H_2 + OH$	$CO_2 \rightleftharpoons CO + 1/2\,O_2$	$1/2\,N_2 + 1/2\,O_2 \rightleftharpoons NO$
298	−164.005	−186.975	−367.480	−92.208	−106.208	−103.762	−35.052
500	−92.827	−105.630	−213.372	−52.691	−60.281	−57.616	−20.295
1000	−39.803	−45.150	−99.127	−23.163	−26.034	−23.529	−9.388
1200	−30.874	−35.005	−80.011	−18.182	−20.283	−17.871	−7.569
1400	−24.463	−27.742	−66.329	−14.609	−16.099	−13.842	−6.270
1600	−19.637	−22.285	−56.055	−11.921	−13.066	−10.830	−5.294
1800	−15.866	−18.030	−48.051	−9.826	−10.657	−8.497	−4.536
2000	−12.840	−14.622	−41.645	−8.145	−8.728	−6.635	−3.931
2200	−10.353	−11.827	−36.391	−6.768	−7.148	−5.120	−3.433
2400	−8.276	−9.497	−32.011	−5.619	−5.832	−3.860	−3.019
2600	−6.517	−7.521	−28.304	−4.648	−4.719	−2.801	−2.671
2800	−5.002	−5.826	−25.117	−3.812	−3.763	−1.894	−2.372
3000	−3.685	−4.357	−22.359	−3.086	−2.937	−1.111	−2.114
3200	−2.534	−3.072	−19.937	−2.451	−2.212	−0.429	−1.888
3400	−1.516	−1.935	−17.800	−1.891	−1.576	0.169	−1.690
3600	−0.609	−0.926	−15.898	−1.392	−1.088	0.701	−1.513
3800	0.202	−0.019	−14.199	−0.945	−0.501	1.176	−1.356
4000	0.934	0.796	−12.660	−0.542	−0.044	1.599	−1.216
4500	2.486	2.513	−9.414	0.312	0.920	2.490	−0.921
5000	3.725	3.895	−6.807	0.996	1.689	3.197	−0.686
5500	4.743	5.023	−4.666	1.560	2.318	3.771	−0.497
6000	5.590	5.963	−2.865	2.032	2.843	4.245	−0.341

Source Gorden J. Van Wylen and Richard E. Sonntag, *Fundamentals of Classical Thermodynamics*, English/SI Version, 3rd ed. (New York: John Wiley & Sons, 1986), p. 723, Table A.14. Based on thermodynamic data given in JANAF, *Thermochemical Tables* (Midland, MI: Thermal Research Laboratory, The Dow Chemical Company, 1971)

Introduce the *extent of reaction*, ξ, which may be assigned to be (in this case)

$$d\xi = dN_{CO_2}$$

In terms of $d\xi$, the following relation holds:

$$d\xi = \frac{dN_{CO}}{-1} = \frac{dN_{O_2}}{-1/2} = \frac{dN_{CO_2}}{1} \tag{182}$$

where the initial mixture is defined by $\xi_{CO} = 1$, $\xi_{O_2} = 1/2$, $\xi_{CO_2} = 0$.
The *reaction equation* of the reaction is, therefore,

$$CO + 1/2\,O_2 \rightarrow (1-\xi)CO + (1/2 - 1/2\,\xi)O_2 + \xi CO_2$$

The quasi-static reaction process can be marked in terms of the *extent of reaction*, ξ. At any point of the reaction, the mole number of the mixture is

$$N = \sum_i N_i = (1-\xi) + \frac{1}{2}(1-\xi) + \xi = \frac{3}{2} - \frac{1}{2}\xi$$

and the mole fractions of each component are

$$\begin{aligned} x_{CO} &= \frac{1-\xi}{\frac{3}{2}-\frac{1}{2}\xi} \\ x_{O_2} &= \frac{\frac{1}{2}(1-\xi)}{\frac{3}{2}-\frac{1}{2}\xi} \\ x_{CO_2} &= \frac{\xi}{\frac{3}{2}-\frac{1}{2}\xi} \end{aligned}$$

We may consider the example of the same reaction but with different initial reactant composition with one kmol of CO and one kmol of O_2. The *same* balancing equation applies here, but the reaction equation becomes

$$CO + O_2 \rightarrow (1-\xi)CO + (1 - 1/2\xi)O_2 + \xi CO_2$$

Correspondingly, the mole fractions are

$$x_{CO} = \frac{1-\xi}{2-\frac{1}{2}\xi}$$
$$x_{O_2} = \frac{1-\frac{1}{2}\xi}{2-\frac{1}{2}\xi}$$
$$x_{CO_2} = \frac{\xi}{2-\frac{1}{2}\xi}$$

the demonstration of which is left as an exercise.

We now consider G as a function of ξ of the quasi-static process of constant T and p. The chemical equilibrium condition is

$$0 = (dG)_{equili} = \sum_i \frac{\partial G}{\partial N_i}\frac{\partial N_i}{\partial \xi} d\xi = \left(\sum_i g_i v_i\right) d\xi \tag{183}$$

Recall that

$$\mu_i = g_i(T, p, x_i) = RT\left(\varphi_i(T) + \ln\frac{p}{p_0} + \ln x_i\right) \tag{179}$$

It follows that

$$\begin{aligned} 0 &= \sum_i RT\left(\varphi_i(T) + \ln\frac{p}{p_0} + \ln x_i\right) v_i \\ &= RT\sum_i \varphi_i(T) v_i + RT\ln\left(\frac{p}{p_0}\right)^{\sum v_i} + RT\sum_i \ln x_i^{v_i} \end{aligned}$$

Define the equilibrium constant K_p

$$\ln K_p(T) \equiv -\sum_i \varphi_i(T) v_i \tag{184}$$

and note that

$$\sum_i \ln x_i^{v_i} = \ln \prod_i x_i^{v_i}$$

The equation can be written as

$$\ln K_p(T) + \ln\left(\frac{p}{p_0}\right)^{-\sum \nu_i} = \ln \prod_i x_i^{\nu_i}$$

or

$$\prod_i x_i^{\nu_i} = \left(\frac{p}{p_0}\right)^{-\sum \nu_i} K_p(T) \tag{185}$$

This is known as the *law of mass action.*

Problem

9.1 Show that the internal energy of a material whose equation of state has the form $p = f(V)T$ is independent of the volume and the pressure. That is

$$\left(\frac{\partial U}{\partial V}\right)_T = 0$$
$$\left(\frac{\partial U}{\partial p}\right)_T = 0$$

9.2 Thermodynamicists sometimes refer to the "first *TdS* equation" and the "second *TdS* equation":

$$Tds = Nc_V dT + \left({}^{T\beta}/_{\kappa_T}\right) dV$$
$$Tds = Nc_p dT - TV\beta dp$$

Derive these equations.

9.3 Use the second *TdS* equation to show that

$$c_p - c_V = \frac{Tv\beta^2}{\kappa_T}$$

9.4 Obtain the Helmholtz equation $\left(\frac{\partial U}{\partial V}\right)_T = T^2\left(\frac{\partial}{\partial T}\frac{p}{T}\right)_V$ from

$$\left\{\frac{\partial}{\partial T}\left[\frac{1}{T}\frac{\partial U}{\partial V} + \frac{p}{T}\right]\right\}_V = \left\{\frac{\partial U}{\partial V}\left[\frac{1}{T}\frac{\partial U}{\partial T}\right]\right\}_T$$

9.5 Consider a composite system made of two subsystems of two ideal gases with the same mole number $N^{(1)} = N^{(1)} = 3$ kmol and with $c_v^{(1)} = (5/2)R$ and $c_v^{(2)} = (3/2)R$. The initial temperatures of the two subsystems are $T_i^{(1)} =$

T_1 = 350 K and $T_i^{(2)} = T_2$ = 250 K. The adiabatic wall separated the two subsystems is then replaced with a diathermic wall. Consider a quasi-static process, i.e., after a given amount of heat (e.g., $\Delta U^{(2)}$) transferred from subsystem$^{(1)}$ to subsystem$^{(2)}$ through the diathermic wall, the original adiabatic wall is placed back as separation wall so that the two subsystems individually will "relax" into thermal equilibrium internally within each subsystem. The "replacement with diathermic wall and the placing-back of adiabatic wall" quasi-static step is then repeated a sufficient number of times until the two systems reach thermal equilibrium at the final equilibrium state. Construct the quasi-static process toward thermal equilibrium of $T_f^{(1)} = T_f^{(2)}$; determine the equilibrium states of both subsystems.

9.6 Callen [1:53–54] wrote

The hypothetical problem of equilibrium in a closed composite system with an internal movable *adiabatic* wall is a unique *indeterminate* problem. Physically, the release of the piston would lead it to perpetual oscillation in the absence of viscous damping. With viscous damping, the piston would eventually come to rest at such a position that the pressures on either side would be equal, but the temperatures in each subsystem would then depend on the relative viscosity in each subsystem. The solution to this problem depends on *dynamical* considerations [underline added]. Show that the application of the entropy maximum formalism is correspondingly indeterminate with respect to the temperatures (but determinate with respect to the pressures).

Consider one such composite system with *He* gas in subsystem 1, and N_2 gas in subsystem 2. The initial conditions of the two subsystems are given as $N^{(1)}$ = 0.5 gm-mole and $N^{(2)}$ = 0.75 gm-mole, $T_i^{(1)}$ = 200 K and $T_i^{(2)}$ = 300 K, $V_i^{(1)}$ = 12 L and $V_i^{(2)}$ = 8 L. Remove the stopper to the piston and the composite system will move towards mechanical equilibrium.

Viscous damping is the key to understanding the process. Instead permitting the piston to oscillate, one can construct a quasi-equilibrium process similar to Problem 9.5: imagine a set of large number of stoppers to the piston; remove the first stopper; permitting the piston to move only a small distance to the next stopper; then permitting the two subsystems to relax to internal thermodynamic equilibrium individually; determine the individual equilibrium states; then repeat the next sequence of steps until $p_f^{(1)} = p_f^{(2)}$. Determine the final pressures and temperatures of both subsystems.

HINT: Each quasi-equilibrium step is a two-phase step: the first phase leads to isentropic heating and cooling in subsystems 1 and 2, respectively; then the eddy kinetic energy in each subsystem will dissipate into internal energy in the second phase; the consideration of the second phase requires the auxiliary assumption of how the eddy kinetic energy is distributed in the two subsystems (you can make your own arbitrary auxiliary assumption).

9.7 Two kmols of H_2O are enclosed in a rigid vessel and heated to a temperature of 2000 K and a pressure of 1 MPa.

$$2H_2O \rightleftarrows 2H_2 + O_2$$

Dissociation of H_2O will take place at such a high temperature. What is the equilibrium composition of the system? What is the composition if the temperature remains constant but the pressure is decreased to 10 kPa?

9.8 In the Example of Problem 9.7, what would the final mole fractions be at a pressure of 1 MPa if the vessel initially had contained 1 mol of oxygen as well as 2 mol of water?

9.9 A system containing 0.02 kg of CO and 0.02 kg of O_2 is maintained at a temperature of 3200 K and a pressure of 0.2 MPa. What is the mass of CO_2 at equilibrium?

9.10 Consider the mixing of gas A and gas B. Initially gas A (its mole number is N_A) occupies V_A volume compartment at T and p. Initially, gas B (N_B) occupies V_B volume compartment at T and p. After the valve connecting the two gases opens, the two gases mix and each occupies the whole volume of $V_A + V_B$. Show that the temperature and pressure of the mixture remains the same. Determine the entropy change of the mixing process.

Now construct a reversible transformation of the system from the same initial state of gas A and gas B in separate compartments and the same final mixed state by the following steps: Assume the availability of a vacuum of the volume of $V_A + V_B$. (1) Arrange for the managed expansion of gas A into a vacuum V_A to end in a volume of $V_A + V_B$ in a submerged heat bath of temperature T and the managed expansion of gas B into a vacuum V_B to end in a volume of $V_A + V_B$ in a submerged heat bath of temperature T. Then, explain that by (2) a reversible mixing of the two gases through a Gibbs thought experiment the end result will be the same final mixed state of the original irreversible mixing as well as the production of the vacuum needed for the two expansion processes. Show that the work obtained by the two expansions together equals to the free heat of the system and T reservoir.

$$\Delta S = N_A R \,\text{In}\, \frac{V_A + V_B}{V_A} + N_B R \,\text{In}\, \frac{V_A + V_B}{V_B}$$

$$Q_{spon} = 0$$

$$Q_{rev} = T\Delta S = N_A R \,\text{In}\, \frac{V_A + V_B}{V_A} + N_B R \,\text{In}\, \frac{V_A + V_B}{V_B}$$

$$W_{rev} = Q_{rev} - Q_{spon}$$

References

1. Callen HB (1st edition, 1960; 2nd edition, 1985) Thermodynamics and an Introduction to Thermostatistics. Wiley, New York
2. Gibbs JW (1961) The Scientific Papers of J. W. Gibbs, Vol. 1: Thermodynamics. Dover (p. 354)
3. Kestin J (1966) A Treatise in Thermodynamics, Volume 1. Blaisdell Publishing Co

A Theory of Heat as Prelude to Engineering Thermodynamics

10

Abstract

Profitable handling of extracted energy in accordance with the energy conversion doctrine has served well expediency for industrialization. Despite that success, this disquisition refutes the energy conversion doctrine in favor of entropy-growth-potential-centric (EGP-centric) approach that is necessary for achieving efficiency rather than expediency. With the new theory of heat, this chapter outlines a philosophical prelude to engineering application of thermodynamics. Though decision is made to separate the project of a textbook on engineering thermodynamics from this disquisition so that details of the textbook project will be worked out after further deliberation, such an engineering thermodynamics text will articulate the distinctive characterization of engineering in terms of physical necessity and causal necessity. It is also anticipated that causal necessity addressed in the textbook project will incorporate solutions, in addition to conventional energy solutions, based on natural EGP, i.e., renewable energy sources (RESs).

Keywords

Engineering thermodynamics · Heat transfer equation versus energy equation for open systems involving shaft work · Physical necessity · Determinism · Causal necessity · Causal necessity and engineering for efficiency

10.1 Engineering Thermodynamics

From the very beginning, I have made the point that for understanding heat and energy it is necessary to consider an energy system to be an interacting system (including the possibility of the triad of system, driving force, and a surroundings reservoir in interaction) rather than a mere object and to treat the system in terms of

L.-S. Wang, *A Treatise of Heat and Energy*, Mechanical Engineering Series,
https://doi.org/10.1007/978-3-030-05746-6_10

both the first law and the second law. The conventional approach often falls short of the two requirements. This includes our conventional understanding of the second law itself, e.g., a common example of a second law statement is, "First, heat flows from hot bodies to cooler ones [spontaneously]. Second, it is impossible entirely to convert heat into work—something is always lost in energetic transformation" by Schneider and Sagan [1].

Schneider and Sagan here repeated the error of impossibility of 100% conversion of heat into work. As it was argued in Chap. 8, the error is not just whether 100% conversion is possible (it is) but that the concept of *conversion* of heat to work is nonsensical: there is no purely reverse conversion. Conversion of extracted heat into work is possible only as a triadic process element, an element-part of a managed triadic relation powered by EGP. We call this process a reversible-like, or reversible process, which is fundamentally different from spontaneous energy conversion processes, i.e., *nonreversible* processes (see Sect. 6.5), such as heat transfer and heat production.

Physics describes spontaneous energy conversion processes to be caused by efficient causation, whereas, the managed triadic-relation processes are caused by efficacious causation. While the study of heat transfer and heat production is part of a thermal fluid engineering curriculum, a survey of engineering thermodynamics texts shows that, in typical texts, the topic of engineering thermodynamics focuses on efficacious reversible-like devices.

That is, "engineering thermodynamics" focuses on engineering systems involving *shaft work* as one kind of thermal engineering problems. While other subjects of thermal engineering such as heat transfer, transport phenomena, or thermal fluids topics deal exclusively with heat transfer and heat production (from resistive work or combustion heat release). In the former case, if positive shaft work is produced the system is said to involve heat extraction as shown in Chap. 8; if negative shaft work is consumed the system is said to convert mechanical work into mechanical energy in the working fluid rather than heat or thermal energy as in the case of converting "resistive" work into heat; what distinguishes both from the latter case (other subjects) is that both are reversible-like processes while the events in the latter case are nonreversible processes.

This disquisition essay highlights the fundamental difference between problems of reversible-like processes involving shaft work and problems of nonreversible processes. Shaft work, its involvement or not, is what demarcates between engineering thermodynamics and other subjects of thermal fluid engineering, yet, none of engineering thermodynamics texts acknowledges explicitly this point. Though the decision is made to separate the project of a textbook on engineering thermodynamics from this disquisition so that details of the textbook project can benefit from time spent on its deliberation, this chapter makes some additional points/comments on differences between the two, as preparation and initiation of the textbook project.

10.2 Heat Transfer Phenomena are Described by Governing Equations

Many engineering problems are treated as applied science or applied physics problems. Heat transfer problems are no exception, and are based on *general laws* and *particular laws* (see below for their definitions). A brief outline of which is given in this section. The purpose of this outline is for identifying an unexamined doctrine of applied science problems, the doctrine of determinism. Thereupon, in the next two sections, Sects. 10.3 and 10.4, a characterization of problems involving shaft work will be made by rejecting the doctrine to inform engineering students of the defining nature of engineering thermodynamics that demarcates it from typical applied sciences.

As presented by Yener and Kakac, [2] the theoretical foundations of heat transfer are general laws:

1. The law of conservation of mass,
2. Newton's second law of motion,
3. The first law of thermodynamics, and
4. The second law of thermodynamics (it should be stricken out for reason given below)

In addition to these general laws, certain particular law, which are dependent on the nature of the medium under consideration while general laws are applicable to all media independent of their nature have to be brought into the application of general laws. They are

1. Fourier's law of heat conduction,
2. Newton's law of cooling, and
3. The Stefan-Boltzmann law of radiation.

It should first be pointed out that the customary inclusion of *the second law of thermodynamics* serves no concrete purpose as the general law is not required independently as every particular law is necessarily consistent with the second law, and thus their inclusion in the first law of thermodynamics guarantees universal entropy growth. Thus, for phenomena considered in this section, the above #4 is merely a distraction. Another immediate comment is that Newton's law of cooling is used for one kind of boundary condition for heat conduction problems as an empirical input but otherwise unnecessary if the heat transfer problems are investigated based on first principles (of general laws and two particular laws of heat conduction and heat radiation).

Let us consider the first law of thermodynamics beginning with Eq. (23A),

$$dE = \delta Q - \delta W,$$

which applies to a closed system, we may write it as a rate equation

$$\frac{dE}{dt} = \dot{Q} - \dot{W} \tag{23B}$$

Consider a moving fluid field by denoting the volume of an identifiable fluid to be $V_m(t)$, which occupies at an initial time the volume $V_m(t_0)$. We shall call this volume the *control volume* of an engineering system, $V_m(t_0) \equiv cv$. Its corresponding boundary surface is called *control surface*, *cs*. We now apply Reynolds' transport theorem to the LHS of the equation for the purpose of applying the first law to this control volume

$$\frac{D}{Dt}\int\limits_{V_m(t)} e\rho d\mathcal{V} = \int\limits_{V_m(t_0)} \frac{\partial(e\rho)}{\partial t} d\mathcal{V} + \int\limits_{cs} e\rho\vec{V}\cdot\hat{n}\, dA$$

Substitution of which into Eq. (23B) yields

$$\int\limits_{cv} \frac{\partial(e\rho)}{\partial t} d\mathcal{V} + \int\limits_{cs} e\rho\vec{V}\cdot\hat{n} dA = \dot{Q} - \dot{W} \tag{191}$$

10.2.1 Governing Equation for Heat Transfer Problems

First, with the application of divergence theorem,

$$\int\limits_{cs} e\rho\vec{V}\cdot\hat{n}\, dA = \int\limits_{cv} \nabla\cdot\left(e\rho\vec{V}\right)d\mathcal{V} = \int\limits_{cv}\left[(\nabla e)\cdot\rho\vec{V} + e\nabla\cdot\left(\rho\vec{V}\right)\right]d\mathcal{V}$$

and the use of the continuity equation

$$\frac{\partial\rho}{\partial t} + \nabla\cdot\left(\rho\vec{V}\right) = 0$$

The LHS of the equation becomes

$$\frac{D}{Dt}\int\limits_{cv} e\rho d\mathcal{V} = \int\limits_{cv} \rho\left[\frac{\partial e}{\partial t} + \vec{V}\cdot\nabla e\right]d\mathcal{V}$$

Substitution of which into Eq. (23B) yields

$$\int\limits_{cv} \rho\left[\frac{\partial e}{\partial t} + \vec{V}\cdot\nabla e\right]d\mathcal{V} = \dot{Q} - \dot{W} = -\int\limits_{cv} \nabla\cdot\vec{q}''\,d\mathcal{V} - \dot{W} \tag{192}$$

The term $\dot{W}$ in Eq. (192) has a positive value when work is done by the control volume on the surroundings. This includes the possible positive useful work output, called *shaft* work. The rate of work done by the control volume is typically subdivided into three classifications

$$\dot{W} = \dot{W}_{shaft} + \dot{W}_{surface} + \dot{W}_{resistive} \tag{193A}$$

Note that $\dot{W}_{resistive}$ is negative definite (resistive work can only be done to the control volume). The second class is in turn subdivided into subclasses [3],

$$\begin{aligned} &\dot{W}_{surface} = \dot{W}_{normal} + \dot{W}_{shear} \\ &\left(= -\int_{cs} \left[\hat{n} \cdot \left(-p\bar{\bar{\delta}} + \bar{\bar{\tau}} \right) \right] \cdot \vec{V} dA \right) = \int_{cs} p\vec{V} \cdot d\vec{A} - \int_{cs} \left(\hat{n} \cdot \bar{\bar{\tau}} \right) \cdot \vec{V} dA \end{aligned} \tag{193B}$$

where the term $\left(\hat{n} \cdot \bar{\bar{\tau}} \right)$ may be expressed as $\vec{\tau}$ (the shear stress in the plane of dA), and $\bar{\bar{\tau}}$ is the stress tensor.

Rewrite Eq. (193B) in view of divergence theorem

$$\begin{aligned} \dot{W} &= \dot{W}_{shaft} + \dot{W}_{surface} + \dot{W}_{resistive} \\ &= \dot{W}_{shaft} + \int_{cv} \nabla \cdot \left(p\vec{V} \right) d\mathcal{V} - \int_{cv} \nabla \cdot \left(\bar{\bar{\tau}} \cdot \vec{V} \right) d\mathcal{V} + \dot{W}_{resistive} \end{aligned} \tag{193C}$$

Substitution of Eq. (193C) with zero shaft work into Eq. (192) yields

$$\begin{aligned} \int_{cv} \rho \left[\frac{\partial e}{\partial t} + \vec{V} \cdot \nabla e \right] d\mathcal{V} = &- \int_{cv} \nabla \cdot \vec{q}'' d\mathcal{V} - \left[\int_{cv} \nabla \cdot \left(p\vec{V} \right) d\mathcal{V} - \int_{cv} \nabla \cdot \left(\bar{\bar{\tau}} \cdot \vec{V} \right) d\mathcal{V} \right] \\ &- \dot{W}_{resistive} \end{aligned}$$

With the expression of $\dot{W}_{resistive}$ (which is negative definite),

$$\dot{W}_{resistive} = -\int_{cv} \dot{q}_{ext-heating} d\mathcal{V}$$

($\dot{q}_{ext-heating}$ may represent external nonwork source such as nuclear-reactive heating, which is not part of *internal energy* of the working fluid), the equation becomes

$$\rho\left(\frac{\partial e}{\partial t}+\vec{V}\cdot\nabla e\right)=-\nabla\cdot\vec{q}''-\nabla\cdot\left(p\vec{V}\right)+\nabla\cdot\left(\bar{\bar{\tau}}\cdot\vec{V}\right)+\dot{q}_{ext-heating} \tag{194}$$

The *equation of mechanical energy* is [4]

$$\rho\left[\frac{\partial\left(\frac{V^2}{2}+gz\right)}{\partial t}+\vec{V}\cdot\nabla\left(\frac{V^2}{2}+gz\right)\right]$$
$$=-\nabla\cdot\left(p\vec{V}\right)+p\nabla\cdot\vec{V}+\nabla\cdot\left(\bar{\bar{\tau}}\cdot\vec{V}\right)-\left(\bar{\bar{\tau}}:\vec{V}\right)$$

Subtraction of the equation of mechanical energy from Eq. (194) yields

$$\rho\left(\frac{\partial u}{\partial t}+\vec{V}\cdot\nabla u\right)=-\nabla\cdot\vec{q}''-p\nabla\cdot\vec{V}+\left(\bar{\bar{\tau}}:\vec{V}\right)+\dot{q}_{ext-heating} \tag{195}$$

It is convenient to express internal energy, u, in terms of temperature, T, by using Eq. (176). We have

$$du=\left(c_p-pv\beta\right)dT+v(p\kappa_T-T\beta)dp \tag{176}$$

Note that

$$p\left(\nabla\cdot\vec{V}\right)=p\left(-\frac{1}{\rho}\frac{D\rho(T,p)}{Dt}\right)=-\frac{p}{\rho}\left(\frac{\partial\rho}{\partial T}\right)_p\frac{DT}{Dt}-\frac{p}{\rho}\left(\frac{\partial\rho}{\partial p}\right)_T\frac{Dp}{Dt}$$
$$=p\beta\frac{DT}{Dt}-p\kappa_T\frac{Dp}{Dt}$$

That is, the term in RHS of (195), $-p\nabla\cdot\vec{V}$, becomes $-p\beta\frac{DT}{Dt}$ and $p\kappa_T\frac{Dp}{Dt}$, which cancels with $-pv\beta dT=\frac{p}{\rho}\beta dT$ and $\frac{p}{\rho}\kappa_T dp$ in Eq. (176) when it is substituted into the LHS of Eq. (195). It follows that

$$\rho c_p\left(\frac{\partial T}{\partial t}+\vec{V}\cdot\nabla T\right)=-\nabla\cdot\vec{q}''+T\beta\frac{Dp}{Dt}+\left(\bar{\bar{\tau}}:\vec{V}\right)+\dot{q}_{ext-heating} \tag{196}$$

With the use of Fourier's law of heat conduction,

$$\vec{q}''=-k\nabla T \tag{197}$$

and the prescription of $\dot{q}_{ext-heating}$, Eq. (196) is the governing equation for heat transfer problems. Bird et al. call it, together with the continuity equation and the Navier–Stokes equations, the *equations of change* [4].

Governing equations, or equations of change, are important in engineering sciences and sciences in general as it is described by Bird et al. in the Postface of Ref. [4:805]:

> Of all the messages we have tried to convey in this long text, the most important is to recognize the *key role of the equations of change*...They can be used to determine velocity, pressure, temperature, and concentration profiles, as well as the fluxes of momentum, energy, and mass, even in complicated time-dependent systems. They are applicable to turbulent systems, and even when complete a priori solutions prove infeasible, simplify the efficient use of data through dimensional analysis.

All laws of physics including laws for heat transfer problems as distilled into Eq. (196) are expressed in terms of governing equations.[1] That is, governing equations for a system *determine* changes in the system inclusively. This is the Bird–Stewart–Lightfoot message that their equations of change serve as the organizing principle of the subject of transport phenomena, and solutions to the equations of change determine every detail of the system that they govern.

Moreover, Eq. (196) with the addition of Eq. (197) inclusively determines all details of change requiring no independent use of the second law because of the addition of Eq. (197) guarantees all predictive details to satisfy the second law.

This is the scientific methodology handed down by Galileo and Newton. The methodological program has been so successful in physics and chemistry that the idea that all changes in nature are determined by such governing equations has been ingrained in our worldview. Such presupposition leads to philosophical determinism: if governing equations inclusively describe all changes, then surely all changes are predetermined as Laplace had concluded, "An intellect which at a certain moment would know all forces that set nature in motion, and all positions of all items of which nature is composed, if this intellect were also vast enough to submit these data to analysis, it would embrace in a single formula the movements of the greatest bodies of the universe and those of the tiniest atom; for such an intellect nothing would be uncertain and the future just like the past would be present before its eyes." Such a universe is a causally closed universe. This is what Poincare noted, "...in the deterministic hypothesis there is only a single possibility" of spontaneous nonreversible process.

10.3 Energy Analysis and Exergy Analysis

For problems of reversible-like processes involving shaft work, as Poincare pointed out, there are other possibilities, in fact, infinite possibilities. We can consider the case in terms of energy equation for a control volume to be derived from the first law

$$\int_{cv} \frac{\partial(e\rho)}{\partial t} d\mathcal{V} + \int_{cs} e\rho\vec{V}\cdot\hat{n}dA = \dot{Q} - \dot{W} \tag{191}$$

[1]The important exception from this rule is the first and the second laws of thermodynamics; this exception was first pointed out in [5].

10.3.1 Rate of Work Done by a Control Volume and Energy Balance in Integral Form for a Control Volume

The term $\dot{W}$ in Eq. (191) has a positive numerical value when work is done by the control volume on the surroundings. This includes the possible positive useful work output, called *shaft work*. The rate of work done by the control volume is typically subdivided into the following three classifications:

$$\dot{W} = \dot{W}_{shaft} + \dot{W}_{surface} + \dot{W}_{resistive} \tag{193A}$$

where

$$\dot{W}_{surface} = \dot{W}_{normal} + \dot{W}_{shear}$$
$$\left(= -\int_{cs} \left[\hat{n} \cdot \left(-p\bar{\bar{\delta}} + \bar{\bar{\tau}}\right)\right] \cdot \vec{V} dA \right) = \int_{cs} p\vec{V} \cdot d\vec{A} - \int_{cs} \left(\hat{n} \cdot \bar{\bar{\tau}}\right) \cdot \vec{V} dA \tag{193B}$$

Note, in this consideration of $\left(\hat{n} \cdot \bar{\bar{\tau}}\right) \cdot \vec{V}$ to the surface of the control volume, *cs*, we have the following special results:

Areas of the *cv* corresponding to *stationary* solid surfaces where $\vec{V} = 0$; thus $\dot{W}_{shear} = 0$

Areas of the *control volume* at inlet and outlet are perpendicular to velocity; thus $\dot{W}_{shear} = 0$

It follows, therefore, that $\int_{cs} \left(\hat{n} \cdot \bar{\bar{\tau}}\right) \cdot \vec{V} dA = 0$, and correspondingly,

$$\dot{W} = \dot{W}_{shaft} + \dot{W}_{normal} + \dot{W}_{resistive} = \dot{W}_{shaft} + \int_{cs} p\vec{V} \cdot d\vec{A} + \dot{W}_{resistive} \tag{198}$$

We now consider the form of the first law for a control volume by using Eqs. (191) and (198). Noting that

$$\int_{cs} p\vec{V} \cdot d\vec{A} + \int_{cs} e\rho\vec{V} \cdot d\vec{A} = \int_{cs} \rho(pv + e)\vec{V} d\vec{A}$$

and

$$e = u + \frac{V^2}{2} + gz,$$

we obtain

$$\dot{Q}_{cv} - \dot{W}_{shaft} - \dot{W}_{resistive} = \frac{\partial}{\partial t}\int_{cv} e\rho d\mathcal{V} + \int_{cs} \rho\left(u + pv + \frac{V^2}{2} + gz\right)\vec{V}\cdot d\vec{A}.$$

That is

$$\dot{Q}_{cv} - \dot{W}_{shaft} - \dot{W}_{resistive} = \frac{\partial}{\partial t}\int_{cv} e\rho d\mathcal{V} + \int_{cs} \rho\left(h + \frac{V^2}{2} + gz\right)\vec{V}\cdot d\vec{A} \quad (199/111)$$

Equations (199) is the familiar form of the first law for a control volume, which is used extensively in engineering application.

10.3.2 Exergy Balance in Integral Form for a Control Volume

We now reproduce the part of exergy balance for a control volume discussion from Chap. 7. Rewrite Eq. (119) as a system exergy rate equation

$$\frac{dEx}{dt} = \sum_j \left(1 - \frac{T_0}{T_j}\right)\dot{Q}_j - \left(\dot{W} - p_0\frac{dV}{dt}\right) - \dot{E}x \quad (119\text{A})$$

in which Ex is given as Eq. (118A),

$$Ex = [U - T_0S + p_0V]_A - [U - T_0S + p_0V]_0 + \frac{1}{2}mv^2 + mgz \quad (118\text{A})$$

Applying Reynolds transport theorem to the LHS of (119A) to express the system change in Ex in terms of the control volume change in Ex and exergy flow across the boundary, cs, of the control volume are as follows:

$$\frac{\partial}{\partial t}\int_{cv} ex\cdot\rho d\mathcal{V} + \int_{cs} ex\cdot\rho\vec{V}\cdot\hat{n}dA = \sum_j\left(1 - \frac{T_0}{T_j}\right)\dot{Q}_j - \left(\dot{W} - p_0\frac{dV}{dt}\right) - \dot{E}x$$

where $\dot{W}$ is in the context of exergy analysis

$$W = W_{shaft} + W_{normal}$$

and W_{normal} is the system mass expansion work. In the context of exergy analysis, we consider the useful work portion of this expansion work by dealing with the combination of W_{normal} and $-p_0\Delta V$. That is, the useful work exerted by the normal surface of a system mass is,

$$W_{normal} - p_0\Delta V$$

Now, consider the rate of this exergy transfer accompanying work. Between t and $t+\Delta t$, the change in volume of the system mass is

$$\Delta V = m_e v_e$$

Exergy transfer accompanying work is then

$$W_{normal} - p_0 m_e v_e$$

The rate of exergy transfer accompanying work is

$$\lim_{\Delta t \to 0} \frac{W_{normal}}{\Delta t} - \lim_{\Delta t \to 0} \frac{m_e}{\Delta t} p_0 v_e$$

Noting that $W_{normal} = m_e p_e v_e$, as $\Delta t \to 0$, we can write the "rate of exergy transfer accompanying normal work" as

$$\dot{m}_e(p_e v_e - p_0 v_e)$$

The mass-specific exergy equals

$$ex = (u - u_0) + p_0(v - v_0) - T_0(s - s_0) + \frac{1}{2}v^2 + gz$$

Correspondingly

$$\int_{cs} \rho(ex)\vec{V}d\vec{A} = -\sum_i \dot{m}_i ex_i + \sum_e \dot{m}_e ex_e$$

Note that the combination of $(pv - p_0 v)$ and $[(u - u_0) + p_0(v - v_0)]$ becomes

$$u + pv - (u + pv)_0 = h - h_0$$

The combination of surface force work term and specific exergy term is, therefore, the following, known as *flow specific* exergy, fe,

$$fe = (h - h_0) - T_0(s - s_0) + \frac{1}{2}v^2 + gz \qquad (120)$$

The exergy balance over a control volume is

$$\frac{\partial Ex_{cv}}{\partial t} = \sum_j \left(1 - \frac{T_0}{T_j}\right)\dot{Q}_j - \dot{W}_{shaft} + \sum_i \dot{m}_i fe_i - \sum_e \dot{m}_e fe_e - \underset{D}{\dot{Ex}} \qquad (121)$$

Equations (199/111) and (121) are the fundamental equations for treatment of engineering problems of reversible-like devices. They will be the core analytical tools for the future textbook project. But, they are neither the governing equations nor the equations of change.

10.4 Shaft Work Entails Mechanism for Its Fulfillment

There is a difference between the notion that every change obeys laws of physics and chemistry, the notion of *physical necessity*, and the notion that laws of physics and chemistry determine every change, i.e., *determinism*. The latter presupposes that *all* laws of physics and chemistry are expressed in terms of governing equations. This is just not true as the aforementioned exception of the first and the second laws of thermodynamics from the general rule noted in Footnote 1. Before we consider the issue in the context of this exception, consider the critique of the presupposition, or the objectivist philosophy, made by one of the most profound thinkers of the twentieth century, Michael Polanyi.

Michael Polanyi (1891–1976)

Polanyi was interested in explaining the phenomena of life and living organisms, and how they are different from nonliving things as explained by physics and chemistry. Living organisms have often been explained in terms of machines, and the fact that all machines are subject to physical necessity in accordance with laws of physics and chemistry has been viewed as proof that living things too are governed by the same laws. Polanyi, however, turned this argument on its head.

On the contrary, says Polanyi, it proves just the opposite, since machines, though of physical necessity, cannot be "explained" in terms of physics and chemistry. In his major work *Personal Knowledge* he wrote, "*The complete [physics and chemistry] knowledge of a machine as an object tells us nothing about it as a machine*" [6]. This is because what defines a machine is its *operational principle*, which patent offices of all nations recognize as a different kind of knowledge from scientific knowledge (see what he wrote below). Polanyi reasoned that if living organs function like machines (indeed, their functions can sometimes be performed by man-made machines, for instance, a kidney replaced with a kidney machine), then life transcends as machines do the domain of physics and chemistry. I am interested here not so much in understanding life but in better appreciation of how the application of the first law and the second law to machines involving shaft work is different from the typical application of scientific laws in the practice of applied sciences. (Though, we can learn a great deal from life phenomena on how to manage the planet Earth for its wellbeing rather than as an object from which man extracts resources.) That is, knowledge about a machine that either produces or consumes shaft work is more than mere scientific knowledge, as general engineering knowledge is different from scientific knowledge, which is based on the notion of observing.

Engineering knowledge involves the additional notion of contriving. Even though the notion of contriving applies to all engineering devices including, e.g., heat exchangers, there is a degree of difference in the kind of contriving required for shaft work machines from simpler ones. Polanyi noted on contriving the following:

> …the physical sciences are predominately observational, while biology and [man-made-things]…, in which observation plays but a subsidiary role…The logic of deductive reasoning has been systematically studied for two millennia, and the logic of empirical inference has been a major preoccupation of philosophy for centuries, but the logic of contriving has found its way only into scattered hints…a machine or a technical process is characterized by an operational principle, which differs altogether from an observational statement. The former, if it is new, represents an invention and can be covered by a patent; the latter, if it is new, is a discovery, which cannot be patented. Contrivances are classes of objects which embody a particular operational principle. [6:87]

In a 1968 paper in *Science* [7] he reiterated the matter this way

> If all men were exterminated, this would not affect the laws of inanimate nature. But the production of machines would stop, and not until men arose again, could machines be formed once more. Some animals can produce tools, but only men can construct machines; machines are human artifacts, made of inanimate material… In constructing a machine and supplying it with power, we harness the laws of nature at work in its material and in its driving force and make them serve our purpose. This harness is not unbreakable; the structure of the machine and with it its working can break down. But this will not affect the forces of inanimate nature on which the operation of the machine relies; it merely releases them from the restriction the machine imposed on them before it broke down. So the machine as a whole works under the control of two distinct principles. The higher one is the principle of the machine's design, and this harnesses the lower one, which consists in the physical chemical processes on which the machine relies.

So, in the conception of machines, we are dealing with a different kind of knowledge from naturalistic observational knowledge, one at a higher level than naturalistic science. We call this higher level knowledge contrivance, operational principle, or the principle of the machine's design, the defining characteristic of which is the notion of "serving our purpose," or the notion of "success" as well as the notion of "breakdown", or "failure". All those notions are foreign to naturalistic observational knowledge since it is of physical necessity.

Machines can achieve high performance, and they can fail. This is an entirely new conception not found in the study of observational sciences of inanimate systems. It would be nonsense to set up an enquiry into why thunderstorms go wrong or how stones make mistakes. They cannot be judged in these terms; they are necessary and they just are. This means that in judging machines we necessarily have in mind the idea of achievement or fulfillment. That is a totally foreign notion to physics and chemistry.

Lest we forget that every operational principle is subject to physical necessity. Both Eqs. (196) and (199/111) are, of course, consistent with the first law and indeed they are derived from the first law

$$\rho c_p\left(\frac{\partial T}{\partial t}+\vec{V}\cdot\nabla T\right)=-\nabla\cdot\vec{q}''+T\beta\frac{Dp}{Dt}+\left(\bar{\bar{\tau}}:\vec{V}\right)+\dot{q}_{ext-heating} \tag{196}$$

$$\dot{Q}-\dot{W}_{shaft}-\dot{W}_{resistive}=\frac{\partial}{\partial t}\int_{cv}e\rho d\mathcal{V}+\int_{cs}\rho\left(h+\frac{V^2}{2}+gz\right)\vec{V}\cdot d\vec{A} \tag{199/111}$$

Of the two, only (196) is a governing equation for heat transfer problem involving no heat extraction locally or globally (system-wide). This does not mean that the first law of thermodynamics itself is a law in the form of governing equation (or, equation of change). Equation (196) is derived from the first law as one special case and certainly not identical to it. Equation (199/111) represents another example of the application of the first law and is not a governing equation.

That Eq. (199/111) is not a governing equation is manifested in the term, $\dot{W}_{shaft}$, which—unlike shear stress work, which can be expressed in terms of velocity fields according to Stokes law of stresses, and $\dot{W}_{resistive}$, which can be prescribed in accordance with Joule resistive heating—cannot be reduced to (or not yet formulated into) a law of causally closed form. To put it in another way, Eq. (199/111) only serves to balance heat, work, and energy once the shaft work is measured, not to determine the shaft work. How much the shaft work is will be a matter to be dependent on an operational principle invented or to be invented subject to the limit of maximum possible shaft work in accordance with (121). The Carnot cycle is one example of a theoretical operational principle, albeit a unique example in the history of thermodynamics, as well as a foundational example in thermodynamics.

A great deal has been pointed out with regard to the uniqueness of the first law and the second law and their profundity. But, I think not enough has been made on the point[2] that the two laws in their application to reversible-like processes/systems, unlike their application to heat transfer problems, are unique among the laws of physics in that they are used not as governing laws but as laws in conjunction with operational principles. It is necessary in providing *mechanism* for *reversible-like processes/systems*.

10.5 Determination and Causal Closure

In his book, *The Nature of Thermodynamics*, [8] Bridgman was embarrassed by the anthropomorphic or economic nature of thermodynamics. But, he should not be. As I identified the cornerstone of the mechanical theory of heat to be universal interconvertibility, here I identify the cornerstone of naturalistic observational sciences to be the notion that laws govern every natural process, i.e., the idea of causal closure.

Every process in nature obeys the laws of nature. This is the idea of physical necessity in philosophy. But, what does necessity mean? Skeptics, most famously David Hume, challenged the idea of causal necessity. Hume correctly argued that man should not project the efficacy in conducting human affairs to the natural events of physical objects—bestowing them efficacy, i.e., causal necessity. This is why we say that there is only efficient causation in the event of July leading to August, but not July *efficaciously causes* August (see [10]). However, Hume did not deny the existence of efficacious causation in human; of which he wrote, “we have a clear idea;” he did not write that it was something “with which we are utterly unacquainted.” Precisely because man erroneously misappropriated a quality (that was clearly understood for man and exercised by man) to material objects, man was led astray to a “false philosophy” [9]. He did not deny the idea of power or efficacious causation in man; he was warning the consequence of misappropriating such power to matter, turning physical necessity (that objects are subject to) into causal necessity (anthropomorphic power, which objects do not possess).

What was the takeaway from Hume’s analysis? This is an example of one chooses to see what one wants to see. The clearly stated case by Hume has almost universally been misunderstood to be him denying the existence of efficacious causation that there is only efficient causation. A rebuttal of this interpretation of the Humean doctrine of causality has been made [10].

This mistake with regard to the nature of causality is associated with the prevailing doctrine of materialistic naturalism that only matter exists and that, if matter is deprive of any power or causal efficacy man too is deprived of causal efficacy: Scientific successes made it possible in avoiding anthropomorphic ways of thinking in explaining natural phenomena. That was exactly what Hume was saying that man

[2]As discussed in Sect. 8.3, Poincare and Lotka have made a similar point in their writing.

has power but should not project anthropomorphic power to matter. But, naturalism, i.e., anti-anthropomorphism, has since acquired a different thrust, in an ironic twist of the first order, that the sound naturalistic idea of physical or nomic necessity (we no longer have to imagine the power of wind in *Anemoi*, the wind gods) became physical determination or governing (man has no free will), and all laws became governing laws. Avoiding anthropomorphic thinking in materialism has become, rather than viewing man as master of his home, instead, viewing man to be powerless in his existence. "…in first stripping away from material objects their anthropogenic power as Hume argued, [materialism turned] 180 degrees to impart to material objects that very anthropogenic power so that matter can govern over [man in] a causally closed world, [as] Collingwood (1940) argued, 'The so-called materialism which was the favorite metaphysical doctrine of these anti-metaphysicians was in consequence only in name a repudiation of anthropomorphism; really it was anthropomorphic at the core.' " [10].

In other words, causal closure is the presupposition that the universe is a causally closed world in which all physical laws are governing equation laws. Causal closure is the scientific version of omnipotent God.

My analysis in this chapter shows that this is simply not true. The first law acquires governing equation status only for a restricted class of problems in the form of Eq. (196). The more general equation Eq. (199/111) inferred from the first law is not a governing equation, as $\dot{W}_{shaft}$ cannot be reduced to or not yet formulated into a law of causally closed form. Should a scientist succeed in achieving the "impossible", all patent offices in the world will be closed. Until that should happen, we have to think and operate, as Polanyi and Collingwood did and every common man and woman does, with the presupposition that the world is not causally closed.

10.6 Engineering for Efficiency

Everything under the sun is subject to physical necessity, but the world is not causally closed. Every process is subject to physical necessity, but only natural processes (thermodynamically speaking, spontaneous nonreversible processes) are determined by laws of physics and chemistry. Laws have only the passive power of physical or nomic necessity of efficient causation, not the active power of causal necessity of efficacious causation. Every event in the Poincare range is subject to physical necessity, but only the spontaneous event is governed (more precisely, description of which provided by laws) while all the rest, to greater or smaller extent, are created by contrivance managing actions.

Heat transfer from a hot body to a cold body is a physical necessity, as well as the great discovery of Mayer and Joule [11] that heat and mechanical energy are always correlated, i.e., conservation of energy prevails in their gain–loss: one gain is coupled with the other loss and vice versa. Mayer–Joule's achievement led to the notion of energy as the capacity for doing work. But, that is misleading and the real reason for heat or energy's capacity for doing work is found in the Carnot's earlier realization that in natural heat transfer processes, man can avail himself of useful work via the creation of reversible processes bringing about possibilities provided by the "descriptive power" of laws. With this insight, Carnot discovered in reversible cycles the causal necessity of efficacious causation.

Thus, Carnot gave philosophical meaning to engineering, especially to engineers practicing engineering thermodynamics aiming for achieving efficiency. He defined efficiency in terms of reversibility. Kelvin and Gibbs added the meaning of available energy to reversibility; generation of engineers, Keenan, Rant, Bejan…, developed the theory of exergy, and, thus, gave meaning to energy in terms of its exergy content as the driver that is capable of making things happen. This disquisition identifies that energy is proxy to entropy growth potential, which is the real driver, as well as demonstrates that the essence of reversibility is the triadic framework under which (including the generalized triadic framework) is how we manage the use of stock EGP and natural EGP that will help us find pathways toward true efficiency. Efficiency is not an energy problem alone; it depends on thinking in terms of natural EGP and triadic framework, which can greatly increase the efficiency of stock EGP usage.

One necessary first step is to apply energy *and* exergy analyses to more problems, as well as general efficiency assessment. One instance of applying energy and exergy analyses can be found in an evaluation of how we used energy in the twentieth century, as reported in a study by Ayres and Warr [12], and a tabulated summary of the study is reproduced as Table 10.1. The numbers in Table 10.1 are the exergy efficiencies of five categories of energy uses and progresses made in their practices in the twentieth century; what the numbers show is that efficiency in *low temperature space* heat category was singularly poor, lower by one order of magnitude in comparison with the other four categories.

One may surmise the starkly poor performance of low temperature heating to be due to that while the four categories' use of energy has been guided by the Carnot–Kelvin formula (at least as an approximated guide for contriving some form of reversible-like processes), low temperature heating, without a formula of any

Table 10.1 Exergy efficiencies of five categories of energy applications in the twentieth century

Year	Electric power	Transportation	High temperature industrial heat	Medium temperature industrial heat	Low temperature space heat
1900	3.8%	3	7	5	0.25
1910	5.7	4.4			
1920	9.2	7			
1930	17.3	8			
1940	20.8	9			
1950	24.3	9			
1960	31.3	9			
1970	32.5	8	20	14	2
1980	32.9	10.5			
1990	33.3	13.9	25	20	3%

relevance for such purpose, has been met by straightforward nonreversible boiler combustion processes, in the vast majority of cases today. Intuition and science and engineering suggest, therefore, that a systematic study of operational principles ("logic of contriving" reversible-like processes) in general and searching for reversible-like process for low temperature heat in particular, are worthy pursuits of engineering thermodynamics.

Industrial civilizations have been built upon the extraction of energy for production of heat and power [13, 14]. Instead, ecological civilization, cognizant of the essence of the second law as "the tendency for living systems to become more complex is not somehow in defiance of the second law. In fact, it is more meaningful to characterise the phenomenon described by the second law as [the] fundamental enabler of all activity, and hence of all life," [15] should be and will be built upon the extraction of heat powered by entropy growth potential.

References

1. Schneider E, Sagan D (2005) Into The Cool—Energy Flow, Thermodynamics, and Life. Univ of Chicago Press
2. Yener Y, Kakac S (2008) Heat Conduction, 4th edition. Taylor & Francis
3. Fox RW, Pritchard PJ, McDonald AT (2009) Introduction to Fluid Mechanics, 7th edition. Wiley
4. Bird RB, Stewart WE, and Lightfoot EN (2007) Transport Phenomena, 2nd edition. Wiley
5. Wang LS (2006) The auxiliary components of thermodynamic theory and their non-empirical, algorithmic nature. Physics Essays 19 (2):174–199
6. Polanyi M (1958) Personal Knowledge. Univ. of Chicago Press
7. Polanyi M (1968) Life's irreducible structure. Science 160:1308–1312
8. Bridgman PW (1961) The Nature of Thermodynamics. Harvard Univ Press
9. Hume D (1739–1740; 1969 Penguin Books edition) A Treatise of Human Nature. Penguin Books, London
10. Wang LS (2011) Causal efficacy and the normative notion of sustainability science. SSPP 7 (No.2):30–40
11. Newburgh R, Leff H S (2011) The Mayer-Joule Principle: The foundation of the first law of thermodynamics. The Physics Teacher 49(November):484–487
12. Ayres RU, Warr B (2010) The Economic Growth Engine: How energy and work drive material prosperity. Edward Elgar (p. 127)
13. Smith C, Wise MN (1989) Energy and Empire: A Biographical Study of Lord Kelvin. Cambridge Univ Press
14. Smil V (2017) Energy and Civilization, A History. MIT Press, Cambridge, MA
15. Floyd J (2007) Thermodynamics, entropy and disorder in futures studies. Futures 39 (2007):1029–1044

Glossary

Caloric theory of heat An obsolete ontological theory of heat that heat consists of a self-repellent, weightless fluid called caloric that flows from hotter bodies to colder bodies. Though obsolete in its materiality conception of heat, it remains useful in how we study heat flowing down the temperature gradients and how we measure the heat value of foods.

Causal necessity The idea that agents can start new causal chains that are not pre-determined by the events of the immediate or distant past in accordance to physical necessity. That is, causation is not limited to efficient causation alone; nor, necessity limited to physical necessity. Causal necessity, therfore, is a new metaphysical presupposition breaking away from the the metaphysical presupposition of physical necessity. According to Poincare (Sect. 8.3), without causal necessity, thermodynamic laws would "no longer have any meaning". It is causal necessity and "indeterministic hypothesis" that give meaning to thermodynamic laws and make them unique in physics.

Chemical energy Energy stored within chemical bonds.

Combustion The process of burning organic chemicals to release heat and light.

Efficacious causation See Causal necessity.

Efficiency According to Energy conversion doctrine: ability of a process or machine to convert energy input to energy output. According to EGP-centric doctrine: ability of a triadic process to produce benefitable outcome from the consumption of stock EGP as measured by the ratio of "benefit" in terms of energy to energy of stock EGP.

Efficient causation See Physical necessity.

Electrical energy Energy made available by the flow of electric charge through a conductor.

Energy conversion Transformation of one form of energy into another, usually to convert the energy into a more useful form.

L.-S. Wang, *A Treatise of Heat and Energy*, Mechanical Engineering Series,
https://doi.org/10.1007/978-3-030-05746-6

Energy conversion doctrine The doctrine of MTH—which is based on the *interconvertibility principle* and the *energy principle*, both of which are self-evident propositions—that "everything that happens is the result of energy and its transformation from one form to another." It is the doctrine serving as the cornerstone of the MTH version of *engineering thermodynamics*. As self-evident truth, it captures a part of truth but it turns out to be incoherent and contradictory on the whole.

The energy principle Mechanical energy dissipates universally into heat; high-grade energy dissipates into lower-grade energy without limit; phenomena of entropy growth are exhausted by energy-dissipation phenomena.

Entropy growth potential Entropy growth potential (EGP) is the driver of every event in a given Poincare range (See Sect. 8.5).

Entropy growth potential principle Every event in a given Poincare range shares the same $EGP_{Poincare\text{-}range}$, therefore, $EGP_{Poincare\text{-}range}$ is the driver of every event of the range, therefore, *general* EGP is the driver of *all* events in nature.

EGP-centric doctrine The doctrine of PETH, which refutes *interconvertibility principle* and considers that the *energy principle* is subsumed under the *entropy growth potential principle*. Furthermore, it denies the existence of pure reverse energy transformation asserting that any reverse energy transformation is an element of a triadic process consisting of EGP the driver, heat from heat reservoir, and energy in useful form, e.g., mechanical energy.

The entropy principle Entropy change of an isolated system (universe), which undergoes certain process, is never negative. That is, entropy change is always positive (entropy always growths) or equal to zero (in case of reversible change).

Equilibrium thermodynamics A branch of thermodynamics, also known as Gibbsian thermodynamics, that is based on the entropy principle as its centerpiece rather than the problematic energy principle or the energy conversion doctrine. That being the case, equilibrium thermodynamics, unlike engineering thermodynamics, is a coherent branch-system of theoretical physics.

Exergy Exergy is the portion of energy that is entirely convertible into all other forms of energy, i.e., convertible to mechanical energy (which can be converted into all other forms without limit). This is one definition of exergy, see Sect. 7.3 for the other definition.

First law of thermodynamics Energy can neither be created nor destroyed.

Form of energy Forms of energy include heat (thermal energy), light, electrical, mechanical, nuclear, sound and chemical.

Heat Energy and entropy in transit, or, a high-entropy form of energy related to its temperature (thermal energy).

Input Matter or energy going into a process.

Interconvertibility principle Heat and mechanical work are universally and uniformly interconvertible in all circumstances. That is, the correlation between heat and mechanical work in accordance with MEH means causation between the two: consumption of heat means heat causes work and, vice versa, consumption of mechanical energy means work causes heat.

Internal reversibility Internal reversibility is not a process operation, but a condition that a process satisfies: defined as the condition that transient states in every second-subset of transient states become "at all times infinitesimally near" their corresponding pairs of equilibrium states of the first subset (See Sect. 6.5).

Kinetic energy Energy of motion, influenced by an objects mass and speed.

Mechanical energy Kinetic energy and potential energy, including pure-exergy energies that can be converted 100% into kinetic or potential energy.

Mechanical equivalent of heat (MEH) Mechanical energy is produced from the consumption of heat and, vice versa, heat is produced from the consumption of mechanical energy. In PETH, the "consumption" of heat is replaced by the "extraction" of heat.

Mechanical theory of heat (MTH) An ontological theory of heat, on the basis of which the current "normal science" of thermodynamics has been developed (See Energy conversion doctrine).

Necessity Necessity is the idea that everything that has ever happened and ever will happen is necessary, and cannot be otherwise. Necessity is opposed to chance and contingency, but a distinction can be made between necessity and determination. This definition is closely aligned with Physical necessity. In that case, there is no distinction between necessity and determination. Distinction can be made, however, if one accepts the possibility of causal necessity (see Causal necessity).

Nomic necessity See Physical necessity.

Non-reversible process Spontaneous natural events are non-reversible processes. Non-reversible processes are necessarily introduced to differentiate them from irreversible reversible-like processes since all real processes are irreversible whether they are non-reversible or reversible-like (See below).

Nuclear energy (atomic) Energy produced by splitting the nuclei of certain elements.

Output Matter or energy coming out of a process.

PETH Predicative entropic theory of heat (See EGP-centric doctrine).

Physical necessity Necessary sequence of event or relationship in accordance with laws of physics and chemistry. Physical necessity is a metaphysical presupposition some philosophers consider that, without which, science (scientific suppositions) is not possible. Scientific suppositions predicated on physical necessity are characterized in terms of efficient causation and determinism.

Poincare range A system undergoes from its initial state to a given final state spontaneously. Given the initial and final states, there are infinite number of possible events bookended by the spontaneous event and the reversible event. The set of all possible events including two bookend events is called Poincare range.

Potential energy Energy that is stored and that comes from an object's position or condition.

Pure exergy energy Energy that 100% of its energy can be converted into mechanical energy.

Pure spontaneity Systems that their EGPs are not associated with any energy form change. Typically, isolated systems with spontaneous tendency towards internal equilibrium possess pure spontaneity.

Quasi-static processes A quasistatic process is an idealization of a spontaneous natural process. The technical definition is: The set consists of a subset-series of infinitely dense succession of equilibrium states *and* the corresponding subsets of all spontaneous transient states between each pairs of equilibrium states in the series.

Reversible process Conventional definition: *a reversible process is a process whose direction can be "reversed" by inducing infinitesimal changes to some property of the system via its surroundings, with no increase in entropy. Throughout the entire reversible process, the system is in thermodynamic equilibrium with its surroundings.* PETH definition: perfection of reversal-like processes that are brought about by triadic operation driven by the consumption of EGP with no entropy growth.

Reversible event A reversible process for a specific Poincare range.

Reversible-like process Reversible-like processes are managed processes driven by entropy growth potential in triadic operation, which is fundamentally different from spontaneous processes whereas the former is brought about by causal necessity the latter is accounted for by physical necessity.

Second law of thermodynamics Entropy growths universally; entropy growth potential is the universal driver of all changes in nature.

Spontaneity See Entropy growth potential.

Spontaneous processes Unmanaged processes associated with entropy growth potential, of opposite category to reversible-like processes (see Reversible-like process).

Triadic framework Triad of EGP, heat from heat reservoir, and useful form of energy. General triad of stock EGP, natural EGP, and useful form of energy or useful heat or useful outcome. The dyadic framework of MTH's interconvertibility principle is what tainted the triumph of the mechanical theory over the caloric theory by misplacing heat in its relational category. Triadic framework is an innovation of the predicative theory (PETH) that places heat in its correct categories both ontologically and predicatively.

Index

L.-S. Wang, *A Treatise of Heat and Energy*, Mechanical Engineering Series,
https://doi.org/10.1007/978-3-030-05746-6

MIX
Papier aus verantwortungsvollen Quellen
Paper from responsible sources
FSC® C105338

If you have any concerns about our products,
you can contact us on
ProductSafety@springernature.com

In case Publisher is established outside the EU,
the EU authorized representative is:
Springer Nature Customer Service Center GmbH
Europaplatz 3, 69115 Heidelberg, Germany

Printed by Libri Plureos GmbH
in Hamburg, Germany